OPERATIONAL AMPLIFIERS AND LINEAR INTEGRATED CIRCUITS

OPERATIONAL AMPLIFIERS AND LINEAR INTEGRATED CIRCUITS

Theory and Applications

DENTON J. DAILEY
Butler County Community College
Butler, Pennsylvania

GLENCOE
McGraw-Hill

New York, New York
Columbus, Ohio
Mission Hills, California
Peoria, Illinois

Sponsoring Editor: John J. Beck
Design and Art Supervisors: Nancy Axelrod and Janice Noto
Production Supervisor: Mirabel Flores

Text Design: P. L. K. Graphics Inc.
Cover Designer: Patricia Lowy
Cover Photography: Wolfson Photography, Inc.
Cover Photograph: An LM3080 operational transconductance amplifier mounted on a TO-5 header. This device is described in Chapter 7, Specialized Linear ICs; see Fig. 7-23*a* and *b*.

Library of Congress Cataloging-in-Publication Data

Dailey, Denton J.
  Operational amplifiers and linear integrated circuits

  Bibliography: p.
  Includes index.
  1. Operational amplifiers.   2. Linear integrated circuits.   I. Title.
  TK7871.58.06D34 1989   621.3815'35   88-13011
  ISBN 0-07-039931-X

Operational Amplifiers and Linear Integrated Circuits: Theory and Applications

Imprint 1995
Copyright © 1989 by Glencoe/McGraw-Hill. All rights reserved.
Copyright © 1989 by McGraw-Hill, Inc. All rights reserved. Printed in the United States of America. Except as permitted under the United States Copyright Act, no part of this publication may be reproduced or distributed in any form or by any means, or stored in a database or retrieval system, without prior written permission from the publisher.

3 4 5 6 7 8 9 10 11 12 13 14 15 RRC/PC 00 99 98 97 96

ISBN 0-07-039931-X

CONTENTS

Preface   xi

CHAPTER 1 • DIFFERENTIAL AMPLIFIERS   1

1-1 DC Analysis of the Differential Amplifier   1
Differential Pair Current Relationships / Differential Amplifier Voltage Relationships

1-2 Current Sources   8
Emitter-Biased Current Source / The Current Mirror / The Zener-Biased Current Source / Additional Current Source Configurations

1-3 AC Analysis of the Differential Amplifier   17
Single-Ended-Input–Single-Ended-Output / Single-Ended-Input–Differential-Output / Differential-Input–Differential-Output / Differential-Input–Single-Ended-Output / Common-Mode Response

1-4 Cascaded Differential Amplifiers   34

1-5 Level-Shifting and Output Stages   38
Level Shifters / Final Output Stages

1-6 Additional Differential Amplifier Variations   44
Active Loading / Increasing Input Resistance

Chapter Review   48
Questions   48
Problems   49

CHAPTER 2 • INTRODUCTION TO OPERATIONAL AMPLIFIERS   56

2-1 Operational Amplifiers and Ideal Voltage Amplifiers   56
Ideal Voltage Amplifier Characteristics / Typical Op Amp Architecture

2-2 Negative Feedback   59
The Noninverting Configuration / The Inverting Configuration / The Voltage Follower / Effect of Negative Feedback on Output Resistance

2-3 Bandwidth Limitations   74
2-4 Cascaded Amplifiers   78
Chapter Review   80
Questions   81
Problems   81

CHAPTER 3 • PRACTICAL OP AMP CONSIDERATIONS   85

3-1 Op Amp Error Sources   85
Bias and Offset Currents / Offset Voltage
3-2 Frequency Compensation and Stability   95
Frequency Compensation / Stability and Phase Margin
3-3 Slew Rate   100
Maximum Rate of Change and Slew Rate / Slew Rate and Distortion
3-4 IC Data Sheets, Identification Numbers, and Packaging   105
IC Identification / Data Sheets
Chapter Review   108
Questions   108
Problems   108

CHAPTER 4 • OP AMP APPLICATIONS: PART A   111

4-1 The Summing Amplifier   111
General Summing Amplifier Analysis / Practical Summing Amp Considerations / Averagers / Additional Summing Amplifier Circuits
4-2 Differential and Instrumentation Amplifiers   123
The Op Amp–Based Differential Amplifier / Instrumentation Amplifiers
4-3 Voltage-to-Current and Current-to-Voltage Conversion   134
The VCIS / The ICVS / The ICIS (Current Amplifier)
4-4 A Voltage Amplifier Variation   140
Chapter Review   142
Questions   143
Problems   143

CHAPTER 5 • OP AMP APPLICATIONS: PART B   146

5-1 The Op Amp with Complex Impedances   146
The Inverting Configuration / The Noninverting Configuration

5-2 Differentiators and Integrators   155
The Differentiator / The Integrator

5-3 Nonlinear Op Amp Circuits   170
The Logarithmic Amplifier / The Antilogarithmic Amplifier / Logarithmic Amplifier Applications

5-4 Precision Rectifiers   181
Precision Half-Wave Rectifier / Precision Full-Wave Rectifier

Chapter Review   184
Questions   184
Problems   185

CHAPTER 6 • ACTIVE FILTERS   189

6-1 Filter Fundamentals   189
Filter Terminology / Basic Filter Theory Review

6-2 Active LP and HP Filters   200
Sallen-Key LP and HP Filters

6-3 Bandpass Filters   213
Wideband BP Filters / Multiple-Feedback Bandpass Filters / Some Bandpass Filter Applications

6-4 Bandstop Filters   222

6-5 State-Variable Filters   226

6-6 The All-Pass Filter   229

Chapter Review   230
Questions   231
Problems   232

CHAPTER 7 • SPECIALIZED LINEAR ICs   234

7-1 Comparators   234
The Op Amp as a Comparator / The Zero-Crossing Detector / The Comparator with Hysteresis / The Window Comparator / Comparator ICs

7-2 **Analog Switches and Track-and-Hold Circuits**   243

Analog Switches / Track-and-Hold Circuits

7-3 **Current Difference Amplifiers**   247

CDA Inverting Configuration / CDA Noninverting Configuration

7-4 **Operational Transconductance Amplifiers**   253

Basic OTA Design Considerations / An OTA Modulator Application

7-5 **Balanced Modulators**   264

The LM1496 Balanced Modulator / Balanced Modulator Applications

7-6 **The 555 Timer**   271

555 Operation Principles / The 555 as a Voltage-Controlled Oscillator

7-7 **Phase-Locked Loops**   275

Phase-Locked Loop Operation / PLL Applications / Commercially Available PLLs

Chapter Review   289
Questions   290
Problems   291

CHAPTER 8 • DATA CONVERSION DEVICES   295

8-1 **D/A Conversional Fundamentals**   295

Resolution and Full-Scale Output / Accuracy / Settling Time

8-2 **D/A Conversion Circuits**   303

Weighted Resistor Summing D/A / The R-2R Ladder D/A Converter / Monolithic D/A Converters

8-3 **A/D Conversion**   312

A/D Conversion Fundamentals

8-4 **A/D Conversion Circuits**   319

Ramp Converters / Successive-Approximation A/D Converters / Dual-Slope A/D Converters / Parallel A/D Converters / Tracking A/D Converters / Monolithic A/D Converters

8-5 **Data Acquisition and Digital Signal Processing**   333

Effects of Sampling / The Nyquist Sampling Theorem / Aliasing / Quantization Noise and Reconstruction

Chapter Review   342
Questions   343
Problems   343

CHAPTER 9 • VOLTAGE REGULATORS   346

9-1 Unregulated Supplies   346
9-2 Linear Voltage Regulators   349
Shunt Regulators / Series Regulators / Overcurrent Protection / Foldback Current Limiting
9-3 Monolithic Linear Voltage Regulators   357
The 723 / The 117 / Dual-Tracking Regulators
9-4 Switching Regulators   362
Switching Regulator Operation / The Step-Down Configuration / The Step-Up Configuration / The Inverting Configuration / Additional Switching Regulator Considerations
Chapter Review   365
Questions   365
Problems   365

APPENDIX A • DATA SHEETS   367

APPENDIX B • SUGGESTED REFERENCES   407

APPENDIX C • ANSWERS TO ODD-NUMBERED PROBLEMS   408

Index   415

PREFACE

Operational amplifiers (op amps) and linear integrated circuits (ICs) are encountered in nearly all electronic systems. A staggering array of different types of general-purpose op amps and more specialized linear ICs is available today. These devices are sophisticated and flexible, and yet they are generally quite inexpensive. These factors make such devices attractive to circuit designers in many areas. Even systems that are primarily digital in nature, such as microcomputers and digital test equipment, often contain many linear ICs.

This text is primarily intended for use in an introductory course on op amps and linear ICs in electronics and electrical engineering technology programs. In order to use this book most effectively, the reader should have had a course in basic dc and ac circuit analysis and a general understanding of semiconductor devices. Knowledge of algebra and trigonometry is also required.

The book is structured in such a way that successive chapters use the information that was provided in preceding chapters. For example, circuits that rely on the operation of comparators are not introduced until comparators themselves have been covered. While many topics required a more structured approach, with one chapter building upon the preceding chapter, the text has a considerable amount of flexibility. For example, Chapters 4 through 7 could be covered in just about any order, without a loss in continuity, because they introduce the applications of op amps and linear ICs.

Each chapter concludes with a series of questions and problems. The degree of difficulty builds from problem to problem, with each chapter containing several problems that are rather difficult and will require a particularly good knowledge of the concepts that were presented. Worked-out examples are used extensively throughout the text.

Several chapters in this book cover material that is glossed over in many other books of a similar nature. In particular, the applications of operational transconductance amplifiers and balanced modulars are presented in some detail. These applications lead directly into signal representation in the frequency domain. In programs without separate communications electronics courses, this material is sorely needed. Also, the coverage of D/A and A/D converters includes a section on the basic concepts of data acquisition and sampling. Because of the increasing use of microprocessors and digital signal processing techniques in consumer, industrial, and military applications, these topics should be addressed.

A laboratory manual containing experiments designed to reinforce the concepts presented in the text is available. Answers to all chapter questions and even-numbered problems can be found in an accompanying instructor's guide.

Denton J. Dailey

OPERATIONAL AMPLIFIERS AND LINEAR INTEGRATED CIRCUITS

CHAPTER 1

DIFFERENTIAL AMPLIFIERS

A large portion of the linear integrated circuits that are used in system designs are based on the differential amplifier. This chapter covers the basics of differential amplifier operation and design from the component level. Knowledge of the operation of differential amplifiers and associated current sources, level-shifting circuits, output stages, and protective circuitry will allow a deeper understanding of the nature of most linear integrated circuits and their characteristics. Throughout this chapter, r-parameter circuit analysis methods will be used for all bipolar-transistor-based circuits.

1-1 ◆ DC ANALYSIS OF THE DIFFERENTIAL AMPLIFIER

The basic configuration of the differential amplifier is shown in Fig. 1-1. Perhaps the most obvious feature of the differential pair is the connection of the emitters of Q_1 and Q_2 to a single resistor R_T. The significance of this connection will be made clear in the upcoming analyses. Since this circuit

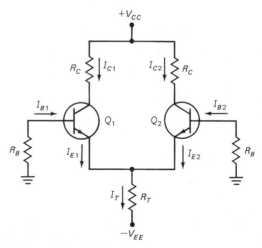

Fig. 1-1 Basic differential amplifier circuit.

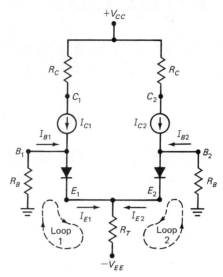

Fig. 1-2 Biasing model for the BJT differential pair.

will ultimately be used to amplify signals, it may be referred to as a differential amplifier. The two transistors and their associated components (in this case resistors) are considered to form a single stage.

Differential Pair Current Relationships

The analysis of the differential amplifier in Fig. 1-1 is relatively straightforward if a few basic assumptions are made at the outset. First, let us assume that Q_1 and Q_2 are matched, that is, that both transistors have identical transconductance curves, β, and temperature tracking characteristics. Let us also assume that both transistors are at the same temperature. Resistors will be assumed to be equal in value if their subscripts are the same. These assumptions are generally valid for practical differential amplifier circuits, although there are some exceptions. For dc analysis purposes, the differential amplifier can be modeled by the circuit in Fig. 1-2. The emitter current for each of the transistors can be determined by applying Kirchhoff's voltage law (KVL) around loops 1 and 2. Analyzing both loops yields the following two equations:

$$0 = V_{EE} - V_{BE} - I_{B1}R_{B1} - R_T(I_{E1} + I_{E2}) \quad \text{for loop 1} \quad \text{(1-1a)}$$

$$0 = V_{EE} - V_{BE} - I_{B2}R_{B2} - R_T(I_{E1} + I_{E2}) \quad \text{for loop 2} \quad \text{(1-1b)}$$

Because of the identical characteristics of the components used in each half of the circuit, Eqs. (1-1a) and (1-1b) differ only in the subscripts that are used in each term. Remember, the values of the terms in each equation are equal because of the matching of the circuit elements in loops 1 and 2. Using this information, we can combine Eqs. (1-1a) and (1-1b) into a single expression.

$$0 = V_{EE} - V_{BE} - I_B R_B - I_E R_T - I_E R_T \qquad (1\text{-}2)$$

where $I_{E1} = I_{E2} = I_E$

$\qquad\quad I_{B1} = I_{B2} = I_B$

$\qquad\quad R_{B1} = R_{B2} = R_B$

In order to produce common factors of I_E in each term of Eq. (1-2), we can define I_B in terms of β through the use of the approximate relation

$$I_B \cong \frac{I_E}{\beta} \qquad \text{because } \beta \cong \beta + 1 \qquad (1\text{-}3)$$

Substituting Eq. (1-3) into (1-2) produces

$$0 = V_{EE} - V_{BE} - \frac{I_E}{\beta} R_B - I_E R_T - I_E R_T$$

Subtracting V_{EE} and V_{BE} from both sides and multiplying by -1 yields

$$V_{EE} - V_{BE} = \frac{I_E}{\beta} R_B + I_E R_T + I_E R_T$$

Factoring the right-hand side of the equation produces

$$V_{EE} - V_{BE} = I_E \left(\frac{R_B}{\beta} + 2R_T \right)$$

Since $I_T = I_E + I_E = I_{E1} + I_{E2}$, we can write

$$V_{EE} - V_{BE} = I_T \left(\frac{R_B}{2\beta} + R_T \right)$$

Finally, solving for I_T yields

$$I_T = \frac{V_{EE} - V_{BE}}{R_B/2\beta + R_T} \qquad (1\text{-}4)$$

We now have a useful analysis equation for the differential amplifier. By definition, under quiescent conditions, the current I_T divides equally between the two emitters. This allows the individual emitter currents to be found using Eq. (1-5):

$$I_E = \frac{I_T}{2} \qquad (1\text{-}5)$$

In a practical circuit, R_T is very much larger than $R_B/2\beta$. This allows further simplification of Eq. (1-4), producing

$$I_T \cong \frac{V_{EE} - V_{BE}}{R_T} \tag{1-6}$$

Regardless of which I_T equation is used [Eq. (1-4) or Eq. (1-6)], the emitter currents for each side of the circuit are assumed to be the same and equal to $I_T/2$.

EXAMPLE 1-1

Determine the (quiescent) values of I_T, I_{E1}, I_{E2}, I_{C1}, I_{C2}, I_{B1}, and I_{B2} for the circuit in Fig. 1-3 based on the use of Eq. (1-4). Assume that $\beta = 100$ and $V_{BE} = 0.7$ V.

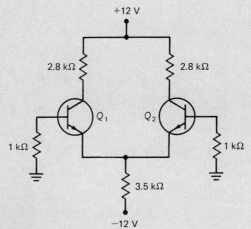

Fig. 1-3 Circuit for Example 1-1.

Solution

(a) $I_T = \dfrac{V_{EE} - V_{BE}}{R_B/2\beta + R_T}$

$= 3.22$ mA

(b) $I_{E1} = I_{E2} = \dfrac{I_T}{2}$

$= 1.61$ mA

(c) $I_{C1} = I_{C2} = I_E \left(\dfrac{\beta}{\beta + 1}\right)$

$= 1.59$ mA

(d) $I_{B1} = I_{B2} = I_E - I_C$

$= 16 \ \mu A$

In Example 1-1, notice that the values of I_C and I_E are nearly equal to one another, and that I_B is relatively small in comparison. In most cases, the analysis of the differential amplifier need not be so exacting. To illustrate this point, let us use Eq. (1-6) to determine I_T and I_E for Fig. 1-3. We shall assume that base current is negligible ($I_B \cong 0$) and $I_C = I_E$.

$$I_T = \frac{V_{EE} - V_{BE}}{R_T}$$

$= 3.23 \text{ mA}$

$$I_C = I_E = \frac{I_T}{2}$$

$= 1.62 \text{ mA}$

Equation (1-6) becomes an even closer approximation for I_T as the ratio of R_T to R_B increases. It will be shown in the next section how the effective value of R_T can be increased drastically. The utility of the more approximate relation of Eq. (1-6) becomes more apparent when we consider that slight variations in transistor V_{BE}, β, and other device parameters can easily negate the small increase in accuracy provided by Eq. (1-4).

Differential Amplifier Voltage Relationships

Once the emitter currents are known, it is possible to determine the voltage drops across the various components in the circuit. When there are equal resistances in both halves of the amplifier, analysis of both sides of the circuit will yield identical values. The following example will apply the values of the currents that were determined in Example 1-1.

EXAMPLE 1-2

Determine the following voltages for Fig. 1-3: V_{CE1}, V_{CE2}, V_{C1}, V_{C2}, V_{CB1}, and V_{CB2}. Use the current values obtained in Example 1-1 to calculate the voltage drops across the appropriate resistors.

Solution

Since both sides of the circuit are identical, we can equate the voltage drops across the transistors and resistors. That is, an analysis of one side yields voltage drops that are applicable to the other side. Using KVL and

Ohm's law, and the current values from the redrawn circuit in Fig. 1-4, we obtain

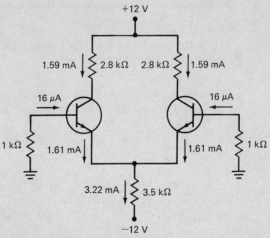

Fig. 1-4 Circuit for Example 1-2.

$$V_{CE1} = V_{CE2} = V_{EE} + V_{CC} - I_C R_C - I_T R_E$$
$$= 24\text{ V} - 4.46\text{ V} - 11.27\text{ V}$$
$$= 8.27\text{ V}$$

$$V_{C1} = V_{C2} = V_{CC} - I_C R_C$$
$$= 12\text{ V} - 4.46\text{ V}$$
$$= 7.54\text{ V}$$

$$V_{CB1} = V_{CB2} = V_{CC} - I_C R_C - I_B R_B$$
$$= 12\text{ V} - 4.46\text{ V} - 16\text{ mV}$$
$$= 7.52\text{ V}$$

Example 1-2 illustrates that the base bias currents produce voltage drops across the base resistors. The effects of unequal bias currents and/or voltage drops across the base resistances can cause significant errors at the output of a differential amplifier. In fact, even when the voltages present at the bases of the transistors are exactly equal, there can be a significant, undesirable effect on the overall circuit output. These errors are discussed in more detail in Sec. 1-3. Also, recall that differences between β and V_{BE} values in the differential pair can cause the circuit to be out of balance. These parameter differences may contribute to the bias current and voltage errors that are present. For now, however, it will suffice for the reader to be aware of the possibility of such errors and their causes.

There are many different variations in the design details of differential amplifiers. For example, in some cases, one side of the differential pair may not include a collector resistor. In such a case, it is apparent that the voltage drops present around the transistors will not be equal to one another. Consider the following example.

EXAMPLE 1-3

Determine the values of I_{C1}, I_{C2}, V_{CE1}, and V_{CE2} for the circuit shown in Fig. 1-5. Use Eq. (1-6) to determine circuit current levels.

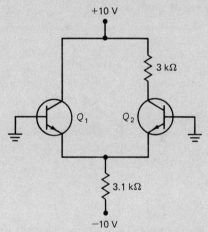

Fig. 1-5 Circuit for Example 1-3.

Solution

From basic transistor analysis theory, we know that unless the transistor is saturated or in cutoff, I_C is essentially independent of the load present between the collector and the power supply. This means that $I_{C1} = I_{C2} = I_T/2$; therefore

$$I_T = \frac{10\ V - 0.7\ V}{3.1\ k\Omega}$$

$$= 3.0\ mA$$

and

$$I_{C1} = I_{C2} = 1.5\ mA$$

> Summing voltage drops from the positive supply to the negative supply across Q_1 yields
>
> $$V_{CE1} = 20\text{ V} - I_T R_T$$
> $$= 20\text{ V} - 9.3\text{ V}$$
> $$= 10.7\text{ V}$$
>
> Likewise, for Q_2 we obtain
>
> $$V_{CE2} = 20\text{ V} - I_T R_T - I_C R_C$$
> $$= 20\text{ V} - 9.3\text{ V} - 4.5\text{ V}$$
> $$= 6.2\text{ V}$$

It is best not to attempt to memorize specific equations for the various differential amplifiers that are presented, as there are too many possible variations. The best analysis approach is to use basic transistor parameter relations, KVL, Kirchhoff's current law (KCL), and Ohm's law to derive the expressions necessary to obtain the desired voltages and currents. Application of these basic techniques becomes even more important as circuit complexity increases.

1-2 ◆ CURRENT SOURCES

The emitter currents in the differential amplifier play the primary role in determining the circuit's operating conditions. Thus, it is advantageous to have a very stable source producing I_T. In other words, I_T should be made independent of the emitter currents flowing in the differential pair. These requirements indicate that it would be desirable to drive the emitters of the differential pair with a current source. A differential amplifier that uses an idealized constant-current source for emitter biasing is illustrated in Fig. 1-6.

Ideally, the internal resistance (and impedance) of a current source is infinite. In order to approximate such an ideal current source, we would need to use an extremely high resistance for R_T. This would invariably result in extremely low quiescent currents in the differential amplifier, which may not be desired in some cases. Fortunately, the collector terminal of a transistor can be used very effectively as a near-ideal current source. Effective collector-to-emitter resistances (r_{CE} or $1/h_{OE}$) can typically range from 100 kΩ to over 1 MΩ. This makes transistorized active current sources nearly ideal candidates for use in differential amplifiers.

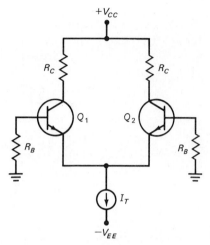

Fig. 1-6 Constant-current source biased differential amplifier.

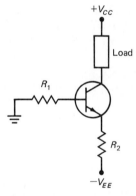

Fig. 1-7 Simple transistor current source.

Emitter-Biased Current Source

Perhaps the simplest bipolar transistor current source that can be analyzed uses emitter biasing. This is illlustrated in Fig. 1-7. Here, the collector current I_T is given by

$$I_T \cong \frac{V_{EE} - V_{BE}}{R_1/\beta + R_2} \tag{1-7a}$$

As usual, if $\beta R_2 \gg R_1$, then Eq. (1-7a) may be reduced to

$$I_T \cong \frac{V_{EE} - V_{BE}}{R_2} \tag{1-7b}$$

A voltage divider could also be used at the base of the current source transistor, as shown in Fig. 1-8a. To determine I_T for this circuit, Thevenin's

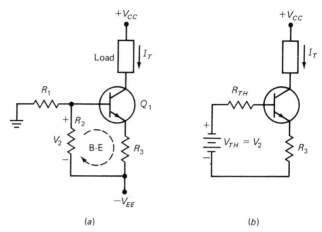

Fig. 1-8 Active current source. (*a*) Voltage divider biased source. (*b*) Circuit redrawn with thevenized biasing circuit.

theorem is applied to convert the voltage divider into an equivalent resistance in series with a voltage source, as shown in Fig. 1-8*b*. The Thevenin voltage (V_{TH}) is equal to V_2 in Fig. 1-8*a*. Notice that in Fig. 1-8*b*, R_3 is not shown connected to V_{EE}. This was done because it is the voltage drop across R_2 that provides bias for the base of Q_1, and V_{EE} is outside the loop in which the bias is developed (dotted line in Fig. 1-8*a*).

EXAMPLE 1-4

Determine I_T for the circuit in Fig. 1-8*a*, given that $-V_{EE} = -12$ V, $R_1 = 10$ kΩ, $R_2 = 2.2$ kΩ, $R_3 = 1.8$ kΩ, and $\beta = 100$.

Solution

$R_{TH} = 10\text{ k}\Omega \parallel 2.2\text{ k}\Omega$

$\quad\quad = 1.8\text{ k}\Omega$

$V_{TH} = V_{EE} \times \dfrac{2.2\text{ k}\Omega}{2.2\text{ k}\Omega + 10\text{ k}\Omega}$

$\quad\quad = 2.16\text{ V}$

Since $\beta R_3 \gg R_{TH}$ and $I_T = I_C \cong I_E$, the effects of the base resistance and base current may be neglected, producing

$I_T = \dfrac{2.16\text{ V} - 0.7\text{ V}}{1.8\text{ k}\Omega}$

$\quad = 811\text{ μA}$

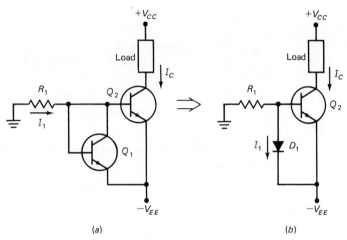

Fig. 1-9 Current mirror. (a) Actual circuit. (b) Equivalent circuit.

The Current Mirror

Figure 1-9 shows the schematic for a BJT current source called a *current mirror*. In Fig. 1-9a, Q_1 is connected as a diode. The effective equivalent of this circuit is shown in Fig. 1-9b. Proper operation of this circuit relies on close matching between Q_1 and Q_2. In designing integrated circuits, it is relatively easy for manufacturers to produce transistors with closely matched characteristics. In order to understand the operation of the current mirror, we assume that the transconductance curves for Q_1 (or D_1) and Q_2 are identical, as shown in Fig. 1-10. Since the base of Q_2 presents a very light load on R_1, compared with diode-connected Q_1, current I_1 may be approximated using

$$I_1 \cong \frac{V_{EE} - V_{BE}}{R_1} \tag{1-8}$$

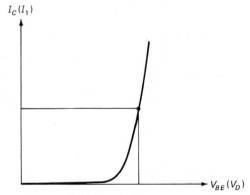

Fig. 1-10 Transconductance curves for matched transistors used to form the current mirror.

Note that V_{BE} appears only once in Eq. (1-8). This is because Q_1 is in parallel with the B-E junction of Q_2, and hence, by definition, the voltage drops are equal. Now, referring back to Fig. 1-10, it can be seen that since the transconductance curves for Q_1 and Q_2 are identical, I_C for Q_2 must equal I_1. The collector current in Q_2 mirrors that of Q_1. Assuming that Q_2 is not saturated, I_C will be independent of the actual collector load. For our purposes, the load represents the emitters of a differential pair.

EXAMPLE 1-5

Determine I_T, I_{C1}, I_{C2}, V_{CE1}, V_{CE2}, and V_{CE3} for the circuit shown in Fig. 1-11.

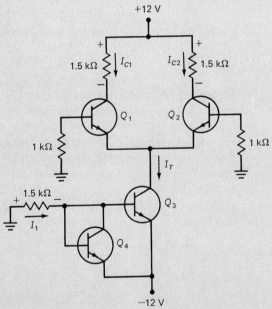

Fig. 1-11 Circuit for Example 1-4.

Solution

The current levels are found using Eqs. (1-7) and (1-5):

$$I_T = I_1 = \frac{12\text{ V} - 0.7\text{ V}}{1.5\text{ k}\Omega}$$

$$= 7.53\text{ mA}$$

$$I_{C1} = I_{C2} = \frac{I_T}{2}$$

$$= 3.77\text{ mA}$$

In order to determine the requested voltages, it is necessary to sum voltages from one supply node to the other. Since both sides of the diff amp are identical, we need only analyze one side.

Writing the KVL equation for the left half of the circuit, we obtain

$$0 = V_{CC} + V_{EE} - I_{C1}R_{C1} - V_{CE1} - V_{CE3}$$

Since I_{C1} is known, the third term in the equation is easily determined. We also can make the usual approximation of $V_{BE} \cong 0.7$ V. This leaves only the C-E voltage terms in the equation as unknowns. The values of these terms are easily determined if we assume that base bias currents for the transistors are negligibly small (that is, that the base resistors produce approximately 0 V drop). This effectively places the bases of Q_1 and Q_2 at ground potential (0 V). In this case,

$$V_{CB1} = V_{CC} - I_C R_C$$
$$= 12 \text{ V} - 5.66 \text{ V}$$
$$= 6.34 \text{ V}$$

We can determine V_{CB3} if we realize that the base of Q_3 is 0.7 V more positive than V_{EE}. This produces $V_{CB3} = 11.3$ V. In order to determine V_{CE} values for the various transistors, we use the fundamental relationship

$$V_{CE} = V_{CB} + V_{BE} \tag{1-9}$$

Applying Eq. (1-9), we obtain

$$V_{CE2} = V_{CE1} = 7.04 \text{ V}$$

Now, solving the original KVL equation for V_{CE3}, we obtain

$$V_{CE3} = 24 \text{ V} - 5.66 \text{ V} - 7.04 \text{ V}$$
$$= 11.3 \text{ V}$$

One of the advantages of the current mirror is that temperature fluctuations that cause changes in Q_2's parameters will cause identical changes in the parameters of Q_1. For example, should the ambient temperature increase, causing a decrease in V_{BE} for Q_2, the value of V_{BE} for Q_1 will also decrease, reducing the bias on Q_2 and hence stabilizing I_T. The thermal stability of the current mirror is greater than that of the circuits in Figs. 1-7 and 1-8. Also, in the manufacture of integrated circuits, it is easier to fabricate transistors and diodes than resistors, especially higher-value resistors ($R > 10-20$ kΩ). This is another reason why active current sources are used in differential amplifier design.

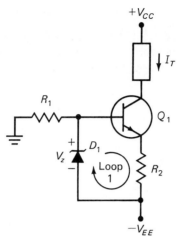

Fig. 1-12 Basic zener-biased current source.

The Zener-Biased Current Source

There are several alternatives to the current mirror. One useful current source uses a zener diode in its biasing circuit, as shown in Fig. 1-12. Zener diodes produce very stable and accurate voltage drops, and they have very low dynamic resistances when operated in reverse breakdown. This means that the bias voltage applied to Q_1 in Fig. 1-12 will remain constant and the Thevenin resistance at the base of Q_1 will be very low. Finally, an added feature of this circuit is that because of the constant voltage drop across the zener diode, regardless of supply voltage variation, I_T will also remain relatively constant if V_{EE} varies. The net effect is an extremely stable current source.

In Fig. 1-12, the two factors that determine I_T are V_z and the value of R_2. Assuming that $V_{EE} > V_z$ (a necessary condition), the following equation for I_T can be developed by summing voltage drops around loop 1:

$$I_T = \frac{V_z - V_{BE}}{R_2} \tag{1-10}$$

Notice that V_{EE} does not enter into Eq. (1-10), because it is outside the E-B junction loop.

For design purposes, we must ensure that the zener diode carries enough current to operate beyond its zener knee. This requirement can be met by choosing R_1 such that $I_z = I_{zT}$, where I_{zT} is the zener test current, as specified by the manufacturer. In equation form, this is written

$$I_{zT} = I_z = \frac{V_{EE} - V_z}{R_1} \tag{1-11}$$

EXAMPLE 1-6

Determine the values for R_1 and R_2 in Fig. 1-13 that are required to give $I_T = 1$ mA. Use standard resistor values.

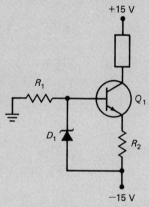

D_1: 1N753; $V_z = 6.2$ V; $I_{zT} = 20$ mA

Fig. 1-13 Circuit for Example 1-5.

Solution

R_1 is chosen to produce $I_z = 20$ mA, using Eq. (1-11).

$$R_1 = \frac{15\text{ V} - 6.2\text{ V}}{20\text{ mA}}$$

$$= 440\ \Omega \quad (\text{use } 470\ \Omega)$$

Now, using Eq. (1-10), the value of R_2 may be found as follows:

$$R_2 = \frac{V_z - V_{BE}}{I_T}$$

$$= 5.5\text{ k}\Omega \quad (\text{use } 5.6\text{ k}\Omega)$$

Additional Current Source Configurations

Often, it is necessary to use multiple current sources in a given circuit design. One possible design for such a circuit is the variation of the current mirror shown in Fig. 1-14. Although this circuit looks formidable, it is really quite easy to analyze. First of all, the solid line bisecting the transistors represents a common connection of the bases of Q_2, Q_3, and Q_4 to diode-connected transistor Q_1. Even though the three bases are in parallel, they still present a light load at the junction of R_1 and Q_1, because of beta multiplication of

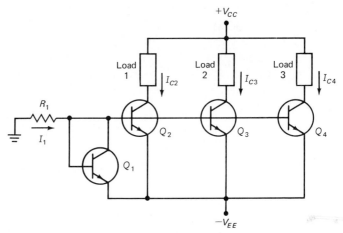

Fig. 1-14 Multiple current sources biased by a single diode-connected transistor.

their emitter resistances. Since this is a current-mirror-type circuit, all of the transistors must be matched for correct operation. This being the case, it can be shown that $I_{C2} = I_{C3} = I_{C4} \cong I_1$, where I_1 is found using Eq. (1-8).

Another useful active current source is shown in Fig. 1-15. In this circuit, the voltage drop across the two series-connected base diodes biases the transistor into conduction. To determine I_T, we apply KVL around the B-E loop, which yields the following equation:

$$0 = 2V_D - V_{BE} - I_T R_2$$

where $V_D = V_{BE} = 0.7$ V

Now, solving for I_T yields

$$I_T = V_D/R_2 = \frac{V_{BE}}{R_2} \tag{1-12}$$

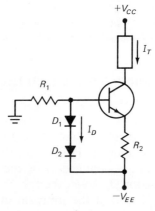

Fig. 1-15 Alternative current source design.

Notice that R_1 does not enter into Eq. (1-12). Again, this is because it is outside the loop that is used to determine the biasing voltage on the transistor. Although the value of R_1 is not critical, it should be chosen such that the diode current is approximately equal to I_T. This helps to ensure that the diodes will be operating at about the same point on their transconductance curves, and at the same junction temperature as the transistor.

EXAMPLE 1-7

Determine the values for R_1 and R_2 in Fig. 1-15 that are required to give $I_T = 5$ mA. Assume that $V_{CC} = 15$ V and $V_{EE} = -15$ V.

Using Eq. (1-12), $R_2 = 0.7$ V/5 mA $= 140\ \Omega$ (use 150 Ω). Now, R_1 is selected such that $I_D = 5$ mA. This resistance is determined by

$$R_1 = \frac{V_{EE} - 2V_D}{I_T} = \frac{13.6\text{ V}}{5\text{ mA}}$$

$\qquad = 2.72\text{ k}\Omega \qquad$ (use 2.7 kΩ)

1-3 ♦ AC ANALYSIS OF THE DIFFERENTIAL AMPLIFIER

There are four differential amplifier configurations: (1) single-ended-input–single-ended-output, (2) single-ended-input–differential-output, (3) differential-input–differential-output, and (4) differential-input–single-ended-output. Each configuration has certain characteristics that make it useful in a given situation. All four types are discussed in this section.

Single-Ended-Input–Single-Ended-Output

Perhaps the easiest of the differential amplifier connections to analyze is that shown in Fig. 1-16, the single-ended-input–single-ended-output configuration. This circuit is redrawn in Fig. 1-17a and b in its ac equivalent forms to make its operation easier to visualize. It can be seen that Q_1 is operated as an emitter follower, while Q_2 is in the common-base configuration. The emitter current source has been replaced with its equivalent internal resistance (ideally), an open circuit. The signal source v_S and its internal resistance R_S are also included in the circuit. For the time being, let us assume that R_S is negligibly small, and hence that $v_{in} = v_S$.

The ac output voltage of the amplifier, under no-load conditions, is given by

$$v_o = i_{C2}R_C \qquad (1\text{-}13)$$

where $\quad i_{C2} = i_{E2}$ (assuming $\alpha = 1$)

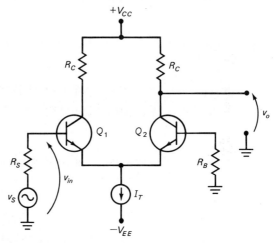

Fig. 1-16 Single-ended-input–single-ended-output differential amplifier configuration.

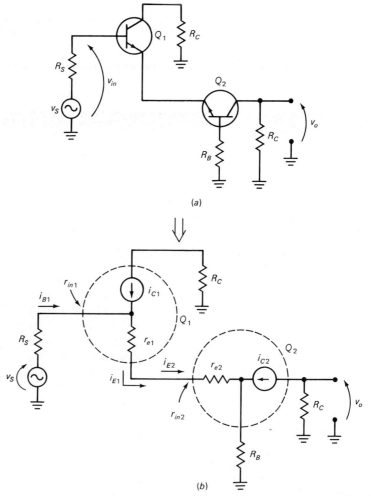

Fig. 1-17 (a) Differential amplifier redrawn to emphasize transistor configurations. (b) Equivalent small-signal model.

The input resistance to common-base amplifier Q_2 is given by

$$r_{in2} = r_{e2} + \frac{R_B}{\beta} \qquad (1\text{-}14a)$$

where r_e, the dynamic resistance of the B-E junction, is given by

$$r_e = \frac{0.026\ V}{I_E}.$$

However, in most circuits, $R_B/\beta \ll r_e$, and therefore we may simplify Eq. (1-14a) to

$$r_{in2} \cong r_{e2} \qquad (1\text{-}14b)$$

For emitter-follower transistor Q_1,

$$i_{B1} = \frac{v_{in}}{r_{in1}} \qquad (1\text{-}15)$$

where $\quad r_{in1} = \beta(r_{e1} + r_{in2}) \qquad (1\text{-}16)$

Substituting Eq. (1-16) into Eq. (1-15) and multiplying by β yields

$$i_{E2} = i_{E1} = i_{C1} = \frac{v_{in}}{r_{e1} + r_{e2}} \qquad (1\text{-}17)$$

Now, substituting Eq. (1-13) into Eq. (1-17) gives

$$v_o = \frac{R_C v_{in}}{r_e + r_{in2}} \qquad (1\text{-}18)$$

The net output voltage would be the algebraic sum of Eq. (1-18) and the quiescent collector voltage V_{OQ}. This will be explored in more detail in a later section.

By definition, $A_v = v_o/v_{in}$; therefore, neglecting the V_{OQ} component for now and dividing both sides of Eq. (1-18) by v_{in} yields

$$A_v = \frac{R_C}{r_e + r_{in2}} \qquad (1\text{-}19)$$

At this point, it may be noted that since both Q_1 and Q_2 of the differential pair are operating at the same quiescent current level, $r_{e1} = r_{e2}$. Hence, we may write

$$A_v = \frac{R_C}{2r_e} \qquad (1\text{-}20)$$

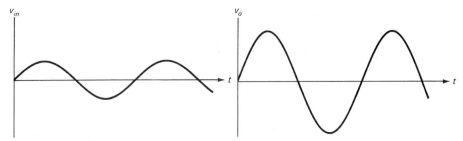

Fig. 1-18 Input-output phase relationship for Fig. 1-16.

The reader should also keep in mind that when a load is applied to the output of the amplifier, the voltage gain equation must be modified accordingly. If we refer to a given load as R_L, Eq. (1-20) would become

$$A_v = \frac{r_C}{2r_e} \qquad (1\text{-}21)$$

where $r_C = R_C \parallel R_L$

The phase relationships between v_{in} and v_o are shown in Fig. 1-18. This amplifier configuration produces an output signal that is in phase with the input. This is reasonable, considering that neither the common-base nor the emitter-follower configuration, both of which comprise this circuit, produces phase inversion.

The overall input resistance, as seen looking into the base of Q_1, can be derived from Fig. 1-17, producing the approximate relationship

$$r_{in} = r_{in1} = \beta 2 r_e \qquad (1\text{-}22)$$

EXAMPLE 1-8

Determine the voltage gain and input resistance of the amplifier in Fig. 1-19, given $R_C = 10 \text{ k}\Omega$ and $I_T = 1.5$ mA. Assume that R_B is negligible and that $\beta = 100$.

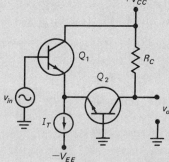

Fig. 1-19 Circuit for Example 1-8

Solution

Under quiescent conditions, the emitter currents are equal, which yields $I_{E1} = I_{E2} = 0.75$ mA; therefore,

$$r_{e1} = r_{e2} = \frac{26 \text{ mV}}{0.75 \text{ mA}} = 34.7 \text{ }\Omega$$

and

$$A_v = R_C/2r_e = \frac{10 \text{ k}\Omega}{69.4 \text{ }\Omega} = 144.1$$

The input resistance is found by applying Eq. (1-22):

$$r_{in} = \beta 2r_e = 100 \times 69.4 \text{ }\Omega = 6.94 \text{ k}\Omega$$

It can now be shown how the input resistance of the amplifier affects the overall output of the circuit. In the circuit of Example 1-8, if the signal source has a relatively low internal resistance, say $R_S = 50\Omega$, with $v_S = 10$ mV, then v_{in} will very nearly equal 10 mV. More precisely, v_{in} is given by Eq. (1-23), which applies to any amplifier.

$$v_{in} = v_S \frac{r_{in}}{r_{in} + R_S} \tag{1-23}$$

Using the values of Example 1-8, we find $v_{in} = v_S \times 0.99 = 9.9$ mV. The actual value of v_o would be given by $A_v \times 9.9$ mV $= 144.1 \times 9.9$ mV $= 1.43$ V. If R_S is high relative to r_{in}, such as in $R_S = 4$ kΩ, then $v_{in} = v_S \times 0.63 = 6.3$ mV. This produces $v_o = 0.91$ V, which is much lower than might be expected.

The net gain of an amplifier accounting for source loading can be calculated, if the source resistance and the input resistance to the amplifier are known, by combining Eq. (1-23) and the gain equation for the given amplifier. For the single-ended-input–single-ended-output circuit, this expression is

$$A_v = \frac{r_{in} r_C}{2r_e(R_S + r_{in})} \tag{1-24}$$

The single-ended-input–single-ended-output configuration is not commonly used in practice. However, the analysis that was just performed is very useful in the study of the other three differential amplifier configurations. This is demonstrated next.

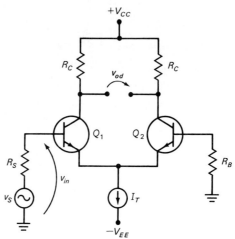

Fig. 1-20 Diff amp connected for single-ended-input–differential-output operation.

Single-Ended-Input–Differential-Output

The single-ended-input–differential-output configuration is shown in Fig. 1-20. In this circuit, the output voltage is sampled across the collectors of the differential pair. This differential amplifier configuration can be used to drive a floating (ungrounded) load.

The ac analysis of Fig. 1-20 will be performed in two parts. First, let us determine the response of v_{C1} to v_{in}. Again, we shall assume that $v_{in} = v_S$, for simplicity. The ac equivalent for Fig. 1-20 is shown in Fig. 1-21a. Since we are only interested in what happens at the collector of Q_1 at this time, we shall use the model in Fig. 1-21b, where Q_2 is replaced by its input resistance as seen by Q_1.

Transistor Q_1 is operating as a common-emitter amplifier. It is shown in basic transistor circuit texts that the voltage gain of the common-emitter amplifier is closely approximated by the equation

$$A_v = \frac{-r_C}{r_e + r_E} \qquad (1\text{-}25)$$

where r_C is the ac resistance seen by the collector, r_e is the dynamic resistance of the emitter, and r_E is the external ac resistance present at the emitter.

It is also important to notice that the common-emitter configuration exhibits phase inversion of v_o relative to v_{in}; therefore, v_{C1} will be 180° out of phase with v_{in}. This is a very important point to remember. Applying Eq. (1-25) to Fig. 1-21b, we find that in this case, $r_E = r_{in2} = r_{e2}$. Recall, however, that it was shown previously that $r_{e1} = r_{e2}$ for the differential amplifier. Hence, we may write

$$A_v = \frac{-r_C}{2r_e} \qquad (1\text{-}26)$$

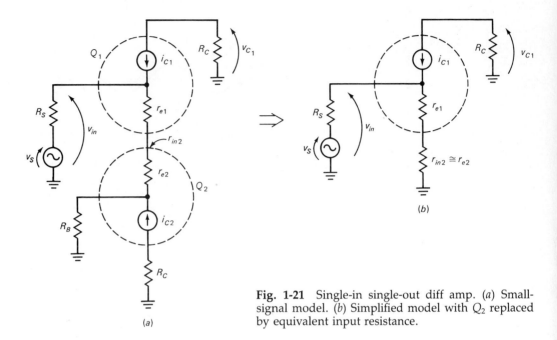

Fig. 1-21 Single-in single-out diff amp. (*a*) Small-signal model. (*b*) Simplified model with Q_2 replaced by equivalent input resistance.

Now, multiplying v_{in} by Eq. (1-26) gives us

$$v_{C1} = \frac{-v_{in}r_C}{2r_e} \tag{1-27}$$

The input resistance, as seen by the source, looking into the base of Q_1 is again given by Eq. (1-22). This should be readily apparent from Fig. 1-21*b*. This completes the first half of the analysis.

To determine the response of the collector of Q_2 to v_{in}, we refer back to Fig. 1-17*b*. Figure 1-17*b* is the model for Fig. 1-20 when we are determining v_{C2}. In fact, the entire second part of the ac analysis of Fig. 1-20 is identical to that of Fig. 1-16. Using this knowledge, we can combine Eqs. (1-20) and (1-26) into a single expression as follows:

Since v_{od} is the difference between v_{C1} and v_{C2}, we may write

$$v_{od} = v_{C2} - v_{C1} \tag{1-28}$$

Substituting Eqs. (1-20) and (1-26) into (1-28),

$$v_{od} = \frac{v_{in}r_C}{2r_e} - \frac{-v_{in}r_C}{2r_e} \tag{1-29}$$

The inversion of v_{C1} makes the right-hand term of Eq. (1-29) negative, which allows us to write

$$v_{od} = \frac{v_{in}r_C}{2r_e} + \frac{v_{in}r_C}{2r_e} \tag{1-30}$$

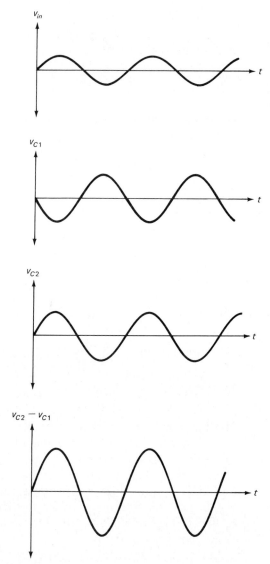

Fig. 1-22 Input-output phase relationships for the single-ended-input–differential-output amplifier.

Dividing both sides by v_{in} gives

$$A_v = \frac{v_{od}}{v_{in}} = \frac{r_C}{2r_e} + \frac{r_C}{2r_e} \tag{1-31}$$

Finally, combining the terms on the right-hand side yields

$$A_v = \frac{r_C}{r_e} \tag{1-32}$$

The sign of the voltage gain depends on which collector terminal is being used as the reference point. The important point to observe for now is that the voltage gain of the single-ended-input–differential-output amplifier is twice that of an equivalent single-ended-input–single-ended-output type. The waveforms present in the circuit of Fig. 1-20 are shown in Fig. 1-22.

EXAMPLE 1-9

Determine the peak-to-peak output voltage for the circuit in Fig. 1-20, given $I_T = 2$ mA, $R_C = 6.8$ kΩ, and $v_{in} = 8 \sin 1000t$ mV.

Solution

The emitter currents for Q_1 and Q_2 are equal, and are determined to be

$$I_{E1} = I_{E2} = \frac{I_T}{2} = 1 \text{ mA}$$

The dynamic resistances of the emitters are found next.

$$r_{e1} = r_{e2} = \frac{26 \text{ mV}}{1 \text{ mA}} = 26 \text{ }\Omega$$

Now, the voltage gain is calculated, using Eq. (1-32). Since no particular load resistance was specified, it is assumed that R_L is infinite (an open circuit).

$$A_v = \frac{r_c}{r_e} = \frac{6.8 \text{ k}\Omega}{26 \text{ }\Omega} = 261.5$$

Since v_{in} was given in trigonometric form, its peak value (8 mV) was specified in the expression. This means that the peak-to-peak amplitude of v_{in} is 16 mV. The peak-to-peak amplitude of v_o is given by $v_o = A_v v_{in} = 261.5 \times 16$ mV $= 4.18$ V$_{P-P}$.

Differential-Input–Differential-Output

The differential-input–differential-output configuration is shown in Fig. 1-23. The output of this curcuit is proportional to the instantaneous difference between v_2 and v_1. The proportionality constant is the differential voltage gain A_d. As can be seen, the output signal is equal to the (instantaneous) difference between v_{C1} and v_{C2}.

In order to derive a useful expression for A_d, we will determine the collector voltages v_{C1} and v_{C2} for each source independent of the other, using the superposition theorem. Let us arbitrarily start the analysis assuming that source 1 is active. The voltage source v_{S2} is "killed" and replaced with its

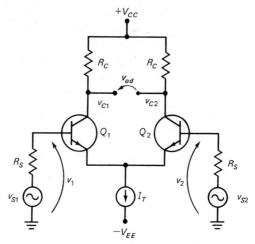

Fig. 1-23 Differential-input–differential-output configuration.

internal resistance R_S. The ac equivalents for these conditions are shown in Fig. 1-24.

The differential output $v_{o(1)}$ due to source 1 is determined in the same manner as was v_{od} for Fig. 1-20. This gives the equation

$$v_{o(1)} = v_{C2} - v_{C1}$$

We may now substitute the identities for v_{C2} and v_{C1}, as determined previously for Eq. (1-31), producing

$$v_{o(1)} = v_1 \frac{R_C}{2r_e} - v_1 \frac{-R_C}{2r_e} \tag{1-33}$$

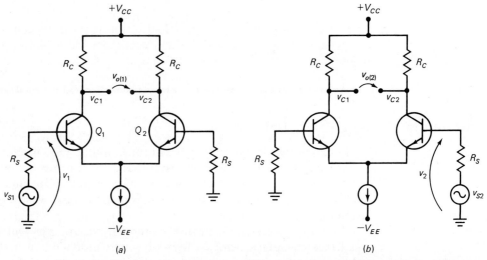

Fig. 1-24 Application of the superposition theorem. (a) Determining response to source v_{s1}. (b) Response to source v_{s2}.

Equation (1-33) could also be expressed as

$$v_{o(1)} = v_1 \frac{R_C}{r_e} \qquad (1\text{-}34)$$

The differential output caused by source 2 is determined in a similar manner. For consistency, the terms on the right-hand side of the equation below are written in the same form as in the derivation of $v_{o(1)}$.

$$v_{o(2)} = v_{C2} - v_{C1}$$

Substituting identities into the right-hand side yields

$$v_{o(2)} = -v_2 \frac{R_C}{2r_e} - v_2 \frac{R_C}{2r_e} \qquad (1\text{-}35)$$

Equation (1-35) can also be expressed as

$$v_{o(2)} = -v_2 \frac{R_C}{r_e} \qquad (1\text{-}36)$$

The overall differential output of the amplifier is the instantaneous algebraic sum of $v_{o(1)}$ and $v_{o(2)}$. That is,

$$v_{od} = v_{o(1)} + v_{o(2)} \qquad (1\text{-}37)$$

Combining Eqs. (1-34) and (1-36) and substituting into Eq. (1-37) yields

$$v_{od} = v_1 \frac{R_C}{r_e} - v_2 \frac{R_C}{r_e} \qquad (1\text{-}38a)$$

$$= (v_1 - v_2) \frac{R_C}{r_e} \qquad (1\text{-}38b)$$

where R_C/r_e is the differential gain A_d of the amplifier.

It is possible that we could have performed this analysis of the amplifier using $v_2 - v_1$ to define the net differential input voltage. Here, the choice is arbitrary. However, it will be seen later that most of the time, in a multiple-stage amplifier, either one form or the other will be correct. The correct form depends on the actual details of the circuit under study. This being the case, let us define the differential input voltage (either $v_2 - v_1$ or $v_1 - v_2$) by the term v_{id}.

The input resistance to the differential-input–differential-output amplifier is defined as the resistance seen by either source while driving the input. Assuming that voltage sources are connected to both inputs, this means that the input resistance may again be approximated by Eq. (1-22).

EXAMPLE 1-10

Determine V_{od}, V_{C1}, V_{C2}, and r_{in} for Fig. 1-25.

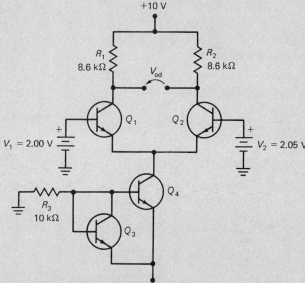

Fig. 1-25 Circuit for Example 1-10.

Solution

Analyzing the current mirror yields

$$I_T = \frac{V_{EE} - V_{BE}}{R_3} = \frac{9.3 \text{ V}}{10 \text{ k}\Omega}$$

$$= 930 \text{ }\mu\text{A}$$

Therefore, $I_{C1} = I_{C2} = 465$ µA. The quiescent collector voltages are now determined, applying the standard technique of killing the input voltage sources:

$$V_{CQ1} = V_{CQ2} = V_{CC} - I_C R_c$$

$$= 10 \text{ V} - 4 \text{ V}$$

$$= 6 \text{ V}$$

Now, calculating emitter resistances,

$$r_{e1} = r_{e2} = \frac{26 \text{ mV}}{465 \text{ }\mu\text{A}}$$

$$= 56 \text{ }\Omega$$

The differential gain is given by

$$A_d = R_c/r_e$$
$$= \frac{8.6 \text{ k}\Omega}{56 \text{ }\Omega}$$
$$= 153.6$$

The differential output is the product of A_d and the difference between the input voltages; therefore

$$V_{od} = A_d \times V_{id}$$
$$= 7.68 \text{ V}$$

It can be shown that the differential output will be divided evenly between the two collectors. That is, V_{CQ1} and V_{CQ2} will change by equal amounts in opposite directions. In this case, since $V_2 > V_1$, V_{CQ2} will decrease by 3.84 V ($V_{od}/2$) and V_{CQ1} will increase by 3.84 V. This produces $V_{C1} = 9.84$ V and $V_{C2} = 2.16$ V. Notice that Q_1 is very close to cutoff.

The collector voltages V_{C1} and V_{C2} in Example 1-10 could have been found by applying Eq. (1-27) to the amplifier in two steps, first for V_1 acting alone and then for V_2 acting alone, then applying superposition to determine the collector voltages. This method would produce outrageously large intermediate collector voltages, however, and therefore was not used in the example. The reader may wish to try this analysis method just to be reassured that it will indeed produce the same results as were found in Example 1-10.

Differential-Input–Single-Ended-Output

Perhaps the most generally useful differential amplifier circuit is the differential-input–single-ended-output configuration. Such an amplifier is shown in Fig. 1-26. Although the collector of Q_2 is used to provide the output voltage, the collector of Q_1 could have been used just as well. The choice is usually somewhat arbitrary.

The ac analysis of Fig. 1-26 is quite easily performed at this time, because the basic groundwork has already been laid in the analysis of the previous three configurations. By inspection, it can be seen that the output voltage that is developed in response to v_1 will be given by

$$v_{o(1)} = v_1 \frac{R_C}{2r_e} = v_1 \frac{R_2}{2r_e} \qquad (1\text{-}39a)$$

The output voltage that is developed in response to v_2 is

$$v_{o(2)} = v_2 \frac{-R_C}{2r_e} = v_2 \frac{-R_2}{2r_e} \tag{1-39b}$$

Using superposition and a little algebraic manipulation, the output voltage is found to be

$$v_o = \frac{R_2}{2r_e}(v_1 - v_2)$$

$$= A_d(v_1 - v_2) \tag{1-40a}$$

However, recall that the collector may well be at some quiescent voltage other than 0 V; therefore, we may write

$$v_o = v_{oQ} + A_d(v_1 - v_2) \tag{1-40b}$$

EXAMPLE 1-11

Refer to Fig. 1-26.

(a) Determine the quiescent output voltage V_{oQ}.
(b) Write the expression for and sketch v_o for $v_1 = 20 \sin \omega t$ mV and $v_2 = V_2 = +15$ mV.

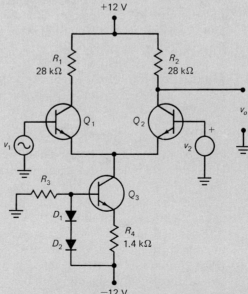

Fig. 1-26 Example of the differential-input–single-ended-output configuration.

Solution

(a) The current source, Q_3, produces $I_T = 500\ \mu A$, which in turn produces $I_{C1} = I_{C2} = 250\ \mu A$. Using KVL, the quiescent collector voltage, which is also V_{OQ}, is found to be

$$V_{OQ} = V_{CC} - I_C R_C$$
$$= 12\ V - 7\ V$$
$$= 5\ V$$

(b) The differential gain of the amplifier is

$$A_d = \frac{R_C}{2r_e}$$
$$= \frac{28\ k\Omega}{208\ \Omega}$$
$$= 134.6$$

Now, since one of the inputs is a dc voltage and the other is a sinusoid, it is convenient to determine the net output voltage using superposition. Starting with v_1 active,

$$v_{o(1)} = A_d \times 20\ \sin \omega t\ mV$$
$$= 2.7\ \sin \omega t\ V$$

The output signal component due to V_2 is

$$V_{o(2)} = -A_d \times 15\ mV$$
$$= -2.0\ V$$

Notice that V_{o2} is inverted because of the common-emitter operation of Q_2 for this input. The net output is now given by

$$v_o = V_{oQ} + v_{o(1)} + V_{o(2)}$$
$$= 5\ V + 2.7\ \sin \omega t\ V - 2\ V$$
$$= 3 + 2.7\ \sin \omega t\ V$$

This voltage is illustrated in Fig. 1-27.

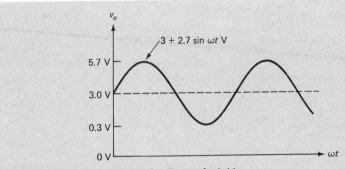

Fig. 1-27 Output voltage for Example 1-11

EXAMPLE 1-12

Determine the value of R_C required in Fig. 1-26 to produce (a) $A_d = 100$, (b) $A_d = 500$.

Solution

Since $A_d = R_C/2r_e$, we may solve for R_C, which gives

(a) $R_c = 2r_e A_d$

$\quad = 2 \times 100 \times 104$

$\quad = 20.8 \text{ k}\Omega$

(b) $R_C = 2 \times 500 \times 104$

$\quad = 104 \text{ k}\Omega$

Solution **a** is valid; however, solution **b** is not realizable. Consider that in order for Q_2 to operate in its active region, its C-B junction must be reverse-biased. If $R_C = 104$ kΩ, then at $I_C = 250$ μA, V_C would be calculated to be -14 V. Obviously, the transistor would be heavily saturated, since with the base at 0 V, the lowest value that V_C could possibly attain (assuming that the transistor is not defective) is about -0.7 V.

Common-Mode Response

It is most important that the reader understand that the output of an amplifier with differential inputs is proportional to the difference between the two input voltages. This implies that if $v_{id} = 0$ (that is, if $v_1 = v_2$), then, ideally, $v_{od} = 0$ V. Such an input condition is termed a *common-mode input voltage* v_{ICM}. Unfortunately, common-mode inputs cause an undesirable change in the output of the amplifier.

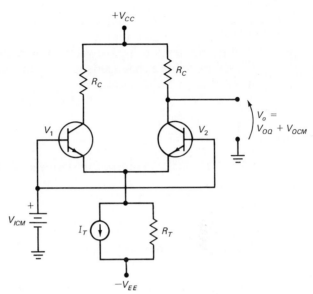

Fig. 1-28 Differential amplifier with common-mode input voltage and nonideal emitter current source.

A reasonable question that might be asked at this point is, what could cause the amplifier to produce an output V_{OCM} in response to a common-mode input? The primary cause of the common-mode response of the differential amplifier is the finite internal resistance of the emitter current source. Ideally, the internal resistance of a current source is infinite; however, this is impossible to attain in practice. A differential amplifier with a common-mode input and nonideal current source is shown in Fig. 1-28.

Common-mode input voltage is defined as the average of the two voltages applied to the differential amplifier:

$$V_{ICM} = \frac{V_1 + V_2}{2} \tag{1-41}$$

The gain of the differential amplifier for common-mode inputs is given by the approximate relationship

$$A_{CM} \cong \frac{-R_C}{2R_T} \tag{1-42}$$

It is desirable to make A_{CM} as low as possible. This is one of the reasons that active current sources are used in differential amplifier circuits. The very high r_{CE} (R_T) values that occur in active current sources make the denominator of Eq. (1-42) very large.

Applying the information just presented to the circuit of Example 1-10, we find that a common-mode input of $+2.025$ V exists. Now, if we assume that Q_3 has $r_{CE} = 50$ kΩ, then the common-mode gain of the circuit is -0.28.

This means that the output will have an additional, unwanted component of −0.57 V summed into it.

A useful relationship between differential gain A_d and common-mode gain is used to define a characteristic called the *common-mode rejection ratio* (CMRR). In symbols, CMRR is given by

$$\text{CMRR} = \left|\frac{A_d}{A_{CM}}\right| \quad \text{where } A_d = \frac{R_C}{2r_e}$$

CMRR is usually expressed in decibels, as defined below:

$$\text{CMRR} = 20 \log \left|\frac{A_d}{A_{CM}}\right| \tag{1-43}$$

In general, it is desirable for CMRR to be very large.

EXAMPLE 1-13

Determine (a) A_d, (b) A_{CM}, (c) V_{OCM}, and (d) CMRR for the circuit in Fig. 1-28, given $R_C = 10$ kΩ, $V_{ICM} = 3.0$ V, $r_e = 75$ Ω, and $R_T = 100$ kΩ.

Solution

(a) The differential gain is $A_d = R_C/2r_e = 10$ kΩ/150 Ω = 66.6
(b) The common-mode gain is $A_{CM} = -R_C/2R_T = -10$ kΩ/200 kΩ = −0.05.
(c) The common-mode is $V_{OCM} = A_{CM}V_{ICM} = 0.05 \times 3.0$ V = −150 mV.
(d) The common-mode rejection ratio is CMRR = $20 \log \left|\frac{66.6}{-0.05}\right|$

$$= 20 \log 1332$$
$$= 20 \times 3.12$$
$$= 62.4 \text{ dB}$$

1-4 ♦ CASCADED DIFFERENTIAL AMPLIFIERS

In many applications, a single-stage differential amplifier cannot meet the requirements demanded. In such cases, multiple-stage amplifiers are necessary. In nearly all instances, differential amplifiers are entirely direct-coupled circuits. This complicates circuit analysis to an extent, but it also has the advantage of allowing the circuit to be used to process dc inputs, as well as ac.

1-4 CASCADED DIFFERENTIAL AMPLIFIERS

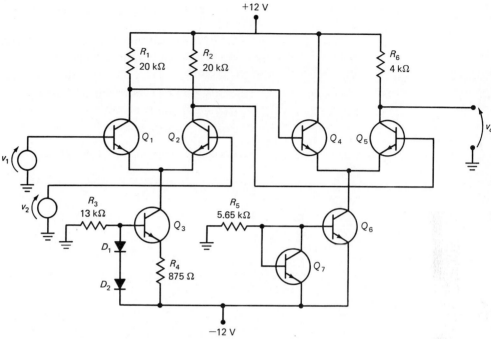

Fig. 1-29 Two-stage differential amplifier.

Figure 1-29 shows a two-stage differential amplifier. The first stage (Q_1, Q_2, Q_3, D_1, D_2, and resistors R_1, R_2, R_3, and R_4) is a differential-input–differential-output section using a two-diode active current source. The second stage is a differential-input–single-ended-output section. Overall, the circuit would be treated as a differential-input–single-ended-output amplifier. This is the most common configuration for such multiple-stage differential amplifiers. An analysis of this circuit is presented in the next example.

EXAMPLE 1-14

Determine the following values for the circuit in Fig. 1-29: I_{C3}, I_{C1}, I_{C2}, I_{C6}, I_{C4}, I_{C5}, V_{CE1}, V_{CE2}, V_{CE4}, V_{CE5}, V_{OQ}, the differential gain of stage 1 A_{d1}, the differential gain of stage 2 A_{d2}, the overall differential gain A_d, and the input resistance r_{in} as seen by either source 1 or source 2. Assume that $\beta = 100$ and $V_{BE} = 0.7$ V.

Solution

The dc analysis begins with stage 1. Calculating I_{C3}, we see that

$$I_{C3} = \frac{0.7 \text{ V}}{875 \, \Omega} = 800 \, \mu\text{A}$$

This current divides between Q_1 and Q_2, producing

$$I_{C1} = I_{C2} = \frac{I_{C3}}{2} = 400\ \mu A$$

Because of the symmetry between both sides of stage 1, we may assume that equal voltage drops will exist around each transistor. Also, the input voltage sources are killed, which places both bases at ground (0 V) potential. This yields

$$V_{CB1} = V_{CB2} = V_{CC} - I_C R_C$$
$$= 12\ V - 8\ V$$
$$= 4\ V$$

Notice here that $V_{CB1} = V_{CB2} = V_{C2}$, because the bases of the transistors are at ground potential.

Now, since $V_{BE} = 0.7\ V$, we may write

$$V_{CE1} = V_{CE2} = V_{CB} + V_{BE} = 4.7\ V$$

The dc analysis of the second stage may now be performed. First, let us determine the collector current of active current source Q_6:

$$I_{C6} = \frac{V_{EE} - V_{BE}}{R_5} = \frac{11.3\ V}{5.65\ k\Omega} = 2\ mA$$

Again, this current divides equally in the differential pair, giving

$$I_{C4} = I_{C5} = \frac{I_{C6}}{2} = 1\ mA$$

Since stage 2 is not symmetrical (the collector resistances differ), Q_4 and Q_5 will have different voltage drops existing across them. Also, note that the bases of these transistors are not at ground potential, since they are directly coupled to the collectors of Q_1 and Q_2. Based on this knowledge, the analysis proceeds as follows.

From the analysis of stage 1, it can be seen that

$$V_{B4} = V_{C1} \quad \text{and} \quad V_{B5} = V_{C2}$$

Now, since $V_{C1} = V_{C2} = 4\ V$,

$$V_{CB4} = V_{CC} - V_{B4} = 12\ V - 4\ V = 8\ V$$

and

$$V_{CB5} = V_{CC} - I_{C5} R_6 - V_{B5} = 12\ V - 4\ V - 4\ V = 4\ V$$

Once again, because all V_{BE} values are known, we may write

$$V_{CE4} = V_{CB4} + V_{BE} = 8 \text{ V} + 0.7 \text{ V} = 8.7 \text{ V}$$
$$V_{CE5} = V_{CB5} + V_{BE} = 4 \text{ V} + 0.7 \text{ V} = 4.7 \text{ V}$$

The quiescent output voltage is the collector voltage of Q_5; therefore we obtain

$$V_{OQ} = V_{C5} = V_{CC} - I_{C5}R_6 = 12 \text{ V} - 4 \text{ V} = 8 \text{ V}$$

This concludes the dc analysis of Fig. 1-29. Since stage 2 is loading stage 1, the ac analysis is best performed starting with stage 2. We shall begin by determining the dynamic emitter resistances of Q_4 and Q_5.

$$r_{e4} = r_{e5} = \frac{0.026 \text{ V}}{I_E} = \frac{0.026 \text{ V}}{1 \text{ mA}} = 26 \text{ }\Omega$$

Now, since stage 2 is a differential-input–single-ended-output circuit, the voltage gain is

$$A_{d2} = \frac{R_6}{2r_e} = \frac{4000 \text{ }\Omega}{52 \text{ }\Omega} = 76.9$$

The input resistance of stage 2 is

$$r_{in2} = 2\beta r_e = 200 \times 26 \text{ }\Omega = 5.2 \text{ k}\Omega$$

With the stage 2 ac analysis complete, a similar process is applied to stage 1.

$$r_{e1} = r_{e2} = \frac{0.026 \text{ V}}{I_E} = \frac{0.026 \text{ V}}{400 \text{ }\mu\text{A}} = 65 \text{ }\Omega$$

In determining the voltage gain of stage 1, we must take into account the loading effect of stage 2. This means that

$$r_{C1} = r_{C2} = R_C \parallel r_{in2}$$
$$= 20 \text{ k}\Omega \parallel 5.2 \text{ k}\Omega$$
$$= 4.13 \text{ k}\Omega$$

Hence

$$A_{d1} = \frac{r_C}{r_e} = \frac{4.13 \text{ k}\Omega}{65 \text{ }\Omega} = 63.5$$

The input resistance to stage 1 is now determined.

$$r_{in} = r_{in1} = 2\beta r_e = 200 \times 65\ \Omega = 13\ k\Omega$$

Finally, the overall voltage gain of the amplifier is given by the product of the individual stage gains. This produces

$$A_d = A_{d1} \times A_{d2}$$
$$= 76.9 \times 63.5$$
$$= 4883$$

The reader should be able to generalize on the analysis presented in this example and apply these techniques to other cascaded differential amplifier circuits.

There are a few more important points to consider concerning Fig. 1-29. First, the output resistance of the amplifier is approximately equal to the value of R_4. This should not be surprising, as the collector of Q_5 is acting as a current source with extremely high internal resistance. Second, the reader should verify that all C-B junctions are reverse-biased. This ensures that all transistors are operating in their active regions.

Since Fig. 1-29 is a single-ended-output diff amp, one of the inputs will act as an inverting (−) input and the other will act as a noninverting (+) input. Determining which input is which is relatively easy to do. The general procedure is as follows:

1. Assume that one input is grounded and that a positive voltage exists at the remaining input. The positive half-cycle of a sine wave or an arrow pointing up are both good symbols to use at the active input.
2. Determine the direction in which the collector voltages of the input transistors would be driven in response to the input and draw half-cycles or arrows designating this response.
3. Repeat steps 1 and 2 for the remaining stages.

These three steps are applied to the simplified two-stage differential amplifier in Fig. 1-30, providing identification of the inverting and noninverting inputs.

1-5 ♦ LEVEL-SHIFTING AND OUTPUT STAGES

It was stated earlier that differential amplifiers are nearly always direct-coupled, in order to allow the processing of dc voltage levels. Since this is the case, it is also desirable to have the quiescent output of the final stage of the differential amplifier at 0 V. That is, for $V_{id} = 0$ V, $V_o = 0$ V. Meeting this requirement necessitates the use of a level shifter.

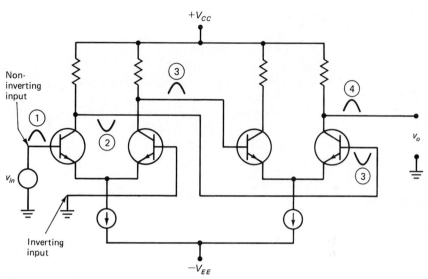

Fig. 1-30 Phase relationships for a two-stage differential amplifier identifying inverting and noninverting inputs.

Level Shifters

The idea behind a level shifter is to shift or translate one voltage level (typically V_c of a single-ended differential amplifier) to another level (typically 0 V). Perhaps the simplest level-shifting circuit is a simple emitter-follower circuit, as shown in Fig. 1-31.

The emitter current of Q_1 is given by

$$I_{E1} = \frac{V_{EE} + V_B - V_{BE}}{R_E} \tag{1-44}$$

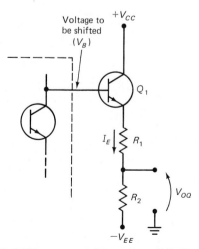

Fig. 1-31 Basic level-shifting stage using emitter-follower.

where $R_E = R_1 + R_2$

It can also be seen that

$$V_{OQ} = V_{EE} - I_E R_2 \qquad (1\text{-}45)$$

For design purposes, a desired value of I_E would be chosen, and the values required for R_1 and R_2 would be determined as follows:
Solving Eq. (1-44) for R_E,

$$R_E = \frac{V_{EE} + V_B - V_{BE}}{I_E} \qquad (1\text{-}46)$$

Now, solving Eq. (1-45) for R_2,

$$R_2 = \frac{V_{EE} + V_{OQ}}{I_E} \qquad (1\text{-}47a)$$

Normally, $V_{OQ} = 0$ V; therefore

$$R_2 = \frac{V_{EE}}{I_E} \qquad (1\text{-}47b)$$

With R_2 now known, R_1 may be determined by

$$R_1 = R_E - R_2 \qquad (1\text{-}48)$$

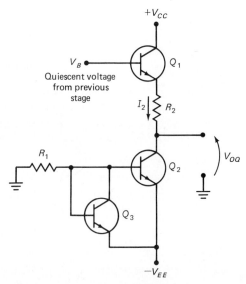

Fig. 1-32 Level-shifting stage using emitter-follower and active current source.

1-5 LEVEL-SHIFTING AND OUTPUT STAGES

Although the level shifter of Fig. 1-31 will perform adequately in many applications, it is common practice to employ an active current source in place of R_2. The circuit in Fig. 1-32 illustrates one such possibility. Even though a current mirror is used in this example, any of the other current sources that were discussed could be used in the design.

An analysis of Fig. 1-32 begins with the determination of the current I_2 that is forced by Q_2. From the designer's perspective, the current source would be designed to produce some desired value of I_2. Now, given a known value for I_2, V_{OQ} is found using

$$V_{OQ} = V_B - V_{BE} - I_2 R_2 \tag{1-49a}$$

Again, in general, $V_{OQ} = 0$ V; therefore

$$V_{OQ} = 0 \text{ V} = V_B - V_{BE} - I_2 R_2 \tag{1-49b}$$

For design purposes, we solve Eq. (1-49b) for R_2 (I_2, V_B, and V_{BE} are known), producing

$$R_2 = \frac{V_B + V_{BE}}{I_2} \tag{1-50}$$

A practical application of the preceding equations is presented in the following example.

EXAMPLE 1-15

Given $V_B = 5$ V, $V_{CC} = +12$ V, and $V_{EE} = -12$ V, determine the values necessary to produce $V_{OQ} = 0$ V for the circuit in Fig. 1-32. Choose $I_2 = 5$ mA.

Solution

The current mirror is designed such that $I_{R1} = 5$ mA using $R_1 = (12 \text{ V} - 0.7 \text{ V})/5 \text{ mA} = 2.26$ kΩ. Now, applying Eq. (1-50), $R_2 = (V_B - V_{BE})/I_2 = 4.3 \text{ V}/5 \text{ mA} = 860$ Ω.

The voltage gain of the level-shifting circuitry must be taken into account when determining the overall gain of the amplifier. Under loaded output conditions, the gain of Fig. 1-31 would be given by

$$A_v = \frac{r_2}{r_2 + R_1 + r_{e1}} \tag{1-51}$$

where $r_2 = R_2 \parallel R_L$

For the circuit of Fig. 1-32, the voltage gain is given by

$$A_v = \frac{R_L}{R_2 + r_{e1} + R_L} \tag{1-52}$$

Since level-shifting circuits are emitter followers, their voltage gains will always be somewhat less than unity. However, in general, the voltage gain of the active current source level shifter will be much closer to unity, because of the high r_{CE} of the current source transistor.

Final Output Stages

Some differential amplifier designs include an additional stage that provides a low output resistance and higher current source/sink capability. This allows relatively low-impedance loads to be driven effectively. The complementary push-pull configuration is often used in such designs.

Figure 1-33a illustrates the basic push-pull design. Transistors Q_3 and Q_4 are biased slightly into conduction by diode-connected transistors Q_1 and Q_2 and resistors R_1 and R_2. The emitter resistors provide negative feedback, stabilizing the Q point of the amplifier. Notice that the components in the upper half (Q_1, R_1, and Q_3) and the lower half (Q_2, R_2, and Q_4) form symmetrical current-mirror-type circuits, such that $I_{R1} = I_{R2} = I_{C3} = I_{C4}$. The following relationship describes these currents.

$$I_{R1} = I_{R2} = \frac{V_{EE} + V_{CC} - 2V_{BE}}{R_1 + R_2} \tag{1-53}$$

When matched transistors are used (the usual case), the total supply voltage will be divided equally between the top and bottom halves of the circuit, producing $V_{OQ} = 0$ V. This assumes that the input of the amplifier is 0 V under quiescent conditions. This condition is met if a level shifter precedes the push-pull stage.

Figure 1-33b shows a modified version of the basic push-pull amplifier. The circuit has been modified to provide overcurrent protection. Here, Q_3 and Q_4 are used to sense the load current that is being supplied by the amplifier. The currents that are shown would be produced if the input were driven positive. Under normal conditions, $I_{B5} = I_1$ and $I_{E3} = 0$. Of course the load current would be approximately $\beta \times I_{B5}$. Notice that I_L causes a voltage drop across R_E that tends to bias Q_3 into conduction. Again, under normal conditions, this voltage drop will be well below the barrier potential of Q_3. Now, should I_L increase to an unsafe level (perhaps because the output is shorted to ground), the voltage drop across R_E will increase, causing Q_3 to be driven into conduction. This shunts the base current away from Q_5, limiting I_L to a safe value. Transistor Q_3 will go into conduction when $I_L R_E \cong 0.7$ V. The lower half of the circuit operates in exactly the same manner for current flow in the opposite direction. Since Q_3 and Q_4 are in cutoff under normal conditions, their presence does not affect the biasing of the amplifier.

1-5 LEVEL-SHIFTING AND OUTPUT STAGES

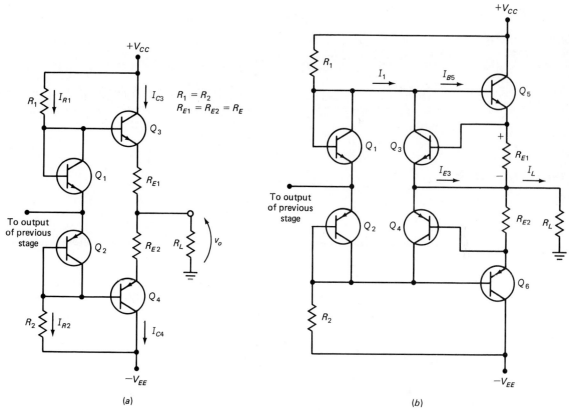

Fig. 1-33 Complementary-symmetry push-pull output stage. (a) Basic circuit. (b) Circuit with current limiting.

EXAMPLE 1-16

Determine the values for R_1, R_2, and R_E for Fig. 1-33, given $V_{CC} = 15$ V and $V_{EE} = -15$ V such that $I_{CQ} = 2$ mA, and $I_{L(\max)} = \pm 15$ mA.

Solution

The values for R_1 and R_2 are found by rearranging Eq. (1-53), giving

$$R_1 + R_2 = \frac{V_{CC} + V_{EE} - 2V_{BE}}{I_{CQ}}$$

$$= \frac{30 \text{ V} - 1.4 \text{ V}}{2 \text{ mA}}$$

$$= 14.3 \text{ k}\Omega$$

Since both halves of the circuit are symmetrical, $R_1 = R_2$; therefore

$$R_1 = R_2 = \frac{14.3 \text{ k}\Omega}{2} = 7.15 \text{ k}\Omega$$

The maximum load current is of the same magnitude for both positive and negative output excursions; therefore $R_{E1} = R_{E2}$. Now, from Ohm's law, at maximum load current (using the top half of the circuit),

$$V_{BE3} = I_{L(max)} R_E$$

Since V_{BE3} will equal 0.7 V at the current limit point,

$$R_{E1} = R_{E2} = \frac{0.7 \text{ V}}{15 \text{ mA}} = 46.6 \text{ }\Omega$$

The push-pull amplifier is a variation of the emitter follower, with a voltage gain slightly less than unity. The gain of the amplifiers in Fig. 1-33 is approximately

$$Av = \frac{R_L}{R_E + r_e + R_L} \qquad (1\text{-}54)$$

Because only one half of the circuit is active at any given time, the transistors in the opposite side will be in cutoff, presenting near-open circuit conditions to the active side of the circuit.

1-6 ◆ ADDITIONAL DIFFERENTIAL AMPLIFIER VARIATIONS

For certain applications, modifications must be made to the basic differential amplifier designs. This section will discuss a few of the more important differential amplifier variations that are used routinely.

Active Loading

An examination of the differential amplifier gain equations [Eqs. (1-21) and (1-32)] indicates that the collector resistance(s) of a given differential pair must be made very large in order to yield high gain. Using high-value resistors is often impractical for several reasons. First, high collector resistance necessitates low collector currents, which in turn increases r_e (the denominator of the gain equations). Second, recall that it was stated that high-value resistors are difficult to fabricate. All this leads to the use of active loading in the collector circuit of the differential amplifier.

Figure 1-34 presents active loading, as applied to differential amplifier design. Under quiescent conditions, Q_1 and Q_2 carry equal collector currents

1-6 ADDITIONAL DIFFERENTIAL AMPLIFIER VARIATIONS

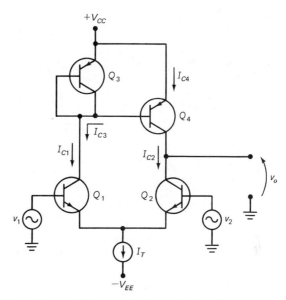

Fig. 1-34 Differential amplifier employing active loading.

($I_{C1} = I_{C2}$). Current I_{C1} is forced through the left side of a current mirror (diode-connected Q_3, and Q_4). The reference current (I_{C1}, I_{C3}) forces an equal collector current to flow in Q_4. All the collector currents are equal, and the circuit is in equilibrium.

In order to clearly see how the circuit works under signal conditions, let us assume that $v_2 = 0$ V and that V_1 is driving the base of Q_1 to some small positive potential. This base drive tends to increase I_{C1} and I_{C3}. Since Q_3 is in the reference side of the current mirror, the collector current of Q_4 will also tend to increase. However, at the same time, because I_T is constant, the collector current of Q_2 tends to decrease. Thus, Q_4 and Q_2 are acting in opposite directions. Eventually (nearly instantaneously in practice), the circuit will regain equilibrium in response to the input signal. The net result of this action is a very large voltage change at the output; in this case, v_o will change in the positive direction (the normal phase relationships still hold).

The actual amount of voltage change for a given input will depend on many factors, such as the relative junction area of the transistors. However, in order to gain an appreciation of the gain magnitudes that may be achieved using active loading, let us apply Eq. (1-20) to Fig. 1-34, assuming the following conditions: $I_T = 2$ mA and $r_{CE4} = 100$ kΩ. These conditions produce $r_e = 26$ Ω (for Q_1 and Q_2).

$$A_d = \frac{r_{CE4}}{2r_e}$$

$$= \frac{100 \text{ k}\Omega}{52 \text{ }\Omega}$$

$$= 1920$$

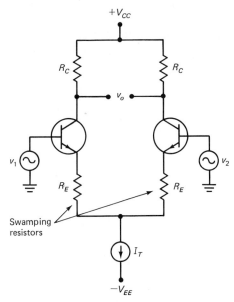

Fig. 1-35 Swamped differential amplifier.

This is an extremely high gain for a single stage of amplification. Such high stage gains are only realizable using active loading. In practice, output loading will reduce the voltage gain to a lower level than that given in the previous calculations. Also notice that the application of Eq. (1-42) would suggest that the common-mode gain of an actively loaded differential amplifier could be relatively high. While this is true, optimization of the I_T current source resistance and the reduction of r_C due to loading will offset this effect to a large extent.

Increasing Input Resistance

It is usually desirable to make the input resistance of a differential amplifier as high as possible. To this end, there are several design alternatives that are used. One way to increase input resistance is through the use of emitter swamping resistors, as shown in Fig. 1-35. An analysis of this circuit shows that

$$r_{in} = 2\beta(r_e + R_E) \tag{1-55}$$

The addition of swamping resistors also affects voltage gain, as may be quantified in the following equations:

$$A_d = \frac{R_C}{2(r_e + R_E)} \quad \text{for single-ended output} \tag{1-56}$$

$$A_d = \frac{R_C}{r_e + R_E} \quad \text{for differential output} \tag{1-57}$$

Two other effective methods of increasing input resistance are through the use of Darlington pairs and field-effect transistors (FETs), as shown in Fig. 1-36. Assuming that all transistors are matched, the input resistance of Fig. 1-36a is

$$r_{in} = 2\beta^2 r_e \tag{1-58}$$

The input resistance of the FET differential amplifier will be the r_{gs} value for the particular devices used. Typically, r_{gs} is around $10^{10}\ \Omega$. Notice also that although JFETs are shown here, MOSFETs could also be used. Darlington pairs are usually used only in the first stage of the differential amplifier, to provide high input resistance. When FETs and bipolar transistors are mixed in a design, the FETs will generally be used in the first stage, again to provide high input resistance, although some IC amplifiers are constructed using all FETs.

The choice between FET and BJT usage in differential amplifiers is determined by the characteristics that are needed in particular applications. For example, bipolar transistors can be matched very closely in V_{BE} characteristics (typically within < 1 mV), whereas the pinchoff voltages of JFETs in a differential pair may easily differ by more than 10 mV. This means that differential amplifiers with BJTs in the first stage will normally produce a differential output V_{od} that is nearer 0 V than that produced by an equivalent FET-type amp. On the other hand, since FETs require nearly zero gate bias current (typically I_G is around 10^{-12} A, which is usually several orders of magnitude lower than I_B for a BJT), the effects of slight inequalities between these bias currents and external gate resistances will be very low. This makes

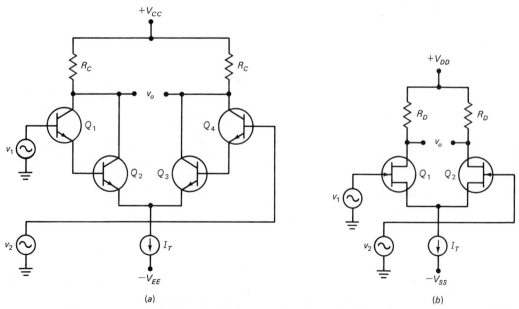

Fig. 1-36 Increasing amplifier input resistance. (a) Using darlington pairs. (b) Using JFETs.

FET input differential amplifiers better suited for operation with very-high-resistance sources. There are other pros and cons associated with each device type; however, these are two of the more important characteristics to consider.

The dc analysis of the Darlington-based differential amplifier is performed in the same manner as analysis of any other BJT differential amplifier, using Ohm's law, KVL, KCL, and basic device theory. Since this is the case, a detailed analysis of differential amplifiers using Darlington pairs is not presented, as the student should be able to apply these techniques to typical circuits. Detailed analysis of the JFET-based differential amplifier will also not be discussed here. The interested reader may wish to refer to a text on semiconductor devices for an explanation of FET operation, particularly in differential amplifier circuits.

Chapter Review

This chapter has introduced the reader to the fundamentals of differential amplifier circuit operation. The information presented provides a solid background which will enable the reader to gain deeper insight into the operation and design of linear IC-based circuits.

It was shown that differential amplifiers must be constructed using components that have closely matched characteristics. Most differential amplifiers are also designed to operate from bipolar power supplies, and direct coupling is used in their design. There are four basic configurations of the differential amplifier: single-ended-input–single-ended-output, single-ended-input–differential-output, differential-input–differential-output, and differential-input–single-ended-output.

Amplifiers that have differential inputs exhibit common-mode rejection. Common-mode rejection is a desirable characteristic. Differential amplifiers often use active current sources to provide emitter biasing. This results in a very stable Q point, and also enhances common-mode rejection. Level-shifting circuitry is normally required in the final stage of a differential amplifier to obtain a quiescent output voltage of zero. The level-shifting circuitry is normally a variation of the emitter-follower-type amplifier.

To achieve very high stage gains, active loading is often used. Active loads are current-mirror-type circuits, and they can be used effectively only in stages that use the single-ended-output configuration. FETs and Darlington pairs are also used in differential amplifier design, to produce high input resistances.

Questions

1-1. How may the voltage gain of a differential amplifier be increased without the use of very-high-value collector resistors?

1-2. State two advantages of using active current sources, as opposed to passive (resistive) current sources, to provide emitter biasing in differential amplifier design.

1-3. Given the amplifier in Fig. 1-16, state two ways in which the gain of this circuit could be increased.

1-4. Of the four basic differential amplifier configurations, which one(s) will not exhibit common-mode rejection?

1-5. Why are push-pull amplifiers sometimes used to provide the output of differential amplifiers?

1-6. For the circuit in Fig. 1-16, assuming fixed values for R_C and transistor parameters β and V_{BE}, what effect would a reduction of I_T have on the circuit?

1-7. Under ideal conditions, what is the voltage gain of an emitter-follower-type level shifter?

1-8. What identifying terms are given to the inputs of a differential-input–single-ended-output amplifier?

Problems

1-1. Determine the following values for the circuit in Fig. 1-37: I_T, I_{C1}, I_{C2}, V_{CE1}, V_{CE2}, A_d, and r_{in}. Assume that $\beta = 100$.

1-2. In Fig. 1-37, if $V_1 = 30$ mV and $V_2 = 55.8$ mV, what is the magnitude of V_{od}? Determine the values of V_{C1} and V_{C2}.

1-3. Determine A_{CM} and CMRR (in decibels) for the circuit of Fig. 1-37.

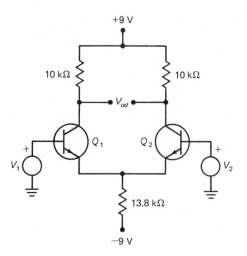

Fig. 1-37

1-4. Refer to Fig. 1-38. Determine I_{C1}, I_{C2}, V_{CE1}, V_{CE2}, V_{CE3}, V_{C1}, and V_{C2}.

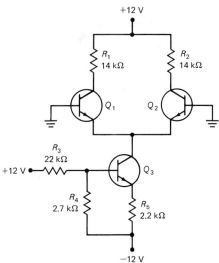

Fig. 1-38

1-5. Refer to Fig. 1-39. Determine I_{C3}, I_{C1}, I_{C2}, V_{CE1}, V_{CE2}, V_{CE3}, V_{C1}, V_{C2}, r_{in}, and A_d. Assume that $\beta = 100$.

1-6. Refer to Fig. 1-39. Given $v_1 = 200 \sin 5000t$ mV and $v_2 = V_2 = +200$ mV, sketch the waveforms that would exist at the collectors of Q_1 and Q_2. Label the peak and mean values of the waveforms.

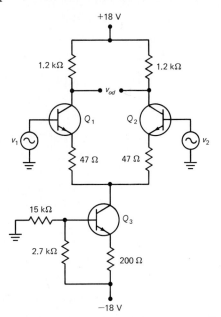

Fig. 1-39

1-7. Calculate the following values for the circuit in Fig. 1-40: I_{C4}, I_{C1}, I_{C2}, V_{CE1}, V_{CE2}, V_{CE4}, r_{in}, and A_d. Assume that $\beta = 100$.

1-8. Refer to Fig. 1-40. Determine the peak-to-peak amplitude of v_{od}, given $v_S = 60$ mV$_{P-P}$.

1-9. In Fig. 1-40, what peak-to-peak value must v_S have to produce v_{od} = 1 V_{P-P}?

1-10. Given the conditions of Prob. 1-9, sketch the waveform for v_{o2} in Fig. 1-40.

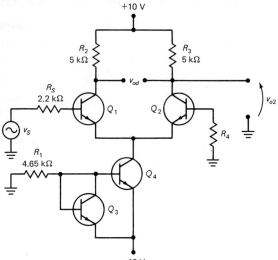

Fig. 1-40

1-11. Refer to Fig. 1-41. Determine the required values for R_1 and R_2 such that the zener current equals I_{zT} and $I_{C1} = I_{C2} = 3.0$ mA.

1-12. Refer to Fig. 1-41. Given the conditions of Prob. 1-11, what value must the collector resistors R_C have in order to produce $V_{CE1} = V_{CE2} = 5$ V?

1-13. Refer to Fig. 1-41. Given the values of R_1 and R_2 as determined in Prob. 1-11 and the collector resistor values shown, sketch the waveforms that will be present at v_{o1} and v_{o2} if $v_1 = -10 \sin 500t$ mV and $v_2 = 10 \sin 500t$ mV.

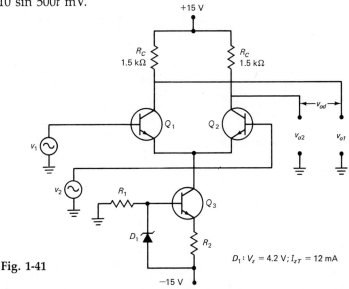

Fig. 1-41

1-14. Refer to Fig. 1-26. Assuming that all transistors have $r_{CE} = 150\ \text{k}\Omega$, $v_1 = +3.60\ \text{V}$, and $v_2 = +3.68\ \text{V}$, determine the following values: A_{CM}, V_{ICM}, V_{OCM}, and CMRR (in decibels).

1-15. Determine I_z and I_L for the circuit in Fig. 1-42.

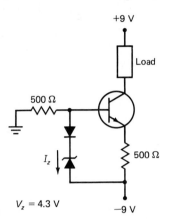

Fig. 1-42 $V_z = 4.3\ \text{V}$

1-16. Determine the following for the circuit in Fig. 1-43: I_{C3}, I_{C1}, I_{C2}, I_{C6}, I_{C4}, I_{C5}, V_{CE1}, V_{CE2}, V_{CE3}, V_{CE4}, V_{CE5}, V_{OQ}, r_{in2}, r_{in1}, A_{d2}, A_{d1}, and A_d (overall). Assume that all transistors have $\beta = 100$.

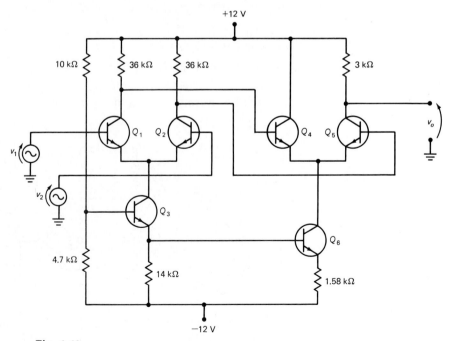

Fig. 1-43

1-17. Determine the following for the circuit of Fig. 1-44: I_{C3}, I_{C1}, I_{C2}, I_{C6}, I_{C4}, I_{C5}, V_{CE1}, V_{CE2}, V_{CE3}, V_{CE4}, V_{CE5}, V_{CE6}, V_{OQ}, r_{in2}, r_{in1}, A_{d2}, A_{d1}, and A_d (overall). Assume that all transistors have $\beta = 100$.

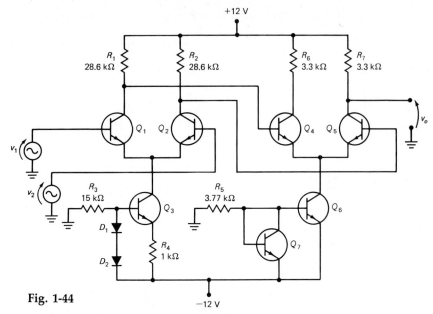

Fig. 1-44

1-18. Analyze the circuit in Fig. 1-45 and determine the following values: I_{C1}, I_{C2}, I_{C3}, I_{C5}, I_{C6}, I_{C7}, V_{CE1}, V_{CE2}, V_{CE3}, V_{CE5}, V_{CE6}, V_{CE7}, V_{OQ}, r_{in}, and A_d. Assume that all transistors have $\beta = 100$.

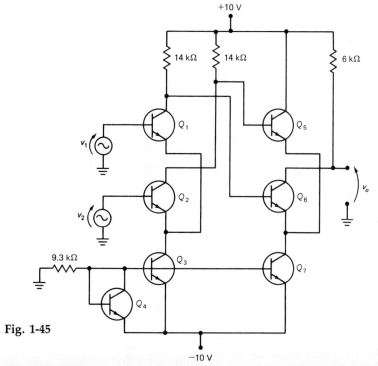

Fig. 1-45

1-19. Refer to Fig. 1-46. Determine the value for R_3 required to give $V_{OQ} = 0$ V. What is the voltage gain of the level shifter when driving a 500-Ω load?

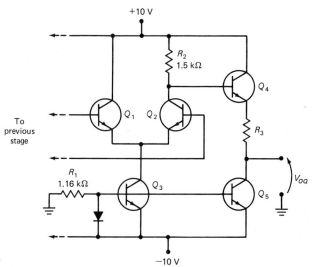

Fig. 1-46

1-20. Refer to Fig. 1-47. Determine the following values: I_{C1}, I_{C2}, I_{C5}, I_{C3}, I_{C4}, V_{CE1}, V_{CE2}, V_{CE3}, V_{CE4}, and V_{C4}. Determine the values for R_7 and R_8 required to give $V_{OQ} = 0$ V with $I_{C8} = 5.0$ mA. Determine the overall differential voltage gain (no load on output) and the input resistance r_{in}. Assume that all transistors have $\beta = 100$.

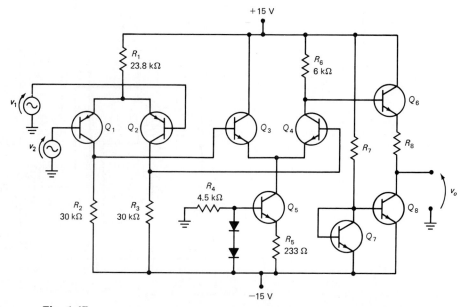

Fig. 1-47

1-21. Refer to Fig. 1-48. Determine I_{C5}, I_{C1}, I_{C2}, I_{C3}, I_{C4}, V_{CE2}, V_{CE3}, V_{CE5}, V_{OQ}, r_{in}, and A_d. Assume that $\beta = 100$ for all transistors.

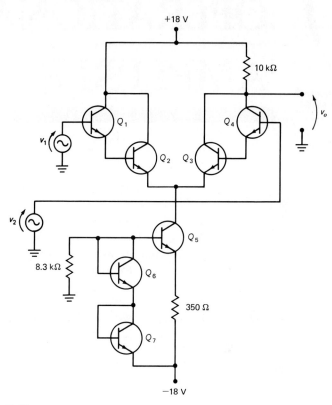

Fig. 1-48

CHAPTER 2
INTRODUCTION TO OPERATIONAL AMPLIFIERS

The differential amplifier theory that was presented in Chapter 1 provides the foundation for a deeper understanding of a large portion of the linear integrated circuits that are available. The most fundamentally important linear integrated circuit is the operational amplifier. Operational amplifiers (op amps) are truly remarkable devices that find use in nearly all areas of linear circuit design. This chapter presents the basic concepts that are necessary for consideration of the operational amplifier as a fundamental linear circuit building block. The basic operation and construction of the typical op amp and the effects of negative feedback on the operational amplifier shall receive primary consideration.

2-1 ♦ OPERATIONAL AMPLIFIERS AND IDEAL VOLTAGE AMPLIFIERS

Realization of the ideal voltage amplifier is an impossible task. A voltage amplifier is an amplifier that accepts a voltage as an input and produces a mathematically related voltage at its output. Since an input voltage causes the generation of the output voltage, the term voltage-controlled voltage source (VCVS) is often applied to the voltage amplifier.

Ideal Voltage Amplifier Characteristics

The ideal voltage amplifier (VCVS) would have the following characteristics:

Voltage gain	A_v	$= \infty$
Bandwidth	BW	$= \infty$
Input resistance	R_{in}	$= \infty$
Output resistance	R_o	$= 0$

In practice, such ideal characteristics cannot be achieved. However, in many practical situations, operational amplifiers can approximate these characteristics. Based on the circuits that were analyzed in the previous chapter, it can be seen that at least a few of these characteristics can indeed be approached, for practical purposes. The following is a list of typical op amp characteristics.

Voltage gain	A_v	=	100,000
Bandwidth	BW	=	1 MHz (unity gain)
Input resistance	R_{in}	=	2 MΩ
Output resistance	R_o	=	50 Ω

Although not all of these typical specifications would be considered outstanding in all applications, there are many different op amps available that are optimized in terms of one or more of these areas. The relative importance of a given performance parameter depends on the application under consideration. In some cases, the circuit designer can alter some performance parameters to meet a given requirement through external component selection.

Typical Op Amp Architecture

Operational amplifiers are nearly always designed using the differential amplifier configurations that were presented in Chap. 1. Figure 2-1 shows a

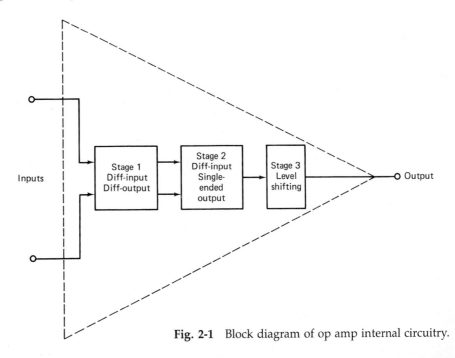

Fig. 2-1 Block diagram of op amp internal circuitry.

block diagram of the internal circuitry that is typical of many op amps. The input stage is a differential-input–differential-output section. Normally, this stage is optimized for high input resistance and high common-mode rejection. The second stage is a differential-input–single-ended-output differential amplifier. This stage is usually designed to provide high voltage gain. The third stage is the level shifter. As was discussed in Chap. 1, the level shifter translates the output of the differential amplifier to 0 V. The level shifter also presents a lower output resistance to the load. This also increases the current sinking and sourcing capabilities of the amplifier. Sometimes a fourth, push-pull stage is added, to further reduce output resistance and increase current drive.

Because of the need for very closely matched component parameters in differential amplifier circuits, op amps are usually produced as monolithic ICs. Because of the nature of monolithic IC fabrication, resistors are more difficult to produce than transistors and diodes. This is especially true of higher resistance values. The difficulties associated with resistor fabrication makes it most practical to replace resistors with transistors wherever possible in the design of a given device.

Resistors that are produced on monolithic ICs are normally quite closely matched to one another (within 1 percent). However, it is very difficult to control the absolute value of resistors in monolithic IC fabrication. In designs in which resistors are required and the absolute value of the resistors is critical, hybrid IC fabrication techniques are used. Hybrid ICs usually have the semiconductor components (diodes and transistors) fabricated on a single silicon chip (die), while resistors (and possibly capacitors) are fabricated separately and are not a homogeneous part of the chip itself. The chip is mounted on a substrate (often a ceramic material), along with and interconnected to film resistors. Film resistors may be formed by the application of conductive materials to the substrate (thick-film resistors) or by the removal of vacuum-deposited metal (thin-film resistors) via an etching process. Thick- and thin-film resistors may also be trimmed very close to their desired absolute values with a laser. The laser vaporizes the resistive material, causing the resistor to increase in value. Both monolithic and hybrid IC fabrication techniques are used in the production of linear ICs.

The schematic symbol for the operational amplifier is shown in Fig. 2-2. All operational amplifiers have two input terminals (one inverting and the other noninverting). The vast majority of op amps available have a single output terminal (single-ended output), although some are available with differential outputs. Notice that power-supply connections are not shown on this symbol, although obviously they would be required. As is the case with

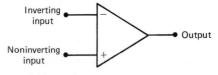

Fig. 2-2 Schematic symbol for the op amp.

digital ICs on logic diagrams, op amp power-supply connections are not shown on functional schematics, but rather are implied. Normally, a separate power-supply connection diagram will contain this information.

2-2 ◆ NEGATIVE FEEDBACK

Even though a real operational amplifier does not have infinite gain, the gain that is available could easily be too high for many applications. Gain may be reduced to any desired value through the use of negative feedback. As will be shown, the use of negative feedback can also provide improvements in other amplifier characteristics.

The Noninverting Configuration

The block diagram in Fig. 2-3 illustrates the implementation of negative feedback in an amplifier circuit. The triangle represents an amplifier with open-loop voltage gain A_{OL}, with v_{id} as its input and v_o as its output. The output voltage is multiplied by the feedback factor β ($\beta < 1$) and subtracted from v_{in}. The sigma represents summation. In this case, we have the summation of v_{in} with $-v_f$, which is effectively a subtraction operation.

Effects of Negative Feedback on Voltage Gain

Inspection of Fig. 2-3 reveals the following relationships:

$$v_o = v_{id} A_{OL} \tag{2-1}$$

$$v_f = \beta v_o \tag{2-2}$$

$$v_{id} = v_{in} - v_f \tag{2-3}$$

Substituting Eq. (2-2) into Eq. (2-3),

$$v_{id} = v_{in} - \beta v_o \tag{2-4}$$

Now, substituting Eq. (2-4) into Eq. (2-1),

$$v_o = A_{OL}(v_{in} - \beta v_o) \tag{2-5}$$

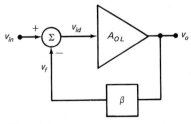

Fig. 2-3 Generalized block diagram for a noninverting amplifier with negative feedback.

Rearranging Eq. (2-5) yields

$$v_o = \frac{A_{OL}v_{in}}{1 + \beta A_{OL}} \qquad (2\text{-}6)$$

Finally, dividing Eq. (2-5) by v_{in} yields

$$A_v = \frac{v_o}{v_{in}} = \frac{A_{OL}}{1 + \beta A_{OL}} \qquad (2\text{-}7)$$

Equation (2-7) defines the feedback, or closed-loop gain (sometimes designated as A_f or A_{CL}) of the amplifier. From this equation it is seen that the feedback factor β determines the overall gain of the amplifier.

The block diagram of Fig. 2-3 is the model of an operational amplifier that is connected in the noninverting configuration, as shown in Fig. 2-4. The noninverting op amp with feedback is usually drawn as in Fig. 2-4a; however, it is redrawn in Fig. 2-4b to make the operation of the circuit clearer. Resistors R_F and R_1 form the feedback network, responsible for the feedback factor β. Let us now idealize the op amp of Fig. 2-4 by assuming that $R_{in} = \infty$ and $A_{OL} = \infty$. As a consequence of the infinite input resistance, $i_{in} = 0$ and $i_2 = 0$. Also, if A_{OL} is infinite, then $v_{id} = 0$ (from $v_{id} = v_o/A_{OL}$). Under these conditions, we see that

$$v_f = v_{in} \qquad (2\text{-}8)$$

The feedback factor is determined by R_F and R_1. From Fig. 2-4b, we see that

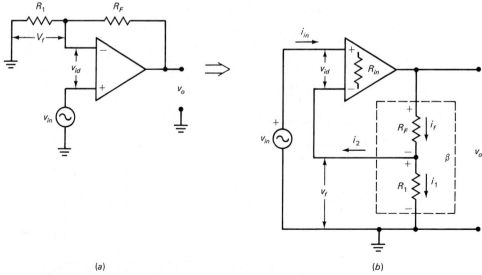

Fig. 2-4 Noninverting op amp with feedback. (a) Standard representation. (b) Circuit redrawn to emphasize feedback network.

these resistors perform voltage division on v_o. Applying the voltage divider equation, we obtain

$$v_f = v_o\beta = v_o \frac{R_1}{R_1 + R_F} \qquad (2\text{-}9)$$

Applying the identity from Eq. (2-8) to Eq. (2-9),

$$v_{in} = v_o \frac{R_1}{R_1 + R_F} \qquad (2\text{-}10)$$

Equation (2-10) may be rearranged, producing

$$A_v = \frac{v_o}{v_{in}} = \frac{R_F + R_1}{R_1} \qquad (2\text{-}11a)$$

An alternative form that is often used is

$$A_v = 1 + \frac{R_F}{R_1} \qquad (2\text{-}11b)$$

Remember that Eqs. (2-11a) and (2-11b) apply to an ideal op amp in the noninverting configuration. The following example will demonstrate the validity of the ideal gain expressions.

EXAMPLE 2-1

Assume that the op amp in Fig. 2-4 is ideal. Determine A_v and V_o given $R_F = 100 \text{ k}\Omega$, $R_1 = 1 \text{ k}\Omega$, and $V_{in} = +20 \text{ mV}$.

Solution

Applying Eq. (2-11), the voltage gain is

$$A_v = 1 + \frac{100 \text{ k}\Omega}{1 \text{ k}\Omega} = 101$$

The output voltage may now be found:

$$V_o = A_v V_{in} = 101 \times 20 \text{ mV} = 2.02 \text{ V}$$

The next example will demonstrate the effect of finite open-loop gain on the noninverting op amp with feedback.

EXAMPLE 2-2

Use Eq. (2-7) to determine the voltage gain and V_o for the circuit of Fig. 2-4, given the resistances of Example 2-1 and $A_{OL} = 100{,}000$.

Solution

$$\beta = \frac{1000}{1000 + 100{,}000} = 9.9 \times 10^{-3}$$

$$A_v = \frac{A_{OL}}{1 + \beta A_{OL}} = \frac{100{,}000}{1 + 990} = 100.9$$

$$V_o = A_v V_{in} = 20 \text{ mV} \times 100.9 = 2.018 \text{ V}$$

The preceding examples indicate the validity of using the ideal op amp as an approximation, at least for the component values and op amp parameters that were given. In order to obtain good accuracy using the ideal gain expression, the following relationship is suggested as a general guide:

$$A_{OL} > \frac{100}{\beta} \tag{2-12}$$

That is, if the condition of inequality (2-12) is met, the gain error (the difference between the closed-loop gain A_v, A_f, or A_{CL} with finite A_{OL} and that with infinite A_{OL}) will be less than 1 percent.

In the case of Example 2-2 (finite A_{OL}), there will be a small differential input voltage. This voltage is given by

$$v_{id} = \frac{v_o}{A_{OL}} \tag{2-13}$$

Often, v_{id} is small enough to ignore (20.18 µV for Example 2-2); however, it is a useful quantity, as is demonstrated in the next discussion.

Noninverting Amplifier Input Resistance

Although the input resistance of the typical op amp is rather high (typically > 1 MΩ) in the noninverting configuration, input resistance can be dramatically increased through the use of negative feedback. This is shown in the following analysis, where R_{in} is the input resistance of the op amp without feedback and R_{inF} is the input resistance with negative feedback. That is, R_{inF} is the effective resistance that is seen by the signal source v_{in}.

In Fig. 2-4b, the effective input resistance is defined by Ohm's law as

$$R_{inF} = \frac{v_{in}}{i_{in}} \qquad (2\text{-}14)$$

Also from Ohm's law, it can be seen that

$$i_{in} = \frac{v_{id}}{R_{in}} \qquad (2\text{-}15)$$

Substituting Eq. (2-15) into (2-14) produces

$$R_{inF} = \frac{v_{in} R_{in}}{v_{id}} \qquad (2\text{-}16)$$

Substituting Eq. (2-13) into (2-16), defining v_o from Eq. (2-6), and performing some algebraic manipulation yields

$$R_{inF} = R_{in}(1 + \beta A_{OL}) \qquad (2\text{-}17)$$

It is apparent that negative feedback increases the effective input of the op amp. This is another desirable occurrence.

The Inverting Configuration

The operational amplifier may be used in the inverting configuration, with feedback, as well as the noninverting configuration. The schematic diagram for such an inverting amplifier is shown in Fig. 2-5. This circuit will first be analyzed using the generalized negative feedback block diagram.

Inverting Amp Voltage Gain with Negative Feedback

Figure 2-6 models the behavior of an inverting amplifier with feedback. The negative sign shown at the v_{in} side of the summation symbol indicates that

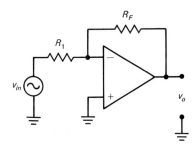

Fig. 2-5 The inverting op amp with negative feedback.

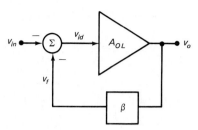

Fig. 2-6 Generalized block diagram for an inverting amplifier with negative feedback.

v_{in} has the opposite polarity from the amplifier's output (180° phase reversal). It can be shown that the voltage gain of such a closed-loop system is given by

$$A_v = \frac{A_{OL}}{1 + \beta + \beta A_{OL}} \qquad (2\text{-}18)$$

where $\beta = R_1/R_f$ for the inverting amplifier

In a practical circuit, $\beta \ll \beta A_{OL}$; therefore Eq. (2-18) may be simplified to

$$A_v = -\frac{A_{OL}}{1 + \beta A_{OL}} \qquad (2\text{-}19)$$

Although Eq. (2-19) is somewhat easier to work with than Eq. (2-18), further simplification is possible, as is shown in the following steps. First, inverting Eq. (2-19) and multiplying by -1 yields

$$-\frac{1}{A_v} = \frac{1 + \beta A_{OL}}{A_{OL}}$$

$$= \frac{1}{A_{OL}} + \frac{\beta}{1}$$

$$= \frac{1}{A_{OL}} + \frac{R_1}{R_F}$$

Now, since A_{OL} is normally very high (on the order of 100,000 or more), the left-hand term may be disregarded, producing

$$-\frac{1}{A_v} \cong \frac{R_1}{R_F}$$

Inverting and changing signs a second time yields the voltage gain A_v:

$$A_v = -\frac{R_F}{R_1} \qquad (2\text{-}20)$$

The validity of Eq. (2-20) is demonstrated in the following example.

EXAMPLE 2-3

(a) Using Eq. (2-18), determine the voltage gain of the circuit in Fig. 2-5, given $R_1 = 1$ kΩ, $R_F = 100$ kΩ, and $A_{OL} = 100{,}000$.

(b) Repeat part **a** using Eq. (2-20), under the assumption that A_{OL} is infinite.

Solution

(a) $\beta = \dfrac{R_1}{R_F} = 0.01$

$A_v = \dfrac{-A_{OL}}{1 + \beta + \beta A_{OL}}$

$= \dfrac{-100{,}000}{1001.01}$

$= -99.899$

(b) $A_v = \dfrac{-R_F}{R_1}$

$= \dfrac{-100 \text{ k}\Omega}{1 \text{ k}\Omega}$

$= -100$

The generalized feedback model is very useful in other areas of electronic circuit and system design and analysis, in addition to the study of op amps. However, a direct analysis of the inverting op amp with feedback will provide some extremely valuable insight into the characteristics of the op amp. For convenience, the inverting op amp with feedback is shown again in Fig. 2-7, with all pertinent voltages and currents included. Given the high degree

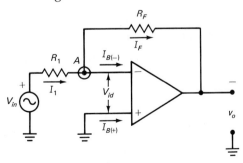

Fig. 2-7 Inverting op amp illustrating currents, differential input voltage, and the virtual ground point.

of accuracy obtained previously under the assumption that the op amp is ideal, the same assumption is made in the following analysis. Specifically, let us assume that $A_{OL} = \infty$ and $R_{in} = \infty$.

The initial conditions are that V_{in} is at some positive level and $V_o = 0$ V. The presence of V_{in} causes a differential voltage to appear between the inverting and noninverting terminals of the op amp. Since the inverting terminal is more positive, the output is driven negative. Ideally, the negative-going excursion of V_o will continue until V_{id} is 0 V. This can be shown as follows.

The fundamental expression for the output of a differential amplifier is

$$V_o = A_d (V_{in(+)} - V_{in(-)}) \tag{2-21}$$

In this case, $A_d = A_{OL} = \infty$ and $V_{in(+)}$ is 0 V, because the noninverting terminal is connected to ground. Now, substituting $V_{id} = V_{in(+)} - V_{in(-)}$ and solving for V_{id} produces

$$V_{id} = \frac{V_o}{A_{OL}}$$

$$= \frac{V_o}{\infty}$$

$$= 0 \text{ V}$$

The important point to recognize here is that the inverting terminal will (ideally) be driven to the same potential as the noninverting terminal, forcing $V_{id} = 0$ V. In this case, the noninverting terminal is at 0 V; hence the inverting terminal of the op amp is also driven to ground potential. Under these conditions, the inverting input (node A) is effectively a simulated ground point. Node A is referred to as a *virtual ground*. The virtual ground concept is central to understanding the operation of many op amp circuits. Now, with node A established as a virtual ground point, it can be seen that the voltage source will supply a current to node A that is given by

$$I_1 = \frac{V_{in}}{R_1} \tag{2-22}$$

Assuming that the op amp inputs have infinite internal resistance, $I_{B(-)} = I_{B(+)} = 0$. Applying Kirchhoff's current law to node A produces

$$I_F = -I_1 \tag{2-23}$$

These results indicate that the output of the op amp is at a negative potential with respect to ground (in order to draw $-I_1$ through R_F). Again, since the inverting terminal is at 0 V, the voltage present at the output, V_o, is the drop across R_F. In equation form,

$$V_o = I_F R_F \tag{2-24}$$

Substituting Eq. (2-23) into Eq. (2-24) yields

$$V_o = -I_1 R_F \qquad (2\text{-}25)$$

Now, substituting Eq. (2-22) into Eq. (2-25),

$$V_o = \frac{-V_{in} R_F}{R_1} \qquad (2\text{-}26)$$

Finally, dividing both sides of Eq. (2-26) by V_{in} and multiplying by -1 produces

$$A_v = -\frac{V_o}{V_{in}} = -\frac{R_F}{R_1} \qquad (2\text{-}27)$$

This is the same gain expression that was derived from the block diagram representation of Fig. 2-6 for the ideal amplifier.

EXAMPLE 2-4

Refer to Fig. 2-7.

(a) Determine the value for R_1 required to give $A_v = -25$, given $R_F = 55\ \text{k}\Omega$.

(b) If $V_{in} = 50\ \text{mV}$, determine I_1, I_F, and V_o.

Solution

(a) $R_1 = \dfrac{R_F}{25} = 2.2\ \text{k}\Omega$

(b) $I_1 = \dfrac{V_{in}}{R_1} = \dfrac{50\ \text{mV}}{2.2\ \text{k}\Omega} = 22.73\ \mu\text{A}$

$I_F = -I_1 = -22.73\ \mu\text{A}$

$V_o = A_v V_{in}$

$\quad = -25 \times 50\ \text{mV}$

$\quad = -1.25\ \text{V}$

The output voltage may also be found using Eq. (2-24):

$V_o = I_F R_F$

$\quad = -22.73\ \mu\text{A} \times 55\ \text{k}\Omega$

$\quad = -1.25\ \text{V}$

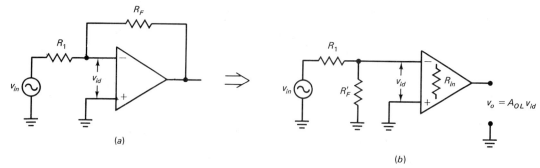

Fig. 2-8 Inverting amplifier. (a) Basic circuit. (b) Equivalent circuit showing the Miller equivalent of R_F at the input.

Inverting Amplifier Input Resistance

The effective input resistance of the inverting op amp with feedback is quite low, compared with that of an equivalent noninverting circuit. We can take an intuitive approach to the determination of an equation to approximate R_{inF} by applying the virtual ground concept and assuming an ideal op amp. Referring to Fig. 2-7, the following relationship is indicated, because node A is (virtually) at ground potential:

$$R_{inF} = R_1 \qquad (2\text{-}28)$$

Thus, ideally, the value of the input resistor R_1 sets the load resistance that must be driven by the V_{in} source.

The input resistance for the inverting amplifier with feedback is found by application of Miller's theorem. Millerizing the input of the amplifier of Fig. 2-8a produces the equivalent circuit in Fig. 2-8b. The Miller input resistance R'_F is given by

$$R'_F = \frac{R_F}{A_{OL} + 1} \qquad (2\text{-}29)$$

In a practical circuit, R'_F will be several orders of magnitude smaller than the input resistance of the op amp; therefore R_{in} may be neglected. It can also be seen from Eq. (2-29) that if A_{OL} is large, the effective overall input resistance of the amplifier is indeed closely approximated by Eq. (2-28).

EXAMPLE 2-5

The circuit in Fig. 2-5 has the following specs: $R_1 = 1.5$ kΩ, $R_F = 30$ kΩ, $A_{OL} = 150{,}000$, $R_{in} = \infty$. Determine A_v, the Miller input resistance, and the overall input resistance.

Solution

$$A_v = \frac{-R_F}{R_1} = \frac{-30\text{ k}\Omega}{1.5\text{ k}\Omega} = -20$$

$$R'_F = \frac{R_F}{A_{OL} + 1}$$

$$= \frac{30\text{ k}\Omega}{150{,}000}$$

$$= 0.2\text{ }\Omega$$

$$R_{inF} = R_1 + R'_F$$

$$= 1.5\text{ k}\Omega + 0.2\text{ }\Omega$$

$$= 1500.2\text{ }\Omega$$

$$\cong 1.5\text{ k}\Omega$$

Most of the time, it is sufficient to approximate the input resistance of the inverting amplifier as being equal to the value of the input resistance R_1. Generally, the noninverting configuration is used in applications that require that the amplifier have very high input resistance. This is so because there are practical limits on the value of R_1 that may be used in the inverting amplifier. These limitations are discussed in Chapter 3.

The Voltage Follower

A special form of the noninverting amplifier is the voltage follower. A voltage follower is an amplifier whose output voltage is equal in amplitude and phase with its input voltage. Such a circuit is shown in Fig. 2-9. Let us first apply an intuitive approach to the analysis of this circuit, again based on the assumption of an ideal op amp.

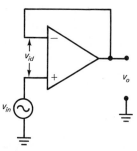

Fig. 2-9 Op amp voltage follower.

Assume that v_{in} is at some positive level. Since A_{OL} is infinite, the output of the op amp will be driven positive until $v_{id} = 0$ V. Since there is no resistance in the feedback loop, there is no attenuation of the output voltage; therefore, when $v_{id} = 0$, $v_o = v_{in}$. This is a case in which the feedback factor β is unity. That is, 100 percent of the output voltage is fed back to the input. It can also be shown that even if resistance is present in the feedback loop, the voltage gain of Fig. 2-9 (and the feedback factor) is unity. The following analysis proves the validity of this statement. This analysis may be skipped without loss of continuity by those readers without a calculus background.

Derivation of the Ideal Voltage-Follower Gain Expression

The basic noninverting amplifier with feedback is shown in Fig. 2-10a. For the noninverting configuration, we know that $A_v = 1/\beta$, where

$$\beta = \frac{R_1}{R_1 + R_F}$$

In the transition from Fig. 2-10a to 2-10b, R_1 is increased from some finite value to infinite resistance (an open circuit). In order to evaluate this change mathematically, we take the limit of Eq. (2-11a) as R_1 approaches infinity.

$$\lim_{R_1 \to \infty} \frac{R_1}{R_1 + R_F} = \frac{\infty}{\infty + R_F} = \frac{\infty}{\infty}$$

Now, since in the limit we obtain the indeterminate form ∞/∞, L'Hospital's rule may be applied. Taking the derivatives of the numerator and the individual terms of the denominator produces

$$\beta = \frac{R_1'}{R_1' + R_F'} = \frac{1}{1 + 0} = 1$$

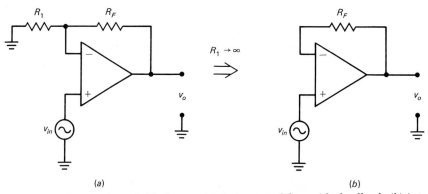

Fig. 2-10 Conversion of (a) the noninverting amplifier with feedback (b) into a voltage follower.

The voltage gain may now be defined as

$$A_v = \frac{1}{\beta} = \frac{1}{1} = 1$$

Input Resistance of the Voltage Follower

Since the voltage follower is basically a noninverting amplifier with feedback, the previous analysis of the effective input resistance of the noninverting op amp configuration, given by Eq. (2-17), still holds.

$$R_{inF} = R_{in}(1 + \beta A_{OL}) \tag{2-17}$$

Effect of Negative Feedback on Output Resistance

The output resistance of an ideal voltage amplifier (or VCVS) is 0 Ω. This is logical, as the output of the amplifier is equivalent to a (controlled) voltage source. Typically, the open-loop output resistance R_o of an op amp will be around 50 Ω or so, although some types have higher R_o and some others lower. In any case, in some applications this resistance would be too high to adequately model the op amp's output as an ideal voltage source. Fortunately, negative feedback provides the beneficial effect of reducing the effective output resistance of the op amp significantly. It can be shown that for both inverting and noninverting amplifiers that employ negative feedback, the effective output resistance is given by

$$R_{oF} = \frac{R_o}{1 + \beta A_{OL}} \tag{2-30}$$

where $\beta = \dfrac{R_1}{R_F}$ for the inverting amplifier

$\beta = \dfrac{R_1}{R_1 + R_F}$ for the noninverting amplifier

Again, it is rather easily seen that the effective output resistance may be quite small, since A_{OL} is normally very large.

EXAMPLE 2-6

Determine A_v, R_{inF}, R_{oF}, and V_{id} for the amplifier in Fig. 2-11, given

(a) $A_{OL} = 75{,}000$, $R_o = 200\ \Omega$, and $R_{in} = 100\ \text{k}\Omega$.
(b) $A_{OL} = 150{,}000$, $R_o = 30\ \Omega$, and $R_{in} = 20\ \text{k}\Omega$.

Use the exact analysis equations.

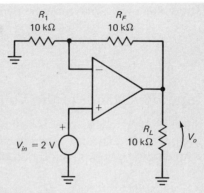

Fig. 2-11 Circuit for Example 2-6.

Solution

Since the feedback factor is the same for both parts of the problem, we will begin with its calculation.

$$\beta = \frac{R_1}{R_1 + R_F} = \frac{10 \text{ k}\Omega}{20 \text{ k}\Omega} = 0.5$$

(a) $A_v = \dfrac{A_{OL}}{1 + \beta A_{OL}}$

$= \dfrac{75{,}000}{37{,}501}$ (2-7)

$= 1.999 \cong 2$

$R_{inF} = R_{in}(1 + \beta A_{OL})$

$= 100 \text{ k}\Omega(1 + 0.5 \times 75{,}000)$ (2-17)

$= 100 \text{ k}\Omega \times 37{,}501$

$= 3.75 \times 10^9 \ \Omega \ (3.75 \text{ G}\Omega)$

$R_{oF} = \dfrac{R_o}{1 + \beta A_{OL}}$

$= \dfrac{200 \ \Omega}{37{,}501}$ (2-30)

$= 5.3 \text{ m}\Omega$

To find V_{id}, we must first determine V_o. Because the output resistance of the circuit is extremely low in relation to the load resistance, it is appropriate to disregard the loading effect on the op amp; hence

$$V_o = A_v V_{in}$$
$$= 2 \times 2 \text{ V}$$
$$= 4 \text{ V}$$

We can now calculate V_{id}:

$$V_{id} = \frac{V_o}{A_{OL}}$$

$$= \frac{4 \text{ V}}{75{,}000} \qquad (2\text{-}13)$$

$$= 53.3 \ \mu\text{V}$$

(b) The second analysis is carried out in the same manner.

$$A_v = \frac{A_{OL}}{1 + \beta A_{OL}}$$

$$= \frac{150{,}000}{75{,}001}$$

$$= 1.999 \cong 2$$

$$R_{inF} = R_{in}(1 + \beta A_{OL})$$

$$= 20 \text{ k}\Omega \times 75{,}001$$

$$= 1.5 \times 10^9 \ \Omega \ (1.5 \text{ G}\Omega)$$

$$R_{oF} = \frac{R_o}{1 + \beta A_{OL}}$$

$$= \frac{30 \ \Omega}{75{,}001}$$

$$= 400 \ \mu\Omega$$

$$V_o = A_v V_{in}$$
$$= 2 \times 2 \text{ V}$$
$$= 4 \text{ V}$$

$$V_{id} = \frac{V_o}{A_{OL}}$$

$$= \frac{4 \text{ V}}{150{,}000}$$

$$= 26.7 \ \mu\text{V}$$

74 CHAPTER 2 INTRODUCTION TO OPERATIONAL AMPLIFIERS

In Example 2-6, the op amp parameters varied substantially between the two analyses, yet the resulting gain values were, for all practical purposes, identical. In both cases, the effective input resistances were high enough to approximate infinite resistance, and the output resistances were low enough to be considered zero. Such predictable behavior is attributable to the use of negative feedback. In this example, a rather high degree of feedback (0.5) was used. If the feedback factor had been lower, the voltage gains and effective input and output resistance would have been less close to the ideal, but it is more than likely that the differences would still have been negligible. A similar analysis using the inverting configuration is left as an exercise at the end of the chapter.

2-3 ♦ BANDWIDTH LIMITATIONS

Some of the basic amplifier analysis equations were derived based on dc sources producing the input, while others were derived based on the application of sinusoidal voltages. To preserve generality, all the equations could have been derived for ac conditions, using impedances instead of resistances. When this is done, purely resistive components are just a special case, in which the imaginary parts of the impedances are zero. Ideally, for the purely resistive case, it should make no difference whether the op amp is amplifying dc levels or ac signals. This would imply that the op amp can process signals at any frequency, which would require the ideal amplifier characteristic of infinite bandwidth.

The real-world op amp can indeed behave nearly ideally, provided all signal components are restricted to a certain range of frequencies. Such a range of frequencies is termed the *midband*. Because op amps use direct

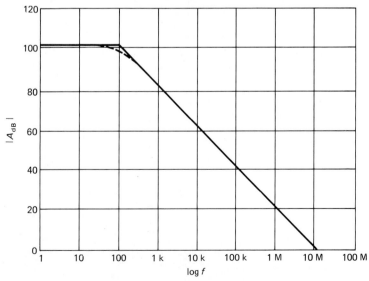

Fig. 2-12 Typical op amp Bode plot.

coupling, the midband ranges from dc (zero frequency) to some upper limit, called the *corner frequency*, *break frequency*, or *critical frequency*. These three terms are used synonymously. A representative plot of A_{OL} versus frequency for an op amp is shown in Fig. 2-12. This is referred to as a *Bode plot*. The dashed curve is the actual response of the amplifier. The solid line represents a piecewise-linear approximation of the true response. Notice that the gain axis is scaled in decibels (dB). The break frequency f_c is defined as the frequency at which the gain of the amplifier has dropped 3 dB from the passband gain.

EXAMPLE 2-7

A certain op amp has $A_{OL} = 120{,}000$ and $f_c = 100$ Hz. Determine the midband gain of the amplifier A_{mid}, in decibels, and the numerical gain of the amplifier at f_c.

Solution

The midband decibel gain of the op amp is

$$A_{mid(dB)} = 20 \log |A_{OL}|$$
$$= 20 \times 5.08$$
$$= 101.6 \text{ dB}$$

The decibel gain at $f = f_c$ is

$$A_{c(dB)} = A_{mid} - 3 \text{ dB}$$
$$= 101.6 \text{ dB} - 3 \text{ dB}$$
$$= 98.6 \text{ dB}$$

This is converted into numerical gain as follows:

$$A_c = \log^{-1} \frac{A_{c(dB)}}{20}$$
$$= \log^{-1} 4.93$$
$$= 85{,}000$$

The numerical gain at the break frequency can also be found by multiplying the numerical value of the midband gain ($A_{mid} = A_{OL}$) by 0.707. This is equivalent to a -3-dB change in gain.

$$A_c = 0.707 \, A_{mid}$$
$$= 84{,}800$$

The difference between the two break frequency gain figures is due to slight rounding errors and slight inaccuracies in the assumed break frequency gain reduction figures. The actual numerical gain is $A_c = A_{mid}/\sqrt{2}$, which in decibels is approximately a change of -3.0103 dB.

In general, the range of frequencies for which the gain of the op amp is within 3 dB of maximum is called the *bandwidth* of the amplifier. Bandwidth is a general term that also applies to other frequency-sensitive systems. The open-loop bandwidth of the op amp whose Bode plot is shown in Fig. 2-12 is 100 Hz. Here, we are assuming that A_{OL} remains essentially constant from 1 Hz (the lowest frequency on the Bode plot) to 0 Hz.

The Bode plot in Fig. 2-12 crosses unity gain (0-dB gain) at a frequency of slightly greater than 10 MHz, and the slope of the gain curve for frequencies above f_c is -20 dB/decade (a decade is a tenfold increase or decrease). This is termed a *first-order response*. The product of the closed-loop gain and bandwidth is constant for op amps that have a single break frequency and a gain rolloff of -20 dB/decade. This constant is called the *gain-bandwidth product* (GBW).

EXAMPLE 2-8

Determine GBW for the op amp whose Bode plot is given in Fig. 2-12.

Solution

GBW can be found by multiplying the frequency and numerical gain values that intersect on the curve on or to the right of the break frequency. It is most convenient to use 0 dB and 10 MHz in this case. The numerical gain for 0 dB is 1; therefore

GBW = 1 × 10 MHz
    = 10 MHz

This result is also obtained using any other intersection. For example, for $f = 1$ kHz, $A_v \cong 80$ dB $= 10,000$.

GBW = 10,000 × 1 kHz
    = 10 MHz

Since the product of gain and bandwidth is a constant, the use of negative feedback, which reduces gain, also increases the bandwidth of the amplifier. If GBW is known, the bandwidth at a given closed-loop gain can easily be determined. If a Bode plot for the op amp is available, the bandwidth may also be determined graphically.

EXAMPLE 2-9

The op amp in Fig. 2-13 has the Bode plot of Fig. 2-12. Determine the closed-loop gain A_v and the bandwidth of the amplifier.

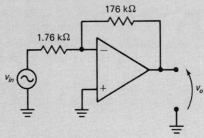

Fig. 2-13 Circuit for Example 2-9.

Solution

Because the open-loop gain of this op amp is very high, the ideal gain equation may be used, with negligible error.

$$A_v = 1 + \frac{176 \text{ k}\Omega}{1.76 \text{ k}\Omega}$$

$$= 101$$

$$\cong 40 \text{ dB}$$

The gain-bandwidth product of this op amp was determined to be approximately 10 MHz; therefore

$$BW = \frac{GBW}{|A_v|}$$

$$= \frac{10 \text{ MHz}}{101}$$

$$= 99 \text{ kHz}$$

The bandwidth can also be estimated graphically, by plotting across from 40 dB to the gain curve. Using this technique, it is seen that the intersection occurs at approximately 100 kHz.

For closed-loop gains of ±10 or greater, both inverting and noninverting op amps will have very nearly the same bandwidth, assuming that they are constructed using similar op amps. At low gains, however, the inverting op amp will exhibit lower bandwidth than a noninverting circuit with the same gain magnitude. This conclusion is drawn from the following relationship, which holds for *both* inverting and noninverting feedback amplifiers:

$$f_c = \frac{f_{\text{unity}}}{1 + R_F/R_1} \tag{2-31}$$

where f_{unity} is the frequency at which A_{OL} of the op amp is unity. Once again, this equation applies to both inverting and noninverting op amp configurations. In the next chapter, we shall call $1 + R_F/R_1$ the noise gain A_N of the amplifier.

The denominator is the usual A_v expression for the noninverting op amp with feedback, but this factor is 1 greater than the gain of the inverting amp with feedback. To illustrate the significance of this relationship, an evaluation of Eq. (2-31) for inverting and noninverting op amps at various gains is presented below. Here, we are assuming that the op amp has $f_{\text{unity}} = 1$ MHz. Notice that the bandwidth of an inverting amp with $A_v = -1$ is 500 kHz, versus 1 MHz for a noninverting op amp with $A_v = 1$.

A_v	Bandwidth ($f_{\text{unity}} = 1$ MHz)
100	10 kHz
10	100 kHz
1	1 MHz
−1	500 kHz
−10	90.9 kHz
−100	9.9 kHz

In general, the bandwidth of the inverting op amp with feedback may be approximated as being the same as that of an equivalent noninverting op amp if its closed-loop gain magnitude is greater than 10.

Just as the open-loop gain of the typical op amp is too high for most applications, the open-loop bandwidth is usually too low. Negative feedback increases the bandwidth of the amplifier to a useful level. It will be seen in the next chapter that many op amps have much lower open-loop bandwidths than the hypothetical op amp that was used in this discussion.

2-4 ♦ CASCADED AMPLIFIERS

The amplification requirements of many applications cannot be met using a single op amp. For example, assume that the op amp whose Bode plot is

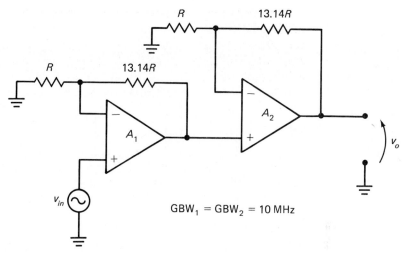

Fig. 2-14 Cascaded op amps.

shown in Fig. 2-12 is to be used in a design that requires a closed-loop gain of 200 (noninverting) and a bandwidth of at least 100 kHz. Obviously, a single op amp circuit will fall short of one specification or the other, as this application requires an overall gain-bandwidth product of 20 MHz. This is twice the GBW of the available op amp. If we set $A_v = 200$, then BW = 50 kHz, and conversely, if we allow for a bandwidth of 100 kHz, then $A_v = 100$.

Let us see whether the design requirements can be met by cascading two op amps. If both stages are set up for a closed-loop gain of approximately 14.14, as shown in Fig. 2-14, the overall gain is 200. For each stage, the bandwidth will be about 707 kHz. However, the overall bandwidth of the circuit is not 707 kHz. For amplifiers using cascaded identical stages, the overall bandwidth (BW_T) is given by the following equation:

$$BW_T = BW_s \sqrt{2^{1/n} - 1} \tag{2-32}$$

where BW_s is the closed-loop bandwidth of each of the stages and n is the number of stages cascaded. Applying Eq. (2-32) to the circuit of Fig. 2-14, we find $BW_T = 455$ kHz. This is substantially lower than the bandwidth of the individual op amps (707 kHz), but it also meets the stated design requirements.

It should be mentioned that maximum bandwidth will be obtained when gain is divided equally between identical amplifier stages. Also, beyond the common critical frequency, the overall gain rolloff will be twice as rapid for the two-stage amplifier as for a single-stage amplifier (-40 dB/decade versus -20 dB/decade). In general, the overall bandwidth of a multiple-stage amplifier will be lower than that of the stage with the lowest bandwidth.

EXAMPLE 2-10

The circuit of Fig. 2-15 is to be used in place of the circuit in Fig. 2-14. Determine the overall voltage gain, BW for each stage, and BW_T.

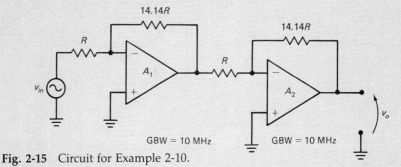

Fig. 2-15 Circuit for Example 2-10.

Solution

Each op amp has $A_v = -14.14$; therefore $A_{vT} = 200$. Each op amp will have the same closed-loop bandwidth. Applying Eq. (2-31), we obtain $BW_1 = BW_2 = BW_s = 660.5$ kHz. The overall bandwidth is found using Eq. (2-32). $BW_T = 425.1$ kHz.

Chapter Review

This chapter has introduced the fundamentals of negative feedback, primarily as it relates to operational amplifiers. In general, negative feedback enhances the performance of an amplifier because it reduces voltage gain to practical levels, increases effective amplifier input resistance, decreases effective amplifier output resistance, and increases amplifier bandwidth. Negative feedback also tends to make an amplifier insensitive to various amplifier parameters, such as open-loop gain and input resistance. In other words, negative feedback makes the op amp act more like an ideal amplifier.

The noninverting op amp configuration allows very high input resistance to be realized. Input impedance increases as the feedback factor increases toward the maximum of unity. The inverting configuration will invariably have lower input resistance than an equivalent noninverting amplifier, because the inverting input terminal is a virtual ground point. The concept of virtual ground is very important in the study of op amp circuits.

The Bode plot is used to determine the bandwidth of an op amp at various voltage gain settings. Normally, Bode plots are straight-line approximations of the actual op amp response, with a maximum error of $+3$ dB occurring at the break frequency. The Bode plot applies to the op amp regardless of whether it is in the inverting or noninverting configuration. When amplifier stages are cascaded, the stage with the lowest critical frequency determines the overall bandwidth of the amplifier.

Questions

2-1. What user-determined parameter controls the closed-loop gain of an amplifier?

2-2. What term is given to the inverting terminal of an op amp configured as an inverting amplifier?

2-3. Based on Eq. (2-17), which particular amplifier circuit will provide the highest effective input resistance for a given op amp?

2-4. What is the minimum closed-loop gain that may be achieved through feedback for a noninverting amplifier?

2-5. Given two different two-stage amplifiers, constructed using identical op amps, if amplifier A has $A_{v1} = 20$ and $A_{v2} = 50$, and amplifier B has $A_{v1} = 100$ and $A_{v2} = 10$, which amplifier will have the higher bandwidth?

2-6. What is the rolloff rate of a first-order system?

Problems

2-1. The amplifier in Fig. 2-3 has $A_{OL} = 10,000$ and $\beta = 0.005$. What is the closed-loop gain of the circuit?

2-2. Based on the conditions stated in Prob. 2-1, given $v_{in} = 10 \sin \omega t$ mV, write the expressions for v_o and v_{id}.

2-3. Refer to Fig. 2-3. If $A_{OL} = 100,000$, what feedback factor is required to produce $A_v = 80$?

2-4. Determine β and A_v for the circuit in Fig. 2-16.

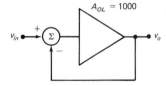

Fig. 2-16

2-5. Refer to Fig. 2-17. Assuming that the op amp is ideal, determine A_v, i_{R1}, i_{RF}, i_L, and v_o, given that $R_F = 68$ kΩ, $R_1 = 27$ kΩ, $R_L = 27$ kΩ, and $v_{in} = 600 \sin 10^3 t$ mV.

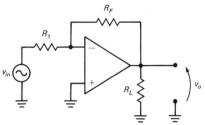

Fig. 2-17

2-6. Using the component values stated in Prob. 2-5, write the expression for v_{in} if $v_o = 6.8 \sin \omega t$ V.

2-7. Refer to Fig. 2-18. Given that $R_F = 330$ kΩ, what value should R_1 have to result in $A_v = 20$? Assume that the op amp is ideal.

2-8. Refer to Fig. 2-18. Given $A_{OL} = 25{,}000$, $R_{in} = 10$ kΩ, $R_o = 100$ Ω, $R_1 = 2$ kΩ, $R_F = 120$ kΩ, $R_L = 1$ kΩ, and $v_{in} = V_{in} = -100$ mV, determine β, R_{inF}, R_{oF}, A_v, I_{R1}, I_{R2}, I_L, and V_o.

2-9. Refer to Fig. 2-18. If the op amp has GBW = 1 MHz, what gain is required (via negative feedback) to produce $f_c = 10$ kHz? Assume that response is flat from dc to f_c.

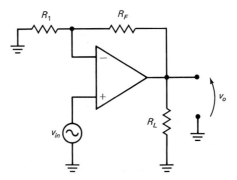

Fig. 2-18

2-10. A certain noninverting amplifier has BW = 20 kHz when the closed-loop gain is 100. Determine GBW for the op amp. At what frequency will this op amp exhibit $A_{OL} = 0$ dB?

2-11. Determine the overall voltage gain of the circuit shown in Fig. 2-19. Assume that the op amp is ideal.

2-12. Refer to Fig. 2-19. What value of v_{in} will result in $v_o = 5 \sin 2\pi 1000t$ V?

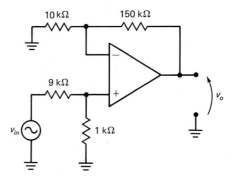

Fig. 2-19

2-13. Refer to Fig. 2-20. Given $A_{OL} = 1000$, $R_{in} = 1000\ \Omega$, $R_F = 1000\ \Omega$, and $V_{in} = 1$ V, determine V_{id}, I_F, and V_o.

2-14. What is the feedback factor for Fig. 2-20?

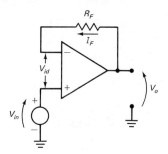

Fig. 2-20

2-15. Assuming that an ideal op amp is used in the circuit of Fig. 2-21, if $R_F = 18\ \text{k}\Omega$, $R_1 = 1.8\ \text{k}\Omega$, and the input voltage source has an internal resistance of 200 Ω, find I_1, I_F, and V_o.

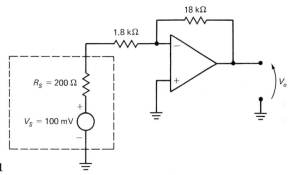

Fig. 2-21

2-16. Determine V_o for the circuit of Fig. 2-22, given $V_{in} = 70$ mV.

2-17. Suppose that the op amp in Fig. 2-22 has GBW = 5 MHz. What is the closed-loop bandwidth of the circuit?

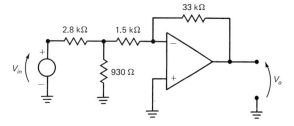

Fig. 2-22

2-18. The op amps in Fig. 2-23 have GBW = 1 MHz. Determine the required ratios for R_2 to R_1 and R_4 to R_3 such that maximum bandwidth is obtained, with an overall voltage gain of 400. Determine the overall bandwidth of the circuit.

2-19. Refer to Fig. 2-23. Each of the op amps has GBW = 1 MHz. Determine the ratios for R_2 to R_1 and R_4 to R_3 such that A_{vT} = 50, with maximum bandwidth.

2-20. The op amps in Fig. 2-23 have GBW = 1 MHz. If R_1 = 10 kΩ, R_2 = 47 kΩ, R_3 = 18 kΩ, and R_4 = 383 kΩ, determine the overall gain A_{vT} and the corner frequencies f_{c1} and f_{c2} for each stage.

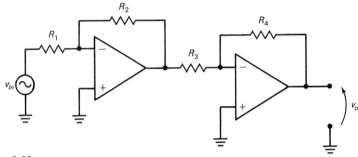

Fig. 2-23

2-21. Refer to Fig. 2-24. What is the overall gain of the amplifier in decibels?

2-22. Given an ideal op amp with R_1 = 10 kΩ, design a single-stage, non-inverting amplifier for A_v = 20 dB.

2-23. Repeat Prob. 2-22 for A_v = 38 dB and R_F = 100 kΩ.

2-24. A certain op amp has a midband A_{OL} of 100 dB and f_c = 5 Hz. Determine the numerical and decibel values of A_{OL} at the following frequencies: **(a)** f = 5 Hz, **(b)** f = 50 Hz, **(c)** f = 500 Hz.

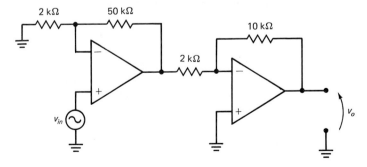

Fig. 2-24

PRACTICAL OP AMP CONSIDERATIONS

In the previous chapter, the op amp was examined primarily in terms of the effects of negative feedback on its behavior. In this chapter, we shall examine some additional nonideal op amp characteristics, and their effects on circuit operation and performance. Additional information on op amp frequency response characteristics, error sources, and the interpretation of op amp data sheets is also presented.

3-1 ♦ OP AMP ERROR SOURCES

Up to this point, the effects of the following nonideal op amp characteristics have been examined:

1. Finite open-loop gain
2. Finite input resistance
3. Nonzero output resistance
4. Finite bandwidth

The reader should be quite familiar with the limitations imposed by these op amp characteristics; however, a few points should be stressed a final time before we move on. In most cases, the open-loop gain of the op amp is high enough to be considered infinite. As a consequence of this very high open-loop gain, negative feedback must be used in nearly all op amp circuit designs. The use of negative feedback improves the op amp's effective input and output resistance and bandwidth characteristics. The point to be made here is that in terms of the parameters listed above, most op amps can be made to behave quite ideally, through the use of negative feedback. There are, however, other nonideal characteristics of the op amp that are not directly affected by negative feedback.

Bias and Offset Currents

Recall that internally, the heart of the op amp consists of one or more differential amplifier sections. The input stage of an op amp will always be a differential-input-type section. Whether bipolar transistors or FETs are used in the input-stage design, a dc path to ground must be provided for the bias currents of these input-stage transistors. This is most obvious in the case of an op amp with a BJT input stage, which we shall consider first.

Generally, the first stage of an op amp is designed to operate at very low collector currents. This has two major effects. First, it keeps the dynamic resistance of the differential pair high, which in turn provides the op amp with very high input resistance. Second, and most important for this discussion, the low input-stage collector currents keep the transistor base bias currents very low (in the nanoampere range, or less). For analysis purposes, the input of an op amp can be modeled as shown in Fig. 3-1. The bases of the differential pair transistors are represented by current sources.

It is standard practice to define the op amp's input bias current I_B as the average of the individual base bias currents:

$$I_B = \frac{I_{B(+)} + I_{B(-)}}{2} \tag{3-1}$$

Ideally, if both transistors in the differential pair have exactly the same β's, both bias currents will be equal ($I_{B(-)} = I_{B(+)}$). In reality, some small difference will exist between the two bias currents. This difference is called *input offset current* I_{io}, and it gives rise to the third current source (I_{io}) shown in Fig. 3-1. Since which base bias current will actually be greater is essentially random, input offset current is defined as the absolute magnitude of the algebraic difference between $I_{B(+)}$ and $I_{B(-)}$. This relationship is defined in Eq. (3-2).

$$I_{io} = |I_{B(+)} - I_{B(-)}| \tag{3-2}$$

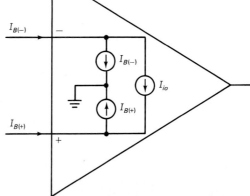

Fig. 3-1 Representation of op amp input circuitry.

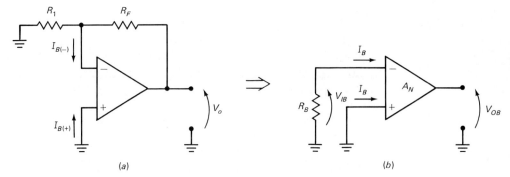

Fig. 3-2 (a) Op amp bias currents. (b) Equivalent circuit with noise gain A_N.

For op amps with BJT inputs, I_B is typically in the range from 50 to 100 nA, whereas I_{io} will usually be around 10 to 20 nA. These currents are relatively low, but they still can have a significant effect on the output of the op amp. Consider the circuit shown in Fig. 3-2a. This could represent either an inverting or a noninverting amplifier in which the input signal (voltage) source has been replaced by its internal resistance, a short circuit. As far as the bias current sources are concerned, the external resistors at the inverting input node appear in parallel, as the output of the op amp is effectively a ground point. This is shown in Fig. 3-2b. The effective resistance seen by the op amp's inverting terminal is given by

$$R_B = R_1 \| R_F \tag{3-3}$$

In Fig. 3-2b, it is assumed that the bias currents are equal, and they are both represented by I_B, for generality. Ideally, when there is no input voltage present, the output of the amplifier is zero. However, as seen in Fig. 3-2b, I_B will cause a small voltage drop to occur across the effective external resistance at the inverting input. This produces an effective differential input of

$$V_{IB} = I_B R_B \tag{3-4}$$

Based on the current flow directions used, the noninverting terminal will be more positive than the inverting terminal by the amount V_B. The bias current direction used here indicates the use of NPN transistors at the op amp's input. If PNP transistors were assumed, the bias currents would flow in the opposite direction. The gain of the op amp to the bias-current-induced voltage is called the *noise gain*, which is given by

$$A_N = \frac{R_F + R_1}{R_1} \tag{3-5}$$

The reader should recognize that the noise gain equation is the same as the gain expression for the noninverting amplifier with feedback.

The output voltage that is produced in response to V_{IB} is given by

$$V_{oB} = A_N V_{IB} \tag{3-6}$$

EXAMPLE 3-1

Refer to Fig. 3-2. Determine V_{oB}, given that $R_1 = 1\ \text{k}\Omega$, $R_F = 10\ \text{k}\Omega$, and $I_{B(-)} = I_{B(+)} = 40\ \text{nA}$.

Solution

The effective resistance as seen by the inverting terminal is

$$R_B = R_1 \parallel R_F$$
$$= 909.1\ \Omega$$

The bias-current-induced input voltage is now determined:

$$V_{iB} = R_B I_B$$
$$= 909.1\ \Omega \times 40\ \text{nA}$$
$$= 36.36\ \mu\text{V}$$

The noise gain of the circuit is

$$A_N = \frac{R_1 + R_F}{R_1}$$
$$= \frac{1\ \text{k}\Omega + 10\ \text{k}\Omega}{1\ \text{k}\Omega}$$
$$= 11$$

The output voltage is now determined as

$$V_{oB} = A_N V_{iB}$$
$$= 11 \times 36.36\ \mu\text{V}$$
$$= 400\ \mu\text{V}$$

In order to negate the effects of bias-current-induced output voltage error, the voltage drop across R_B in Fig. 3-2b must be effectively reduced to zero. Reducing either R_F or R_1 to zero would accomplish this, but normally such circuit modifications are impractical, or even impossible. The best approach is to apply a voltage equal to V_B at the noninverting input of the op amp. To produce this voltage, all that needs to be done is to equalize the external

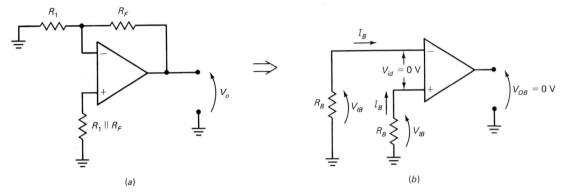

Fig. 3-3 Op amp with bias current compensation resistor. (a) Actual circuit. (b) Equivalent circuit representation.

resistance at each input terminal. This is easily accomplished by placing a resistor R_B in series with the noninverting input terminal. Figure 3-3 shows this modification. Now, if we assume that $I_{B(-)} = I_{B(+)}$, then the voltage drops across the input resistors will be equal in polarity and magnitude, causing $V_{id} = 0$ V, and hence $V_{oB} = 0$ V.

In practice, perfect matching of bias currents cannot be achieved; therefore, even with equal resistances on the op amp inputs, some error voltage may appear at the output. The polarity of the input-offset-current-induced output error voltage depends on which input-terminal bias current is greater, which in turn is usually unknown. This I_{io} error could be eliminated by making R_B adjustable, so that the bias-current-induced voltages on the input terminals could be made equal. However, it will be shown in the next section that there are other error sources that when compensated for will largely eliminate the effects of I_{io} on the op amp. In general, in order to keep bias- and offset-current errors to a minimum, the resistance of the feedback loop of the op amp should be limited to 1 MΩ or less. Also, resistors produce thermal noise, which is proportional to temperature and resistance value. Using low-value resistors will tend to minimize the thermally generated noise. Since R_F is usually the highest-value resistor (R_F must be greater than R_1 for $A_v > 1$), it is usually the resistor whose value is most important in terms of noise and offset-current sensitivity. Keeping R_F as low as practical will likewise result in low values for other resistances in the circuit. Of course, if some minimum input resistance is desired, for example, in the design of an inverting amplifier, then that resistance places a lower limit on the input resistor and hence on R_F. For reasons that will be covered later, for most op amp designs, a lower limit of around 1 kΩ should be imposed on R_1.

Offset Voltage

Another important op amp error source that must often be considered is input offset voltage V_{io}. Input offset voltage is caused primarily by imperfect matching between the V_{BE} (or V_{GS}, for FETs) values of the op amp's input

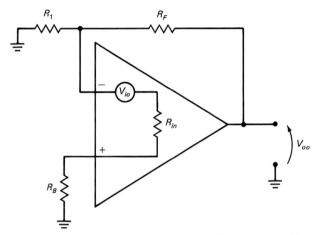

Fig. 3-4 Model of op amp for analysis of input offset voltage effects.

differential pair transistors. What this means is that even though externally the op amp's differential input voltage is zero, a nonzero output voltage will be produced. This is called an *output offset voltage*, V_{oo}. Neglecting other possible error sources, the op amp with input offset voltage may be modeled as shown in Fig. 3-4.

The actual polarity of V_{io} depends on the parameters of components within the op amp, and is usually unknown. The output offset voltage is dependent on the noise gain of the op amp and is given by Eq. (3-7).

$$V_{oo} = A_N V_{io} \tag{3-7}$$

Noninverting Amplifier Offset Compensation

The output offset voltage V_{oo} may be reduced to zero by applying an input voltage that is equal in magnitude but opposite in polarity to V_{io}. This is referred to as *offset null compensation*. The offset null compensation that is used with a given op amp depends on whether the op amp is used in the inverting or noninverting configuration. In either case, an offset voltage is applied to the unused input. For example, consider the noninverting amplifier in Fig. 3-5a. The components in block 1 are used to generate the necessary offset compensation voltage V_{io}, while resistors R_1 and R_F, in block 2, are used to set the gain of the circuit.

The component values necessary for offset compensation in Fig. 3-5a may be determined in several different ways, based on the following requirements. First, block 1 is designed such that the voltage drop across R_Z can match or exceed the maximum V_{io} expected of the op amp in use. Second, R_Y should be very large, relative to R_Z. This makes the effective resistance of the compensation network, R_{eq}, approximately equal to R_Z. The smaller the value of R_Z, the better. Usually, an R_Z value in the range from 10 Ω to 100 Ω or so will be adequate. For example, assume that for the op amp to be used, $V_{io(max)} = 10$ mV, and that symmetrical supply voltages, $\pm V_S$, are

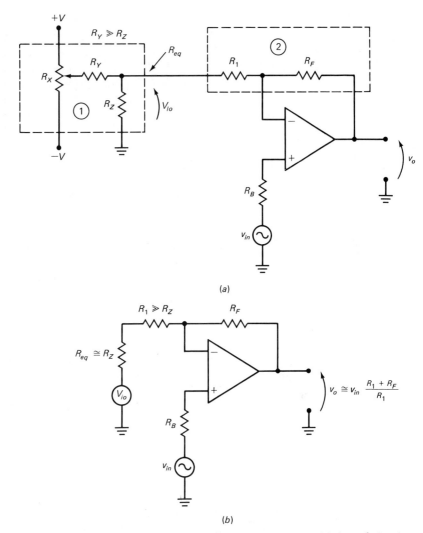

Fig. 3-5 Noninverting op amp with offset compensation. (a) Actual circuit. (b) Equivalent circuit representation.

being used to power the circuit. The wiper of R_X will reach maximum voltage ($\pm V_S$) at the two extremes of its rotation. Setting R_Z to some low value (10 Ω, for example) and applying the voltage divider equation and some algebra, the value for R_Y is determined to be

$$R_Y = \frac{\pm V_S R_Z}{V_{io(\max)}} - R_Z \tag{3-8}$$

The value of the potentiometer, R_X, is chosen to be approximately one-fourth (or less, if possible) of R_Y in value. This helps to prevent the potentiometer from being heavily loaded by the remaining circuitry, and it also improves the linearity of the offset adjustment. The minimum value for the potentiometer is determined by supply loading and potentiometer power dissipation considerations. That is, too low a value for the pot may place an excessive load on the power supply or cause excessive heating of the pot itself. Potentiometer R_X will usually be a trimmer (trim pot) that is adjusted with a small screwdriver or tweaker. These pots are usually rated at $\frac{1}{2}$ W or less. In general, it is advisable to keep the potentiometer current less than 10 mA or so, although this is by no means a set rule.

The remaining resistors, R_1 and R_F, are chosen to set the voltage gain of the amplifier. The main constraint here is that R_1 be much greater in value than R_Z. This allows the standard noninverting op amp gain expression to be used, as illustrated in Fig. 3-5b. A general rule that can be applied to the selection of a value for R_1 is

$$R_1 > 100 R_Z \tag{3-9}$$

If the condition of this inequality is met, the gain error will be less than 1 percent.

EXAMPLE 3-2

The op amp in Fig. 3-5a has $V_{io(max)} = \pm 10$ mV. Design the circuit such that $A_v = 25$. Determine the resistances required for the V_{io} compensating circuitry and the optimum value for R_B. The supply voltages are ± 12 V and $R_Z = 10$ Ω.

Solution

Applying Eq. (3-8),

$$R_Y = \frac{\pm V_S R_Z}{V_{io(max)}} - R_Z$$

$$= \frac{12 \text{ V} \times 10 \text{ }\Omega}{10 \text{ mV}} - 10 \text{ }\Omega$$

$$= 12 \text{ k}\Omega$$

The potentiometer is chosen as 12 kΩ/4 = 3 kΩ. Selecting $R_1 = 1000$ Ω satisfies Eq. (3-9). To obtain $A_v = 25$, the noninverting gain expression is rearranged, producing $R_F = 24$ kΩ.

R_B is determined using Eq. (3-3):

$$R_B = 1 \text{ k}\Omega \parallel 24 \text{ k}\Omega$$

$$= 960 \text{ }\Omega$$

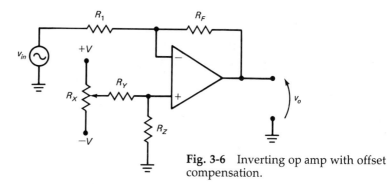

Fig. 3-6 Inverting op amp with offset compensation.

Inverting Amplifier Offset Compensation

The design of the offset compensation circuitry for the inverting amplifier is similar to that for the noninverting configuration. An offset-compensated inverting amplifier is shown in Fig. 3-6. Again, R_Z is chosen first, and R_Y is determined using Eq. (3-8). Since $V_{io(max)}$ is very small in comparison with the supply voltages, R_Y will be much greater than R_Z. This makes the equivalent resistance as seen by the noninverting terminal approximately equal to R_Z. To help compensate for bias-current-induced offset, R_Z is chosen to be approximately equal to $R_1 \| R_F$. The potentiometer is chosen as approximately $R_Y/4$ or less, for the reasons stated previously.

EXAMPLE 3-3

Refer to Fig. 3-6. Assume that the supply voltages are symmetrical at ± 15 V, and the op amp data sheets specify $V_{io(max)} = 20$ mV. Design the circuit such that $A_v = -5$, with input resistance of at least 5 kΩ.

Solution

Resistors R_1 and R_F are determined first. Using 5.6 kΩ (the closest standard value that is greater than 5 kΩ) for R_1 to meet the input resistance requirement, $R_F = 28$ kΩ. The value of R_Z is chosen using Eq. (3-3).

$R_Z = 5.6 \text{ k}\Omega \parallel 28 \text{ k}\Omega$

$ = 4.7 \text{ k}\Omega$

R_Y is found using Eq. (3-8):

$$R_Y = \frac{\pm V_S \, R_Z}{V_{io(max)}} - R_Z$$

$$= \frac{15 \text{ V} \times 4.7 \text{ k}\Omega}{20 \text{ mV}} - 4.7 \text{ k}\Omega$$

$$= 3.5 \text{ m}\Omega$$

Nearly any potentiometer will work here because R_Y is so large. In fact, using a pot that is low in value relative to R_Y is desirable because it makes the offset adjustment very linear. With this in mind, R_X will be chosen somewhat arbitrarily as 10 kΩ. This is much smaller than R_Y, and still presents a negligible load on the supply. Bear in mind that a 25-kΩ, 50-kΩ, or 5-kΩ pot could have been used with little effect on the adjustment.

Additional Offset Compensation Methods

Some op amps have terminals that are available for offset compensation purposes. The 741 op amp, which will be discussed in the next section, is one such device. The offset compensation terminals are typically connected

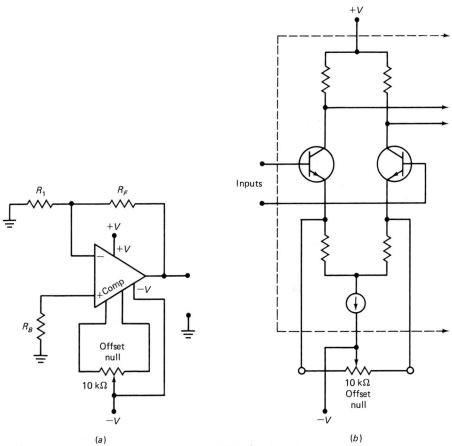

Fig. 3-7 Op amp with offset null compensation terminals. (*a*) Placement of offset null pot. (*b*) Representative internal offset null circuitry.

to the input differential pair, as shown in Fig. 3-7. Moving the wiper of the pot from center causes a slight imbalance in the differential pair, which cancels out the offset voltage that is present.

This offset compensation technique, if available, may be used with both the inverting and the noninverting amplifier configurations. This method is much more convenient than the other two methods that were presented because design effort and parts count are reduced.

Offset compensation is not always required, but if it is implemented, the output offset voltage should be nulled out with the input signal voltage reduced to zero.

3-2 ♦ FREQUENCY COMPENSATION AND STABILITY

Frequency response is a very important consideration in the design and analysis of nearly all electronic circuits and systems. The range of signal frequencies and signal levels that must be processed in a given application dictates the types of devices and components that may be used in the design. Likewise, a knowledge of device frequency characteristics and limitations can be used to predict or explain the behavior of a circuit under various conditions. The major frequency-related characteristics of op amps are examined in this section.

Frequency Compensation

For small-signal applications, the bandwidth of the op amp is usually the most important frequency-related parameter to consider. The term *small signal* is somewhat difficult to define, but in general, a small signal is one that the op amp can amplify in a linear manner.

It was shown in the previous chapter that the closed-loop bandwidth of the op amp is determined by two things: the amount of negative feedback used and the open-loop bandwidth of the op amp. The Bode plot was used to graphically illustrate the relationship between amplifier bandwidth and gain characteristics. Consider the plot of Fig. 3-8, which represents the open-loop characteristics of the industry-standard 741 op amp. The 741 is constructed using internal frequency compensation such that the gain rolls off at a rate of -20 dB/decade, above the break frequency (between 5 and 10 Hz).

Internal frequency compensation is provided to make the op amp easier to use, and to minimize design work and external component count. These are the advantages of internal compensation. On the other hand, internal compensation limits the usefulness of the op amp in some situations. Therefore, many op amps that are available require external compensation for correct operation. An example of an externally compensated op amp is the 709. The Bode plot for this op amp is shown in Fig. 3-9a. In this case, several closed-loop gain curves are shown, along with recommended compensating

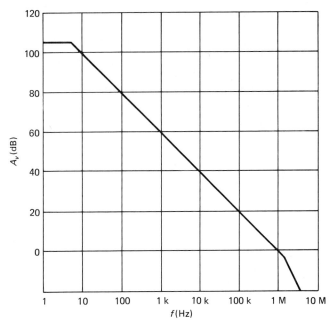

Fig. 3-8 Bode plot for the 741 op amp.

component values. The compensating components are connected to the 709 as shown in Fig. 3-9b. The use of correct compensating component values will help to ensure that the op amp will not oscillate when in operation. Amplifier stability is an important topic, and will be discussed next.

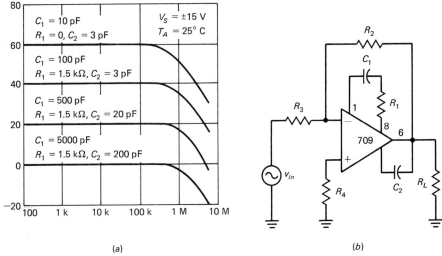

Fig. 3-9 Type 709 op amp. (a) Closed-loop gain frequency response plot. (b) Typical circuit configuration.

Stability and Phase Margin

Most operational-amplifier-based designs rely on negative feedback in order to produce the necessary gain and input and output impedance levels. The actual gain of a given amplifier is a complex quantity that may be expressed in terms of both its magnitude (or modulus) and its phase. At frequencies well below the break frequency, the feedback signal will be about 180° out of phase with the input. (Recall that a fraction of the output is fed back to the inverting input, producing this phase inversion.) If at some higher frequency the feedback signal should happen to become in phase with the input that produces it, the amplifier may break into oscillation. The requirements for oscillation are given by the following relation:

$$A_{OL} \beta > 1 \underline{/0°}, 360° \ldots \quad (3\text{-}10)$$

where the product $A_{OL}\beta$ is termed the *loop gain* of the circuit. Alternatively, the loop gain is equal to the difference between A_{OL} and the closed-loop gain, A_v. In equation form, this is

$$\text{Loop gain} = A_{OL} - A_v \quad (3\text{-}11)$$

In general terms, if an amplifier exhibits a first-order low-pass response (a -20-dB/decade rolloff), to the 0-dB (unity) gain level, it will be stable for any closed-loop gain. In regard to Fig. 3-8, it can be seen that for any closed-loop gain, all the way down to unity gain, the rolloff rate is -20 dB/decade. The phase shift associated with the gain of the amplifier with a first-order low-pass response will increase from 0 at dc (zero frequency) toward $-90°$ asymptotically. At the break frequency, the phase shift will be $-45°$. When this is the case, the feedback signal cannot be in phase with the input, and the circuit will be unconditionally stable.

Some op amps have several break frequencies, each of which occurs before A_{OL} decreases to 0 dB. A similar situation was encountered in Chap. 2, for the amplifier in Fig. 2-15, where two single-break-frequency op amps were cascaded. Each successive break frequency present will contribute an additional -20 dB/decade rolloff to the overall frequency response plot and, just as importantly, an additional increase in phase shift of the feedback signal. The Bode plot for an amplifier with multiple break frequencies is shown in Fig. 3-10. An amplifier that has multiple break frequencies can, under certain conditions, become unstable.

A useful method of evaluating amplifier stability is through the determination of the amplifier's *phase margin*. For an op amp with n break frequencies, the total phase shift of the feedback signal is given by

$$\theta_t = -\tan^{-1}\left(\frac{f}{f_{C1}}\right) - \tan^{-1}\left(\frac{f}{f_{C2}}\right) - \cdots - \tan^{-1}\left(\frac{f}{f_{Cn}}\right) \quad (3\text{-}12)$$

where f is the closed-loop break frequency, as determined on the Bode plot, and $f_{C1}, f_{C2}, \cdots, f_{Cn}$ are the open-loop break frequencies of the op amp.

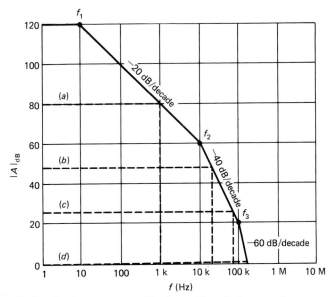

Fig. 3-10 Bode plot for op amp with several break frequencies.

Using Eq. (3-12), the phase margin of the op amp, under closed-loop conditions, is given by

$$\theta_{pm} = 180° - |\theta_t| \tag{3-13}$$

Here are the general rules for determining amplifier stability:

1. If the phase margin is positive, the amplifier will be stable, regardless of the loop gain.
2. If the loop gain is greater than 1 and the phase margin is negative, the amplifier will be unstable (it will oscillate).

In general, a phase margin of about 45° will result in good overall performance in terms of frequency response and pulse response (reasonably low pulse overshoot and ringing). Very low positive phase margins (less than 10°) are to be avoided, as slight changes in circuit parameters could result in a negative phase margin and instability. Also, amplifiers with very low phase margins and high loop gains will tend to overshoot and ring in response to pulse inputs.

Recall that an op amp with a single break frequency (above the unity gain crossing frequency) will be stable for any closed-loop gain. This is also the case for an op amp with multiple break frequencies, as long as the closed-loop break frequency lies on the -20-dB/decade portion of the op amp's Bode plot (curve a in Fig. 3-10). If the closed-loop break frequency lies on the -40-dB/decade portion of the Bode plot, the amplifier may or may not be stable (curves *b* and *c* of Fig. 3-10). Here, the amplifier is said to be conditionally stable. Determination of the phase margin is definitely required for these conditions. Finally, if the closed-loop break frequency is on the -60-dB/decade portion of the Bode plot, as in curve *d* in Fig. 3-10, the amplifier will be unstable. The following example demonstrates these concepts.

EXAMPLE 3-4

Determine the phase margins for the four closed-loop gains shown in Fig. 3-10.

(a) The closed-loop values are $A_v = 80$ dB and $f_c = 1$ kHz. Using Eq. (3-11), the loop gain is found to be 40 dB, which is obviously much greater than 1. The total phase shift is found as follows:

$$\theta_t = -\tan^{-1}\frac{1000}{10} - \tan^{-1}\frac{1000}{10{,}000} - \tan^{-1}\frac{1000}{100{,}000}$$

$$= -\tan^{-1}100 - \tan^{-1}0.1 - \tan^{-1}0.01$$

$$= -89.4° - 5.7° - 0.57°$$

$$= -95.7°$$

The phase margin is now found using Eq. (3-13):

$$\theta_{pm} = 180° - |\theta_t|$$

$$= 180° - 95.7°$$

$$= 84.3°$$

This is a postive phase margin; therefore, the amplifier is stable for this closed-loop gain.

(b) Here, the closed-loop gain is approximately 50 dB, with $f_c = 20$ kHz. Obviously, the loop gain is quite large. In fact, as the closed-loop gain is reduced, loop gain increases; therefore, for the remaining parts of this example, it will not be calculated. The total phase shift is

$$\theta_t = -\tan^{-1}\left(\frac{20{,}000}{10}\right) - \tan^{-1}\left(\frac{20{,}000}{10{,}000}\right) - \tan^{-1}\left(\frac{20{,}000}{100{,}000}\right)$$

$$= -90° - 63.4° - 11.3°$$

$$= -164.7°$$

The phase margin is

$$\theta_{pm} = 180° - 164.7°$$
$$= 15.3°$$

The amplifier is stable, but this is a rather small phase margin. Care should be exercised in circuit layout and component selection to ensure stability.

(c) The closed-loop gain is about 25 dB, and $f_c = 90$ kHz. Because of the coarse scaling on the Bode plot, f_c has been estimated slightly higher than it probably would be. This way, if we err, it will be in the worstcase direction. This is generally good practice. Now, calculating the phase shift,

$$\theta_t = -\tan^{-1}\left(\frac{90{,}000}{10}\right) - \tan^{-1}\left(\frac{90{,}000}{10{,}000}\right) - \tan^{-1}\left(\frac{90{,}000}{100{,}000}\right)$$
$$= -90° - 83.7° - 42°$$
$$= -215.7°$$

The phase margin is

$$\theta_{pm} = 180° - 215.7°$$
$$= -35.7°$$

Because the loop gain is greater than 1 and a negative phase margin exists, the amplifier is unstable. When f_C lies on the -40-dB/decade portion of the Bode plot, the amplifier is conditionally stable.

(d) In this final example, it is obvious that the amplifier will be unstable, as the loop gain is 120 dB and the phase margin is $-60.5°$.

An interesting conclusion may be drawn from the preceding example: Amplifiers tend to become more unstable when greater amounts of negative feedback are applied. This means that unity gain amplifiers may be very troublesome, relative to higher closed-loop gain amplifiers, unless appropriate compensation is provided. Frequency compensation is not required with some op amps, but for those that require it, the manufacturer's recommendations should be observed.

3-3 ♦ SLEW RATE

The slew rate of an amplifier is defined as the maximum rate at which the output of the amplifier can change, usually specified in volts per microsecond. Unlike bandwidth, slew rate is considered a large-signal parameter.

The hypothetical ideal op amp would have infinite slew rate, which of course is impossible to achieve in practice. A possible slew rate specification for an op amp might be 5 V/μs, for example. That is, the output of this particular op amp can change no more than ±5 V in 1 μs. It will be shown that for many applications, this is far from being a good approximation of the ideal.

Maximum Rate of Change and Slew Rate

It is particularly informative to consider the effects of finite slew rate in the case in which the op amp is to produce a sinusoidal waveform at its output. The maximum rate of change of a sinusoidal signal occurs at zero crossing, and is dependent on both the frequency and amplitude of the signal. The maximum rate of change (ROC_{max}) of the sinusoid is given by the following equation:

$$ROC_{max} = 2\pi f V_P \quad \text{V/s} \tag{3-14a}$$

where f is the frequency of the signal in hertz and V_P is the peak amplitude of the signal. Alternatively, this relation may be expressed as

$$ROC_{max} = \omega V_P \quad \text{V/s} \tag{3-14b}$$

where ω is the radian frequency of the sinusoid.

Equations (3-14a) and (3-14b) will be applied to signals that are to be produced at the output of the amplifier. The following examples will provide clarification.

EXAMPLE 3-5

Determine the maximum rate of change, in volts per microsecond, of the following signals:

(a) $v = 5 \sin 2\pi 1000t$ V
(b) $v = 10 \sin 2\pi 10{,}000t$ V
(c) $v = 10 \sin 5000t$ V

Solution

(a) $ROC_{max} = 2\pi 1000 \times 5$ V
$= 314{,}159$ V/s
$= 0.0314$ V/μs

(b) $ROC_{max} = 2\pi 10{,}000 \times 10$ V
$= 628{,}319$ V/s
$= 0.628$ V/μs

(c) $\text{ROC}_{max} = 5000 \times 10 \text{ V}$

$= 50{,}000 \text{ V/s}$

$= 0.05 \text{ V/}\mu\text{s}$

EXAMPLE 3-6

The 741 op amp of Fig. 3-11 has a typical slew rate of 0.5 V/μs. Determine the highest frequency that may be processed that will produce $v_o = 20 \text{ V}_{\text{P-P}}$ without exceeding the slew rate of the op amp.

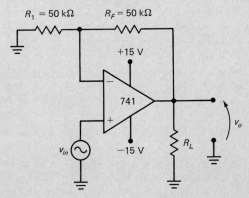

Fig. 3-11 Circuit for Example 3-6.

Solution

At the output, the amplitude and slew rate are given; we must therefore solve Eq. (3-14a) for f and convert the given slew rate in volts per microsecond into volts per second. Converting the slew rate first,

$\text{SR}_{(V/s)} = \text{SR}_{(V/\mu s)} \times 1{,}000{,}000$

$= 0.5 \text{ V/}\mu\text{s} \times 1{,}000{,}000$

$= 500{,}000 \text{ V/s}$

Now, applying Eq. (3-14a) (recall that the peak value of V_o must be used, where $V_P = V_{\text{P-P}}/2$),

$f = \dfrac{\text{SR}}{2\pi V_P}$

$= \dfrac{500{,}000 \text{ V/s}}{62.83 \text{ V}}$

$= 7.96 \text{ kHz}$

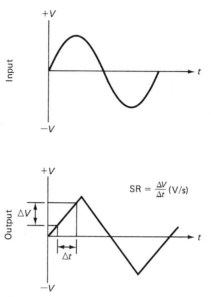

Fig. 3-12 Slew-rate distortion of a sinusoid.

The closed-loop gain of the circuit in Fig. 3-11 is 2. Based on bandwidth considerations alone, if we refer to Fig. 3-8, we would expect the amplifier to be able to handle signals up to about 500 kHz. However, the preceding example shows that the slew rate of the op amp would be exceeded long before the closed-loop break frequency is reached. At least, this is the case for an output amplitude of 20 V_{P-P}. It can easily be determined that in order to realize the full 500-kHz bandwidth of the circuit, the output must be limited to a peak amplitude of 0.159 V. For a given peak amplitude, the maximum rate of change of a sinusoidal signal will increase or decrease in direct proportion with frequency. For example, using the op amp and conditions set forth in Example 3-6, a 250-kHz output signal can have a peak amplitude of 0.318 V before slew limitation will be reached.

The slew rate of an op amp may be determined experimentally by driving it with a sinusoidal signal such that the output becomes triangular in shape, as shown in Fig. 3-12. Such a high degree of distortion results from extreme slew limitation, but it will not harm the op amp.

Slew Rate and Distortion

A sinusoidal signal that exceeds only the bandwidth of an amplifier will come out of the amplifier at a lower amplitude than if it was within the midband, but it will still have the same "sine-wave" shape. Complex signals, such as speech and music, consist of many different sinusoidal components. Those components that exceed f_c will be attenuated, relative to the lower frequencies. This has an effect on the tonal quality, or color, of the music (or speech), but is not objectionable in some applications. Figure 3-13 illustrates the input

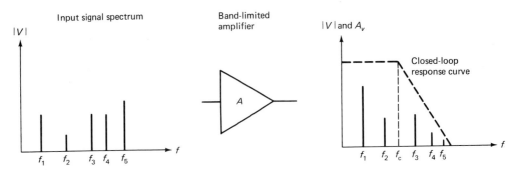

Fig. 3-13 Effects of amplifier band limiting on signal frequency spectrum.

and output of an amplifier under these band-limiting conditions, in the frequency domain. The important thing to notice here is that no new frequency components are created. Only the relative amplitudes of the spectral components have been altered. The relative phase of the components may also be altered somewhat, but this is beyond the scope of this discussion.

A signal with a complex spectrum like that of Fig. 3-13 will be rather difficult to analyze when slew rate is exceeded. Therefore, a pure sine-wave input will be assumed for this discussion. When an amplifier goes into slew limiting, the output signal is no longer exactly sinusoidal in shape. It can be shown that new spectral components will be created under these conditions. As noted earlier, when an amplifier's slew rate is exceeded by a large margin, a triangular waveform will be generated in response to a sinusoidal input. The time and frequency domain representations for this case are shown in Fig. 3-14. Here, frequency components at odd multiples of the fundamental frequency are produced. Such distortion is usually unacceptable. Some pos-

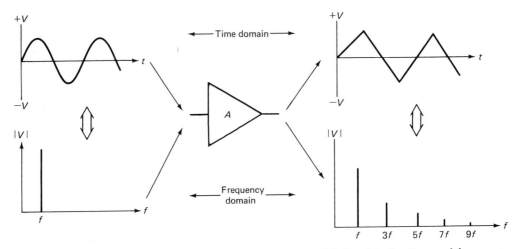

Fig. 3-14 Effects of slew limiting on a sinusoidal signal in the time and frequency domains.

sible cures for this problem are using an op amp with a higher slew rate, limiting the range of frequencies to be amplified through filtering, or limiting the output amplitude to an acceptable level.

A wide variety of op amps with different slew rates are available. For example, National Semiconductor's LM318 has a guaranteed slew rate of 50 V/µs. Another op amp from National Semiconductor, the LH0032, has a slew rate of 500 V/µs. This is an extremely fast, externally compensated op amp, with a useful bandwidth of over 70 MHz.

3-4 ♦ IC DATA SHEETS, IDENTIFICATION NUMBERS, AND PACKAGING

The data sheets for a given op amp contain a great deal of useful information relating to the performance characteristics and limitations, and packaging options. Also, many data sheets provide schematics illustrating typical applications, circuit layout guidelines, and hints for obtaining optimum device performance. Appendix A of this book contains data sheets for several popular op amps and other linear integrated circuits. The reader is urged to refer to this material throughout the following discussion.

IC Identification

Integrated circuits are identified by part numbers that are printed on them. There are several standardized numbering systems in use. One device code system is called the Pro-electron code. Another coding system, used by the military, is the JAN (Joint Army-Navy) numbering system. Components that have a JAN number have met very tight processing and performance requirements, and are of high reliability. The details of these two coding systems are available from many different sources, such as IC manufacturers and the government.

The numbering system that is most often encountered is that of the individual manufacturer of a given IC. Most IC part numbers begin with two or three letters that indicate the manufacturer of that device. A few manufacturers' ID codes are listed below.

ID code	Manufacturer
µA	Fairchild
LM, LH, LF	National Semiconductor
SN	Texas Instruments
CA, CD	RCA
MC, MFC	Motorola
NE, SE	Signetics
BB	Burr-Brown

There are many other possible examples that could be listed. Regardless of the manufacturer, if the device is a 709 op amp, for example, it must behave in the same manner and meet the same specifications as a 709 produced by another company.

An IC's part number may also specify the range of temperatures over which it will operate and the type of package and packaging material used in its construction. For example, the 741 is produced by National Semiconductor in several versions. One possibility is the LM741CN. Here, the designation LM stands for Linear Monolithic, and of course 741 is the op amp type number. The last two characters, C and N, specify the temperature range over which the op amp will operate and the package type, respectively. The temperature range for commercial-grade ICs (this is what the C indicates) is from 0° C to 70° C. Two other temperature ranges apply for industrial-grade devices (0° C to 85° C) and military-grade devices ($-55°$ C to 125° C). The last character, N, means that the IC is housed in a plastic DIP (Dual Inline Package). Another possibility would be an H, as in LM741CH. The H means that the IC is housed in a TO-5 metal can. Figure 3-15 illustrates the most commonly available IC packages.

Most often, an IC will have other numbers aside from the part number printed on it. Usually, these additional numbers are date codes. As might be

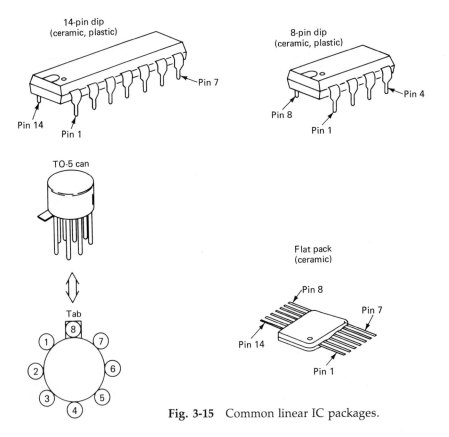

Fig. 3-15 Common linear IC packages.

expected, date codes are used to determine when the IC was manufactured. Finally, a manufacturer's logo and the country of origin will often be found printed on a given IC. These markings are essentially self-explanatory.

Data Sheets

Integrated circuit data sheets are generally divided into five sections. These sections are usually organized as follows.

General Description

This section highlights the main features, options available, and ordering information for the device.

Schematic Diagram and Pin Connections

Many IC manufacturers provide an internal schematic for the device. This can be used by the engineer to determine certain characteristics of the circuit that might not be listed in the data. The schematic can also provide insight into why the IC behaves as it does, and is usually necessary to determine the suitability of the device for various applications. Pin connection diagrams illustrate the package options that are available.

Absolute Maximum Ratings

This section lists such information as the maximum supply voltage, differential input voltage, power dissipation, storage and operating temperature ranges, soldering temperature, and other important data. These specifications should never be exceeded or the device may be destroyed, or its performance permanently impaired.

Electrical Characteristics

This section lists minimum, typical, and maximum values for various device parameters. For example, in the case of an op amp, data relating to offset voltages and currents, CMRR, slew rate, input and output resistance, and temperature coefficient characteristics are provided. Generally, these data are used to determine the suitability of the device to a given application.

Performance Characteristic Curves

This section provides device characteristics in graphical form. In this way, characteristics that vary widely with frequency, temperature, supply voltage, and other parameters can be obtained very quickly.

Typical Applications

This is a very useful section when quick circuit design is necessary. The manufacturer will usually provide examples of standard and rather unconventional device applications that may be helpful to the device user.

Chapter Review

A well-designed op amp-based circuit will usually function quite well for many years if the device limitations are considered and accounted for. Although nearly ideal in many applications, op amps do have some inherent error sources. Offset voltage and offset current, finite gain, finite input resistance, nonzero output resistance, finite bandwidth, and finite slew rate are the main factors that should be considered in most applications.

Offset voltage and current can be nulled out by adding appropriate components to the circuit. Bandwidth can often be altered by varying the closed-loop gain of the op amp or by using external compensating networks. Bandwidth limitations are important in the design of small-signal amplifiers, while slew rate is important in the design of circuits that process large signals.

Device specifications are provided in manufacturers' data sheets. Data sheets provide information on device packaging, performance, and applications. Most ICs are available from several sources; however, a given device will behave in the same way no matter who manufactures it.

Questions

3-1. What term is given to the gain that is associated with offset components produced at the input of an op amp?

3-2. In BJT-based op amp designs, what is primarily responsible for the presence of input offset voltage? Input offset current?

3-3. In general, will op amps with FET input stages normally have higher or lower input bias currents?

3-4. Which frequency-related op amp parameter is of greater significance under large-signal operating conditions?

3-5. What circuit parameters determine the stability of an amplifier?

3-6. Given a Bode plot of an amplifier's frequency response, when will the amplifier be unconditionally stable? Conditionally stable? Unstable?

3-7. Describe two ways to determine the loop gain of an op amp circuit.

3-8. In general, for sinusoidal waveforms, when does the signal undergo its maximum rate of change?

3-9. What type of stability should be expected of an op amp for which the closed-loop gain curve intersects the open-loop gain curve at a point at which gain rolls off at -20 dB/decade?

Problems

3-1. A certain op amp has $I_{B(+)} = 8$ nA and $I_{B(-)} = 12$ nA. Determine I_{Io}.

3-2. Based on the individual bias currents given in Prob. 3-1, determine I_B for the op amp.

3-3. Refer to Fig. 3-4. $R_F = 270$ kΩ, $R_1 = 18$ kΩ. Determine the optimum value for R_B.

3-4. Refer to Fig. 3-2a. Given $R_F = 150$ kΩ, $R_1 = 15$ kΩ, and $I_B = 20$ nA, determine A_N and V_{OB}.

3-5. Refer to Fig. 3-2a. Given $R_F = 100$ kΩ, $R_1 = 50$ kΩ, and $I_B = 15$ nA, determine A_N and V_{OB}.

3-6. Refer to Fig. 3-5a. Given $V_{io(max)} = \pm 20$ mV, $V_s = \pm 15$ V, $R_F = 27$ kΩ, and $R_1 = 4.7$ kΩ; design the offset compensation network using $R_Z = 50$ Ω. Determine the correct value for R_B.

3-7. Repeat Prob. 3-6 given $V_{io(max)} = \pm 8$ mV and $R_Z = 27$ Ω.

3-8. Refer to Fig. 3-6. Given that $V_{io(max)} = \pm 15$ mV, supply voltage $V_S = \pm 10$ V, $R_1 = 3.3$ kΩ, and $R_F = 33$ kΩ, determine the component values required by the offset compensation circuitry.

3-9. Repeat Prob. 3-8, given $R_1 = 2.2$ kΩ and $R_F = 27$ kΩ.

3-10. Refer to Fig. 3-9. Given $R_2 = 180$ kΩ and $R_3 = 2$ kΩ, determine the correct values for C_1, C_2, R_1, and R_4. Determine the bandwidth of the circuit.

3-11. Refer to Fig. 3-9. Given $R_2 = 68$ kΩ, determine the component values required such that $A_v = -10$.

3-12. A certain op amp is characterized by the Bode plot in Fig. 3-16. Determine θ_t, θ_{pm}, and the loop gain for the following closed-loop gain values, and state whether the circuit is stable or not for each case: **(a)** $A_v = 80$ dB, **(b)** $A_v = 50$ dB, **(c)** $A_v = 0$ dB.

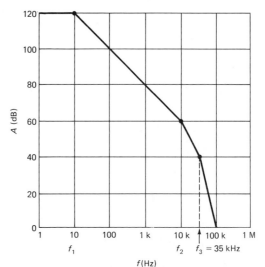

Fig. 3-16

3-13. Refer to Fig. 3-10. Give A_v = 40 dB, determine θ_t, θ_{pm}, and the loop gain of the circuit. Is the circuit stable?

3-14. Determine the maximum instantaneous rate of change for each of the following voltages: **(a)** v_1 = 5 sin $2\pi 20{,}000t$ V, **(b)** v_2 = 0.5 sin $5000t$ V, **(c)** v_3 = 10 cos $(10{,}000t + 0.2)$ V.

3-15. A certain op amp has SR = 2 V/μs. Given A_v = 20, determine the highest frequency (sinusoidal) that can be produced at the output such that V_o = 24 V_{P-P}.

3-16. A certain application requires an op amp to operate over a frequency range from 20 Hz to 20 kHz with an output voltage swing of 20 V_{P-P}. Determine the minimum slew rate required of an op amp as required in this application.

3-17. The National Semiconductor LH0101 op amp has BW = 5 MHz, SR = 10 V/μs. Determine the maximum frequency that may be produced at the output of this op amp if V_o = 26 V_{P-P}.

3-18. Typical high-quality stereo power amplifiers will produce output signals at frequencies up to 100 kHz, 60 V_{P-P}. Determine the slew rate of an amplifier that meets these criteria.

OP AMP APPLICATIONS: PART A

There are literally thousands of applications in which op amps are used. This chapter presents several useful operational amplifier circuits that, for the most part, are extensions of the circuits covered previously. Uses of the op amp in summing amplifiers, voltage-controlled current sources, current-controlled voltage sources, and differential amplifier circuits are presented. Instrumentation amplifier circuits and applications are also introduced.

4-1 ♦ THE SUMMING AMPLIFIER

A summing amplifier is an amplifier whose output is proportional to the sum of the signals applied to its inputs. Summing amplifiers are used in applications in which a linear mixing of several signals is required. One application is in the recording of music, where the signals produced by various musical instruments and voices must be combined and processed to produce a record.

General Summing Amplifier Analysis

An n-input op amp-based summing amplifier is shown in Fig. 4-1. The op amp in the inverting configuration (with feedback) is nearly ideal for use as a summing amplifier. The rationale behind this statement will be made clear in the following analysis. The concept of the virtual ground point is central to the operation of the inverting summing amplifier.

Recall that the inverting op amp with feedback, such as in Fig. 4-1, will drive its output such that $v_{id} = 0$ V, provided that A_{OL} is very large (which is generally true for most op amps). In Fig. 4-1, this causes the inverting input of the op amp to become a virtual ground point. Now, assuming that the current entering the inverting terminal is essentially zero, we may apply Kirchhoff's current law (KCL) to node A, producing

$$-i_F = i_1 + i_2 + i_3 + \cdots + i_n \qquad (4\text{-}1)$$

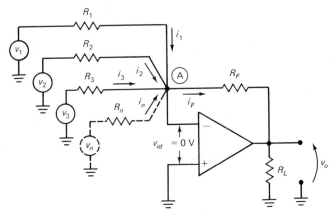

Fig. 4-1 Op amp summing amplifier.

Because of the virtual ground, we may define the current terms of Eq. (4-1) as follows:

$$i_1 = \frac{v_1}{R_1} \qquad i_2 = \frac{v_2}{R_2} \qquad i_3 = \frac{v_3}{R_3} \qquad \cdots \qquad i_n = \frac{v_n}{R_n} \qquad (4\text{-}2)$$

Substituting Eq. (4-2) into (4-1) yields

$$-i_F = \frac{v_1}{R_1} + \frac{v_2}{R_2} + \frac{v_3}{R_3} + \cdots + \frac{v_n}{R_n} \qquad (4\text{-}3)$$

Recall from Chap. 2 that v_o is defined by

$$v_o = i_F R_F \qquad (4\text{-}4)$$

Substituting Eq. (4-3) into (4-4) produces

$$v_o = -R_F \left(\frac{v_1}{R_1} + \frac{v_2}{R_2} + \frac{v_3}{R_3} + \cdots + \frac{v_n}{R_n} \right) \qquad (4\text{-}5a)$$

Alternatively, Eq. (4-5a) may be expanded to produce

$$v_o = -v_1 \frac{R_F}{R_1} - v_2 \frac{R_F}{R_2} - v_3 \frac{R_F}{R_3} - \cdots - v_n \frac{R_F}{R_n} \qquad (4\text{-}5b)$$

We now have two general expressions for the output voltage of a summing amplifier.

There are a few important conclusions that may be drawn, based on the preceding analysis. First, notice that if the number of inputs applied to the summer is reduced to one, Eq. (4-5b) reduces to the familiar inverting amplifier output voltage expression. It is also instructive to note that Eq. (4-5b) indicates that the output of the summer is the superposition of each input voltage, multiplied by a scaling constant (gain). Finally, all inputs are independent of one another. This means that a given input may be adjusted for any desired gain without affecting the other inputs. Because each input source is effectively driving to ground through a resistance, there is no source interaction at the input. This is a very desirable characteristic.

EXAMPLE 4-1

Refer to Fig. 4-1. Assume that only the first three input sources are used, with $v_1 = V_1 = +50$ mV, $v_2 = V_2 = +100$ mV, and $v_3 = V_3 = -80$ mV. The resistor values are as follows: $R_1 = 1$ kΩ, $R_2 = 2.2$ kΩ, $R_3 = 4$ kΩ, and $R_F = 100$ kΩ. Determine the following quantities: I_1, I_2, I_3, I_F, and V_o.

Solution

The input currents are found using Eq. (4-2):

$$I_1 = \frac{50 \text{ mV}}{1 \text{ kΩ}} \qquad I_2 = \frac{100 \text{ mV}}{2.2 \text{ kΩ}} \qquad I_3 = \frac{-80 \text{ mV}}{4 \text{ kΩ}}$$

$$= 50 \text{ μA} \qquad\qquad = 45.5 \text{ μA} \qquad\qquad = -20 \text{ μA}$$

I_F is the sum of the input currents (with opposite polarity), and, applying Eq. (4-3), we obtain

$$I_F = -(I_1 + I_2 + I_3)$$
$$= -(50 \text{ μA} + 45.5 \text{ μA} - 20 \text{ μA})$$
$$= -75.5 \text{ μA}$$

Now, applying Eq. (4-4), we obtain V_o:

$$V_o = I_F R_F$$
$$= -75.5 \text{ μA} \times 100 \text{ kΩ}$$
$$= -7.55 \text{ V}$$

It can easily be shown that Eq. (4-5a) and (4-5b) will also produce the correct value for V_o. This is left for the reader to verify.

The virtual ground point in a summing amplifier is often referred to as a summing point, summing junction, or summing node. Recall that the summation notation was used in the block diagram of the op amp with feedback

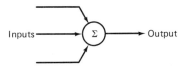

Fig. 4-2 Block diagram representation of a summing amplifier.

in Chap. 2 (Figs. 2-3 and 2-6). Often, subsections of complex systems are represented by functional symbols in block diagram form. A summing amplifier is usually represented as shown in Fig. 4-2.

Practical Summing Amp Considerations

Figure 4-1 was drawn specifically to emphasize the nodal characteristic of the summing junction. In practice, summing amps usually are not drawn this way. Figure 4-3 is more representative of the typical summing amplifier schematic diagram. The usual rules relating to offset voltages and currents apply just as strongly to summing amps (and all other op amp circuits) as to the basic amplifier configurations covered in the earlier chapters. Resistor R_B was included in Fig. 4-3 for bias current compensation purposes. Such a resistor may be required if the bias currents produce an unacceptable offset at the circuit's output. In general, for a summing amp with n inputs, the value of R_B is given by

$$R_B = R_F \| R_1 \| R_2 \| R_3 \| \cdots \| R_n \tag{4-6}$$

This is the effective resistance to ground, as "seen" by the inverting input of the op amp. Equation (4-6) is derived by deactivating all voltage sources driving the inverting input (including the output of the op amp) and finding the equivalent Thevenin resistance.

Ideally, a summing amplifier may have any number of inputs. In practice, the number of inputs that can be used may depend on several factors. The output of the op amp could be driven to one supply rail or the other, if many

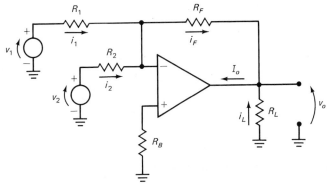

Fig. 4-3 Typical two-input summing amplifier.

inputs of the same polarity are summed. Similarly, if the sum of the input currents exceeds the current source/sink capability of the op amp, the output will be in error. These two problems can occur in any op amp circuit, not just in summing amplifier circuits, but because of the greater number of inputs, it is more difficult to keep track of the ranges of v_o and i_F that are expected to occur. Again, keeping resistor values higher than a few kilohms will help reduce the load on the output by the feedback circuit.

There is one problem that may occur in inverting summing amplifiers that is not obvious, and that may catch the unwary designer off guard. That is, the bandwidth of the summing amplifier will be decreased in proportion to the effective resistance between the inverting terminal and ground. Consider the summing amplifier in Fig. 4-3. Let us assume that the op amp is a 741, with $R_F = 50\text{ k}\Omega$, $R_1 = 10\text{ k}\Omega$, and $R_2 = 1\text{ k}\Omega$. Let us also assume that $v_2 = 0\text{ V}$. Now, if v_1 is a time-varying voltage (a sine wave perhaps), then, applying Eq. (4-5), the gain of the op amp relative to source v_1 is given by

$$A_{v(1)} = -\frac{50\text{ k}\Omega}{10\text{ k}\Omega}$$

$$= -5$$

Based on the fact that the 741 has a gain-bandwidth product of 1 MHz, using previously developed analysis techniques, one might expect the bandwidth to be BW = 167 kHz. For the normal inverting amp the bandwidth of the circuit would indeed be 167 kHz. However, the actual bandwidth of the summer is dependent on the noise gain of the circuit. Noise gain is defined as

$$A_N = 1 + \frac{R_F}{R_{in(eq)}} \tag{4-7}$$

where, in general, $R_{in(eq)} = R_1 \parallel R_2 \parallel R_3 \cdots \parallel R_n$

Let us now apply this information to a typical summing amp. For Fig. 4-3, based on the component values given, we find

$$R_{in(eq)} = R_1 \parallel R_2$$
$$= 10\text{ k}\Omega \parallel 1\text{ k}\Omega$$
$$= 909\text{ }\Omega$$

$$A_n = \frac{R_F}{R_{in(eq)}} + 1$$
$$= \frac{50\text{ k}\Omega}{909\text{ }\Omega} + 1$$
$$= 56$$

Now, the actual bandwidth of the circuit is found by applying the GBW relationship:

$$BW = \frac{GBW}{A_n}$$

$$= \frac{1 \text{ MHz}}{56}$$

$$= 17.9 \text{ kHz}$$

This is much lower than the bandwidth of an equivalent inverting amplifier. In general, the more inputs that are summed, the lower the effective bandwidth will become. If time-varying signals are to be mixed, this is a very important consideration. Keeping the noise gain A_N of the summer as low as possible helps to keep the bandwidth high, but as more inputs are summed, the bandwidth will suffer nevertheless.

EXAMPLE 4-2

Refer to Fig. 4-1. Assume that this circuit is used as a three-input summing amplifier with the following component values: $R_1 = R_2 = R_3 = 1 \text{ k}\Omega$, $R_F = 47 \text{ k}\Omega$, and the op amp is a 741. What is the effective bandwidth of the circuit?

Solution

The equivalent external resistance is

$$R_{in(eq)} = 1 \text{ k}\Omega \parallel 1 \text{ k}\Omega \parallel 1 \text{ k}\Omega$$

$$= 333.3 \text{ }\Omega$$

The noise gain, based on the equivalent resistance, is

$$A_N = \frac{47 \text{ k}\Omega}{333.3 \text{ }\Omega} + 1$$

$$= 142$$

Since GBW = 1 MHz,

$$BW = \frac{1 \text{ MHz}}{142}$$

$$= 7042 \text{ Hz}$$

An input signal containing frequency components above 7042 Hz will suffer a significant amplitude reduction.

There are many possible design techniques used to overcome the summing amplifier bandwidth reduction problem. One method is to use an op amp with a very high bandwidth in the design of the circuit. Another possibility is to scale the input voltages individually through separate op amps, then combine them into a summer with equal-valued input resistors. The use of equal-valued input resistors results in the highest value of $R_{in(eq)}$ for the summing amp. The term *scaling* refers to multiplying the various inputs by different factors (gains or scaling constants). This is also often referred to as weighting the inputs. More will be said about this in the discussion that follows.

Averagers

A special case of the summing amplifier is the averager. The output of an averager is proportional to the average of the input variables. In equation form, for n inputs, this is given as

$$V_{avg} = \frac{V_1 + V_2 + V_3 + \cdots + V_n}{n} \tag{4-8}$$

The op amp averager is designed such that the expression for its output voltage is of the same form as Eq. (4-8). The circuit of Fig. 4-1 could represent an n-input averager, assuming that the feedback and input resistor ratios are correct. To design an averager, we use the rule that given some particular value of R_F, the input resistor values are chosen such that

$$R = nR_F \tag{4-9}$$

The application of Eq. (4-9) results in an averager with equally weighted inputs. Since the op amp summer inverts, the sign of the output will be opposite to that of the true average, but this can be easily accounted for and corrected if necessary.

EXAMPLE 4-3

Design a three-input averager, given $R_F = 10$ kΩ. Determine V_o given $V_1 = 2.5$ V, $V_2 = -1.5$ V, and $V_3 = 5$ V.

Solution

The input resistor values are found using Eq. (4-9):

$$R_1 = R_2 = R_3 = 30 \text{ k}\Omega$$

The circuit appears as shown in Fig. 4-4.

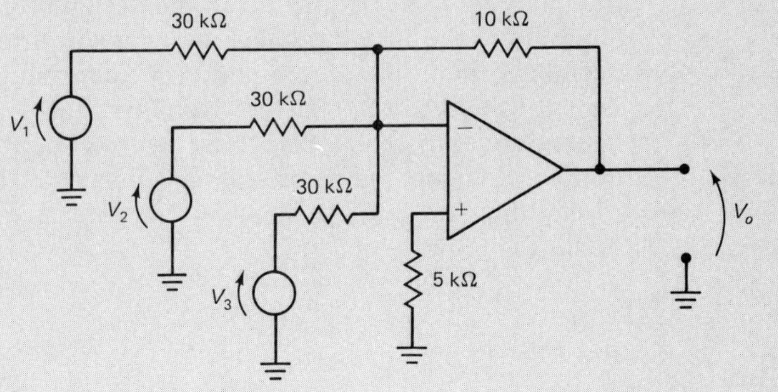

Fig. 4-4 Circuit for Example 4-3.

The output voltage may be determined using standard op amp analysis techniques or using Eq. (4-5a) or (4-5b). However, since this is an averager, it is easiest to use Eq. (4-8), making use of the fact that $V_o = -V_{avg}$.

$$V_{avg} = \frac{2.5 \text{ V} - 1.5 \text{ V} + 5 \text{ V}}{3}$$

$$= 2 \text{ V}$$

$$V_o = -2 \text{ V}$$

It is suggested that the reader verify that $V_o = -2$ V, using the alternative analysis techniques.

Occasionally, a weighted average of input variables is required. That is, one or more of the averager's inputs is to have a greater or lesser weight than the other normally weighted inputs. In such a case, the input resistors on the normally weighted inputs are determined using Eq. (4-9). An input that is to have a greater or lesser weighting will have its associated input resistance decreased or increased in inverse proportion to its weighting factor. For example, if V_1 in Fig. 4-4 was to be weighted twice the normal, its new value would be given by

$$R_W = \frac{R}{W}$$

$$= \frac{30 \text{ k}\Omega}{2}$$

$$= 15 \text{ k}\Omega$$

Conversely, if this input was to be weighted at half the normal, its new value would be

$$R_W = \frac{R}{W}$$

$$= \frac{30 \text{ k}\Omega}{0.5}$$

$$= 60 \text{ k}\Omega$$

From the circuit analysis standpoint, a weighted averager would require the use of standard op amp analysis methods, as the simple average formula would no longer be valid.

Additional Summing Amplifier Circuits

In op amp-based circuits and systems, it is common to find signals being applied to summing amplifiers at both the inverting and noninverting input terminals of a given op amp. The analysis of such amplifiers is relatively straightforward, and builds upon the fundamentals covered thus far. A representative circuit is shown in Fig. 4-5.

One of the more intuitive methods of analyzing Fig. 4-5 relies on the use of superposition. Because the op amp operates in a linear manner (superpositon applies to linear systems), the output of this circuit can be considered to be the superposition, or sum, of three distinct components, as defined below:

$$V_o = V_{o1} + V_{o2} + V_{o3} \tag{4-10}$$

where V_{o1} = output due to source V_1

V_{o2} = output due to source V_2

V_{o3} = output due to source V_3

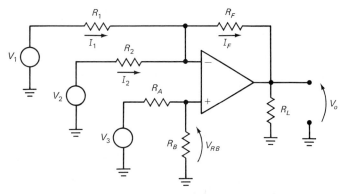

Fig. 4-5 Op amp circuit with signals applied to both inputs.

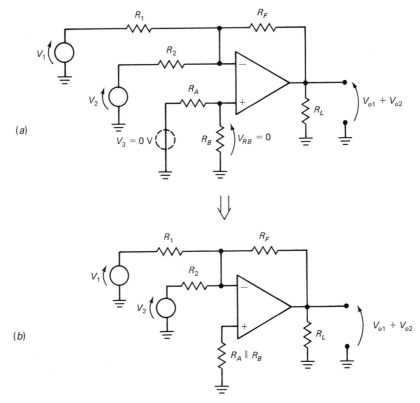

Fig. 4-6 Circuit of Fig. 4-5, (a) with source V_3 inactive, and (b) simplified equivalent circuit.

Output components V_{o1} and V_{o2} are determined using the analysis methods covered in the last section, with source V_{o3} reduced to zero. This replacement is illustrated in Fig. 4-6. Here, it is apparent that the circuit reduces to a simple summing amplifier, whose output is given by Eq. (4-5a) or (4-5b). In mathematical terms,

$$V_{o1} = -V_1 \frac{R_F}{R_1} \qquad V_{o2} = -V_2 \frac{R_F}{R_2}$$

thus

$$V_{o1} + V_{o2} = -\left(V_1 \frac{R_F}{R_1} + V_2 \frac{R_F}{R_2}\right)$$

The third output signal component present is determined by zeroing sources V_1 and V_2 and applying the noninverting amplifier analysis techniques. This replacement is shown in Fig. 4-7. Source V_3 drives the op amp's noninverting input through a voltage divider (V_{RB} is the voltage applied to

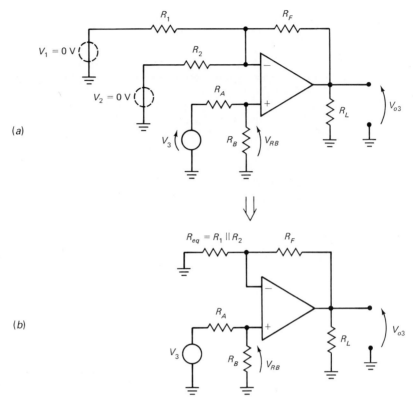

Fig. 4-7 Circuit of Fig. 4-5, (a) with sources V_1 and V_2 inactive, and (b) simplified equivalent circuit.

the input terminal). The output of the voltage divider is given by

$$V_{RB} = V_3 \frac{R_B}{R_A + R_B}$$

The gain of the op amp to inputs on the noninverting terminal is found by applying the usual noninverting gain equation, with the equivalent value of the input resistance being the parallel equivalent of R_1 and R_2. In equation form, the output caused by V_3 is given by

$$V_{o3} = V_3 \left(\frac{R_B}{R_A + R_B}\right) \left(\frac{R_{eq} + R_F}{R_{eq}}\right)$$

The overall output of the amplifier is found by combining the equations just derived, and substituting into Eq. (4-10).

$$V_o = V_3 \left(\frac{R_B}{R_A + R_B}\right) \left(\frac{R_{eq} + R_F}{R_{eq}}\right) - V_1 \frac{R_F}{R_1} - V_2 \frac{R_F}{R_2} \quad (4\text{-}11)$$

Equation (4-11) can easily by expanded to handle additional inputs to the inverting terminal by including more negative terms as required.

EXAMPLE 4-4

Refer to Fig. 4-5. Assume that $R_1 = 2.2$ kΩ, $R_2 = 4.7$ kΩ, $R_A = 10$ kΩ, $R_B = 10$ kΩ, $R_F = 33$ kΩ, $V_1 = -10$ mV, $v_2 = 100 \sin 500t$ mV, and $V_3 = +100$ mV. Determine the expressions for i_F and v_o. Sketch v_o.

Solution

Let us first determine the value of R_{eq}.

$$R_{eq} = R_1 \| R_2 = 2.2 \text{ k}\Omega \| 4.7 \text{ k}\Omega = 1.5 \text{ k}\Omega$$

The expression for v_o is found by direct application of Eq. (4-11).

$$v_o = 100 \text{ mV} \left(\frac{10 \text{ k}\Omega}{10 \text{ k}\Omega + 10 \text{ k}\Omega}\right) \left(\frac{1.5 \text{ k}\Omega + 33 \text{ k}\Omega}{1.5 \text{ k}\Omega}\right)$$
$$- (-10 \text{ mV}) \left(\frac{33 \text{ k}\Omega}{2.2 \text{ k}\Omega}\right) - (100 \sin 500t \text{ mV}) \left(\frac{33 \text{ k}\Omega}{4.7 \text{ k}\Omega}\right)$$
$$= (100 \text{ mV})(0.5)(23) + (10 \text{ mV})(15) - (100 \sin 500t \text{ mV})(7)$$
$$= 1.15 \text{ V} + 0.15 \text{ V} - 0.7 \sin 500t \text{ V}$$
$$= 1.3 - 0.7 \sin 500t \text{ V}$$

The sketch of v_o is shown in Fig. 4-8.

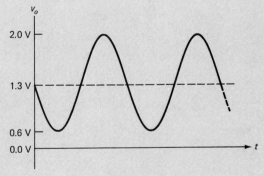

Fig. 4-8 Output waveform for Example 4-4.

The expression for i_F may be found based on the relationship

$$i_F = \frac{V_{RF}}{R_F}$$

The inverting terminal is not at ground potential but is at the same potential as the noninverting terminal; thus $V_{RF} = V_0 - V_{RB}$. This yields

$$i_F = \frac{V_0 - V_{RB}}{R_F}$$

$$= \frac{1.3 - 0.7 \sin 500t - 0.05}{33 \text{ k}\Omega}$$

$$= 37.8 - 21.2 \sin 500t \ \mu A$$

4-2 ◆ DIFFERENTIAL AND INSTRUMENTATION AMPLIFIERS

It stands to reason that since the op amp is designed around the differential amplifier, the op amp itself can be used as a differential amplifier. Op amp differential amplifiers generally behave quite ideally and find use in many applications.

The Op Amp–Based Differential Amplifier

A simple op amp–based differential amplifier is shown in Fig. 4-9. Correct operation of the differential amplifier requires that the op amp amplify V_1 and V_2 by equal but opposite amounts. That is, the gain magnitudes for inverting and noninverting inputs must be equal. This is achieved by meeting the following requirements:

$$R_1' = R_1 \quad \text{and} \quad R_F' = R_F$$

It is apparent that the differential amplifier is a special case of the amplifier that was shown in Fig. 4-5. Therefore, Eq. (4-11) may easily be modified to

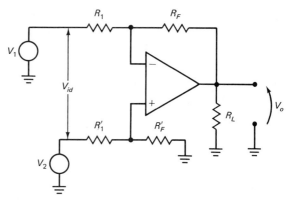

Fig. 4-9 Op amp differential amplifier.

yield the output of the differential amplifier. The required analysis expression is derived as follows:

$$V_o = V_2 \left(\frac{R_F'}{R_F' + R_1'}\right)\left(\frac{R_1 + R_F}{R_1}\right) - V_1 \frac{R_F}{R_1}$$

Now, since $R_F' = R_F$ and $R_1' = R_1$, we may obtain

$$V_o = V_2 \frac{R_F'}{R_1'} - V_1 \frac{R_F}{R_1}$$

Factoring equivalent terms,

$$V_o = (V_2 - V_1)\frac{R_F}{R_1} \tag{4-12}$$

Equation (4-12) indicates that the output of the differential amplifier is proportional to the difference between the two input voltages. The gain of the differential amplifier is easily set to the desired value by choosing the correct ratio of R_F to R_1.

Recall that ideally the output of a differential amplifier is zero when equal voltages exist at the input terminals ($V_o = 0$ V when $V_1 = V_2$). Such conditions produce a common-mode input voltage (V_{iCM}). A real differential amplifier will produce an output in response to the common-mode input. The response of the amplifier to a common-mode voltage is given by

$$V_{oCM} = A_{CM}V_{iCM}$$

The reader should also recall that the ability of a differential amplifier to reject common-mode signals and amplify differential signals is quantified by the common-mode rejection ratio, which is normally expressed in decibels.

$$\text{CMRR} = 20 \log \left|\frac{A_d}{A_{CM}}\right|$$

It is desirable for CMRR to be as large as possible. Op amps have relatively high CMRRs. The 741, for example, has a typical CMRR of 90 dB. Other op amps are available with CMRR ratings of >110 dB. For a circuit such as that in Fig. 4-9, assuming that the op amp itself has a very high CMRR, the main limiting factor in terms of CMRR is the matching between $R_1 - R_1'$ and $R_F - R_F'$. Laser-trimmed precision resistors are often used in applications in which CMRR optimization is of prime importance.

EXAMPLE 4-5

The circuit in Fig. 4-9 has the following component values: $R_1' = R_1 = 10$ kΩ, $R_F' = R_F = 270$ kΩ. Given $V_1 = +2.00$ V and $V_2 = +2.05$ V, determine V_o for the following CMRR values: **(a)** CMRR = ∞, **(b)** CMRR = 80 dB, **(c)** CMRR = 40 dB. Assume that A_{CM} is possible.

Solution

The desired output of the circuit (the output due to V_{id}) is found using Eq. (4-12). For CMRR = ∞, this is the total output of the circuit (assuming that all offsets have also been nulled out).

(a) $V_o = V_{od} = (2.05 \text{ V} - 2.00 \text{ V}) \times 27$

$\qquad = 1.35$ V

(b) The output of the circuit with finite CMRR is composed of V_{od} and a common-mode output component. The common-mode gain of the circuit [for part **(b)**] is

$$A_{CM} = \frac{A_d}{\log^{-1}\left(\frac{\text{CMRR}_{db}}{20}\right)}$$

$\qquad = 2.7 \times 10^{-3}$

The common-mode input is defined as the average of the input voltages [see Eq. (1-40)].

$V_{iCM} = 2.025$ V

The common-mode output is

$V_{oCM} = A_{CM} V_{iCM}$

$\qquad = 2.7 \times 10^{-3} \times 2.025$ V

$\qquad = 5.47$ mV

Now, the total output of the differential amplifier is found:

$V_o = V_{oCM} + V_{od}$

$\qquad = 5.47 \text{ mV} + 1.35 \text{ V}$

$\qquad \cong 1.355$ V

(c) The common-mode gain and output for CMRR = 40 dB is found in the same way as for part b. This analysis yields

$$A_{CM} = 0.27 \quad V_{oCM} = 547 \text{ mV}$$

The overall output voltage in this case is

$$V_o = 1.897 \text{ V}$$

The preceding example demonstrates the importance of high CMRR in the use of differential amplifiers. It is not unusual for common-mode voltages to greatly exceed differential signals, in practical situations. Also, it was assumed that the common-mode gain was positive. Depending on the op amp used, A_{CM} may be negative (inv.), as would occur for a 741 op amp.

Common-mode voltages may also be time-varying in nature. The differential amplifier should also reject this type of interference. Unfortunately, CMRR decreases with frequency. A plot of CMRR versus frequency for the LM301 is shown in Fig. 4-10. Most op amps will exhibit a similar CMRR response.

Instrumentation Amplifers

The instrumentation amplifier is basically a differential amplifier that has been optimized in terms of dc parameters. In particular, CMRR, offset current, offset voltage, bias current, and temperature drift specifications are very

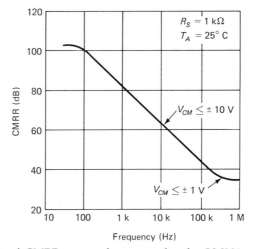

Fig. 4-10 Plot of CMRR versus frequency for the LM301 op amp. (*Courtesy of National Semiconductor Corp.*)

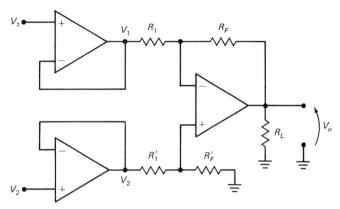

Fig. 4-11 Instrumentation amplifier with voltage-follower buffered inputs.

tightly controlled. In order to minimize loading of the input source(s), the basic instrumentation amplifier will have buffered inputs. This basic circuit is shown in Fig. 4-11. The differential gain of the circuit is given by Eq. (4-12).

One of the most prolific uses of the instrumentation amplifier is in the processing of the output of a bridge circuit. This application is shown in Fig. 4-12. The bridge is composed of four resistors, with R_1 being a thermistor. In this particular case, let us assume that $R_2 = R_3 = R_4 = R_1$ at, say, 25°C. At this temperature, the bridge is said to be in balance, and $V_1 = V_2$; hence $V_{id} = 0$ V. Should the temperature of R_1 change, the value of R_1 will also change, unbalancing the bridge and causing $V_1 \neq V_2$. Such circuits are frequently used to sense temperature changes and provide input to process control systems. Because of the extremely high input resistance of the instrumentation amplifier (R_{in} is typically > 1000 MΩ for voltage followers), loading of the bridge is essentially nonexistent. For operation at 25°C (bridge in balance), $V_1 = V_2 = V_{ref}/2$. This is a common-mode voltage at the input of the amplifier. If the CMRR of the op amp is very large, $V_o \cong 0$ V. This

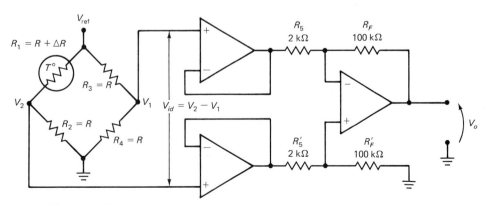

Fig. 4-12 Instrumentation amplifier used to process the output of an active bridge.

shows why CMRR is so important in differential and instrumentation amplifier applications. The following example will illustrate the analysis of the instrumentation amplifier.

EXAMPLE 4-6

Refer to Fig. 4-12. At $0°$ C, $R_1 = R_2 = R_3 = R_4 = 10.000$ kΩ. The resistance of the thermistor varies linearly from 9.000 kΩ at $+10°$C to 11.000 kΩ at $-10°$ C (a negative temperature coefficient). Determine V_o at the following temperatures. Assume $V_{ref} = +10$ V, CMRR $= \infty$. **(a)** $T = +5°$ C, **(b)** $T = +10°$ C, **(c)** $T = -5°$ C, **(d)** $T = -10°$ C.

Solution

First, the differential gain of the instrumentation amplifier is determined:

$$A_d = \frac{R_F}{R_1}$$

$$= \frac{100 \text{ k}\Omega}{2 \text{ k}\Omega}$$

$$= 50$$

For each of the temperatures given, voltages V_1 and V_2 may be determined by simply applying the voltage divider equation to both halves of the bridge. Before finding the voltages, however, we must first determine the resistance of the thermistor at the temperatures of interest. Because the thermistor varies linearly over the stated temperature range, its resistance at a given temperature may be found using the relationship

$$R_1 = R + \Delta R$$

where $\Delta R = -500T$

This yields the following data:

Temperature	ΔR	R_1
$+10°$ C	-1 kΩ	9.00 kΩ
$+5°$ C	-500 Ω	9.50 kΩ
$-5°$ C	$+500$ Ω	10.50 kΩ
$-10°$ C	$+1$ kΩ	11.00 kΩ

Based on these resistance values, the differential input voltages and the corresponding output voltages are found as follows:

(a) $T = +5°\text{ C}$

$$V_1 = V_{ref}\left(\frac{R_4}{R_3 + R_4}\right) \qquad V_2 = V_{ref}\left(\frac{R_2}{R_2 + R_1}\right)$$

$$= 5\text{ V}\left(\frac{10\text{ k}\Omega}{10\text{ k}\Omega + 10\text{ k}\Omega}\right) \qquad = 5\text{ V}\left(\frac{10\text{ k}\Omega}{10\text{ k}\Omega + 9.5\text{ k}\Omega}\right)$$

$$= 2.500\text{ V} \qquad = 2.564\text{ V}$$

$$V_o = A_d(V_2 - V_1)$$
$$= 50(0.064\text{ V})$$
$$= 3.2\text{ V}$$

In a similar manner, the voltages for the remaining temperatures are found.

(b) $T = +10°\text{ C}$

$V_1 = 2.500\text{ V} \qquad V_2 = 2.632\text{ V}$

$V_o = 50(0.132\text{ V})$
$ = 6.6\text{ V}$

(c) $T = -5°\text{ C}$

$V_1 = 2.500\text{ V} \qquad V_2 = 2.439\text{ V}$

$V_o = 50(-0.61\text{ V})$
$ = -3.05\text{ V}$

(d) $T = -10°\text{ C}$

$V_1 = 2.500\text{ V} \qquad V_2 = 2.381\text{ V}$

$V_o = 50(-0.119)$
$ = -5.95\text{ V}$

In addition to the temperature-sensing application just presented, strain could be determined if the thermistor was replaced with a strain gage. A strain gage is a resistor whose value varies in proportion to the strain to

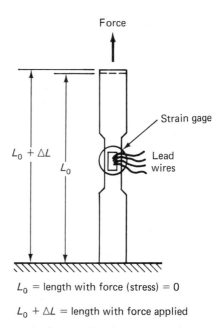

L_0 = length with force (stress) = 0

$L_0 + \Delta L$ = length with force applied

Fig. 4-13 Typical active bridge application.

which it is subjected. For example, in the testing of materials for their mechanical properties, a strain gage may be bonded to a sample of, say, titanium which is to be pulled in a tensile testing machine. This is shown in Fig. 4-13. As the test specimen is pulled, the strain gage stretches and increases in resistance, producing a proportional differential output voltage. This voltage is amplified and processed as required to determine the stress-strain relationships of the sample. The stress is the force per unit cross-sectional area of the test specimen.

In the typical instrumentation amplifier application, the bridge may be located some distance from the amplifier. This is generally the case when the bridge environment is hostile (high temperature, pressure, or vibration) or there is just not enough room for the amplifier and its associated electronics, as in the case of the strain gage. Consequently, long connecting leads must be used between the bridge and the instrumentation amp. Under these conditions, shielding of the leads is a wise precaution. Shielding the conductors helps prevent stray electromagnetic fields from inducing noise voltages onto the signal lines. Of course, shielding is not 100 percent effective, and some induced voltage may well exist. In order to help reduce the effects of this noise further, the lines that are connected to the bridge output will also be twisted with each other. This twisting tends to make the induced noise be of equal amplitude on each line, producing a common-mode noise signal. The ac CMRR is important for rejection of this type of noise. This is illustrated in Fig. 4-14. If the instrumentation amp has a high enough CMRR, the noise that is picked up by the leads will usually cause little error at the output. Some troublesome sources of ac noise are motors, fluorescent lights, and power switching circuits containing SCRs.

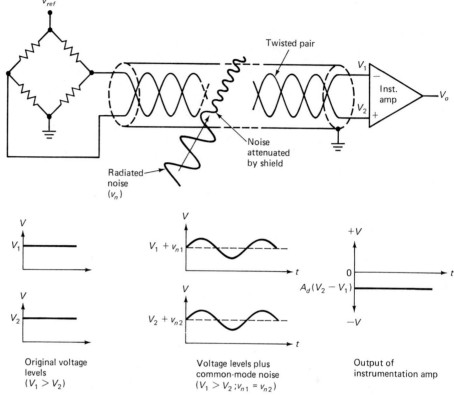

Fig. 4-14 Common noise reduction techniques.

Variable-Gain Instrumentation Amplifiers

The gain of an instrumentation amplifier may be adjusted by making some minor circuit modifications. An adjustable-gain instrumentation amplifier is shown in Fig. 4-15. The expression for the output voltage for this amplifier is

$$V_o = (V_2 - V_1)\left(\frac{R_F}{R_2}\right)\left(1 + \frac{2R_1}{R_G}\right) \tag{4-13}$$

where $R_1' = R_1$

Normally, R_G is chosen to provide the desired voltage gain. If continuously adjustable gain is desired, a potentiometer is inserted for R_G. As usual, for good CMRR, resistor pairs $R_F - R_F'$ and $R_2 - R_2'$ should be closely matched. Slight mismatching between R_1' and R_1 will not affect CMRR, but will cause a deviation in gain, as determined by Eq. (4-13). For minimum offset, the differential amplifier A_3 is designed for low gain, while the input amplifiers A_1 and A_2 are set up to provide higher gains [the third factor in parentheses in Eq. (4-13) is made large].

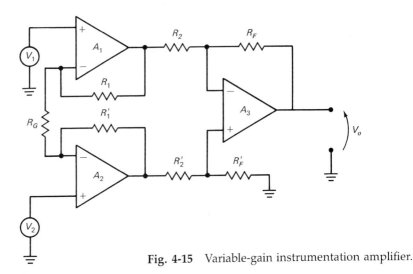

Fig. 4-15 Variable-gain instrumentation amplifier.

Commercial Instrumentation Amplifiers

Instrumentation amplifiers can be constructed using standard op amps and resistors; however, for applications requiring very high performance, it is best to use a dedicated instrumentation amplifier, such as the LH0036/LH0036C from National Semiconductor. The data sheets for this device are presented in the appendix, but for convenience, its equivalent internal circuitry and a few important specifications are given here.

The LH0036's internal circuitry is shown in Fig. 4-16. The resistors that are shown on this diagram are film-type resistors that are fabricated on a ceramic header inside a metal can package. Also mounted on the ceramic header is the silicon chip that comprises the amplifiers that are shown. The LH0036 is a hybrid IC. Among the features of this device are: CMRR = 100 dB, R_{in} = 300 MΩ, adjustable gain, and guard drive.

The gain of the LH0036 is set by placing a resistor of the appropriate value across pins 7 and 4. This gain adjustment technique was shown in Fig. 4-15. In the case of the LH0036, $R_3 = R_4 = R_5 = R_6$, and $R_1 = R_2 = 25$ kΩ; therefore the output voltage equation (4-13) reduces to

$$V_o = (V_2 - V_1)\left(1 + \frac{50 \text{ k}\Omega}{R_G}\right) \tag{4-14}$$

For normal operation, the CMRR preset pin (9) is grounded, and pin 8 (CMRR trim) is left open. This connection will produce CMRR > 80 dB. For higher CMRR, pin 9 is left open and pin 8 is connected to ground through a 5-kΩ potentiometer. The potentiometer is adjusted for minimum V_{oCM}, with a common-mode voltage applied to the amplifier's inputs. Exact procedures are presented in the LH0036 data sheet. In addition to the CMRR, input bias current and bandwidth are also adjustable. Most often, instrumentation am-

4-2 DIFFERENTIAL AND INSTRUMENTATION AMPLIFIERS

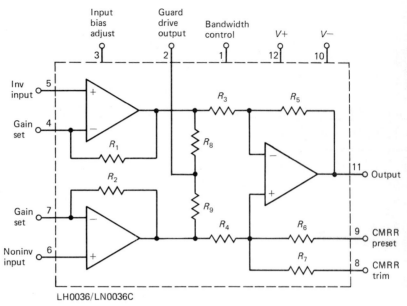

Fig. 4-16 Internal block diagram and pin designations for the LH0036. (*Courtesy of National Semiconductor Corp.*)

plifiers are used to process dc voltages. Therefore, it may be desirable to limit bandwidth in order to decrease the amplification of high-frequency noise.

The final aspect of the LH0036 to be presented is the guard drive output. The guard drive connection is shown in Fig. 4-17. The guard drive output is used to drive the input shielding to the same potential as the common-mode voltage present at the amplifier's input. This reduces current leakage between the input wires and the shield. Also, unequal capacitances between each signal line and the shield can cause unequal attenuation of what would normally be common-mode ac noise (common-mode signals have equal

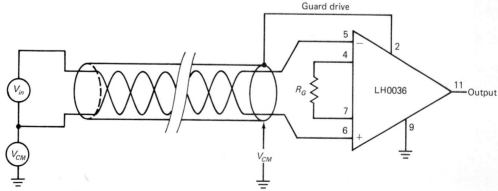

Fig. 4-17 Use of active guarding to reduce common-mode voltage-induced errors.

phase and amplitude). Now, because of the guard drive, the potential difference between the shield and the common-mode noise on signal lines is zero, effectively eliminating the effects of the stray capacitances.

4-3 ♦ VOLTAGE-TO-CURRENT AND CURRENT-TO-VOLTAGE CONVERSION

The circuits covered up to this point have been voltage amplifiers. Recall that such circuits are also referred to as voltage-controlled voltage sources (VCVS). In the most general terms, the quantity that relates the input and output of a system (or circuit) is of primary importance to understanding the behavior and possible uses of that circuit. The primary VCVS or voltage amplifier parameter is v_o/v_{in}, the voltage gain.

In terms of input-output-related classifications, there are three other types of amplifiers: current amplifiers, transconductance amplifiers, and transresistance amplifiers. The term used to describe a given amplifier is taken from the unit obtained when the output unit (V or I) is divided by the input unit (V or I). All four amplifier types and their input-output relationships are listed below.

Output/input	I/O units	Amplifier type
v_o/v_{in}	A_v (unitless)	Voltage amplifier: VCVS (voltage-controlled voltage source)
i_o/v_{in}	g_m (siemens)	Transconductance amplifier: VCIS (voltage-controlled current source)
v_o/i_{in}	r_m (ohms)	Transresistance amplifier: ICVS (current-controlled voltage source)
i_o/i_{in}	A_i (unitless)	Current amplifier: ICIS (current-controlled current source)

This section primarily presents the basics of op amp-based transconductance and transresistance amplifiers. The current amplifier is rarely implemented using op amps, and therefore shall receive cursory treatment here.

The VCIS

The VCIS is used to provide or force a current through a load that is proportional to the VCIS input voltage. The proportionality constant relating output current to input voltage is termed transconductance, g_m. Possibly the most descriptive name given to the VCIS is the term voltage-to-current (V/I) converter. Two simple op amp V/I converters are shown in Fig. 4-18. In both cases, the load is in the feedback loop of the op amp.

4-3 VOLTAGE-TO-CURRENT AND CURRENT-TO-VOLTAGE CONVERSION

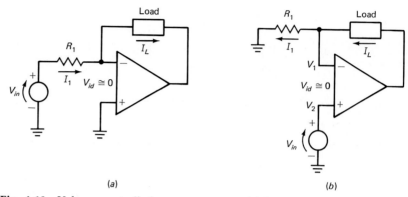

Fig. 4-18 Voltage-controlled current sources. (*a*) Inverting. (*b*) Noninverting.

The analysis of Fig. 4-18*a* is straightforward. In order for the op amp to maintain virtual ground at the inverting input terminal, we know that

$$I_L = -I_1$$

and

$$I_1 = \frac{V_{in}}{R_1}$$

Therefore

$$I_L = \frac{-V_{in}}{R_1} \qquad (4\text{-}15)$$

In terms of transconductance, this is

$$g_m = \frac{-1}{R_1} \qquad (4\text{-}16)$$

Thus we may write

$$I_L = V_{in} g_m \qquad (4\text{-}17)$$

For the V/I converter of Fig. 4-18*b*, because $V_{id} \cong 0$ V and the currents flowing into the op amp input terminals are nearly zero, we may write

$$V_1 = V_2$$

Therefore

$$V_{R1} = V_{in}$$

Applying Ohm's law,

$$I_1 = \frac{V_{in}}{R_1}$$

Now, since $I_L = I_1$, we obtain

$$I_L = \frac{V_{in}}{R_1} \qquad (4\text{-}18)$$

In terms of transconductance,

$$g_m = \frac{1}{R_1} \qquad (4\text{-}19)$$

and

$$I_L = g_m V_{in} \qquad (4\text{-}20)$$

Notice that in both cases, the load current is independent of the load resistance. In order to force a given current through various load resistances, the voltage produced at the output of the op amp will automatically change as necessary.

The maximum current that may be forced through R_L in Fig. 4-18a and b may depend on either the current source/sink capability of the op amp or the supply voltages used. The following example will illustrate these limitations for the inverting V/I converter.

EXAMPLE 4-7

Refer to Fig. 4-18a. Assume that $R_1 = 1\ \text{k}\Omega$, $V_{in} = +2\ \text{V}$, and the op amp has the following limitations: $V_{o(max)} = \pm 12\ \text{V}$, $I_{o(max)} = \pm 10\ \text{mA}$. Determine the following: (a) g_m, (b) I_L, (c) the maximum load resistance that may be used, and (d) the maximum value of V_{in} that may be used, given $R_1 = 1\ \text{k}\Omega$ and $R_L = 0\ \Omega$.

Solution

(a) From Eq. (4-16),

$$g_m = \frac{-1}{1\ \text{k}\Omega} = -1\ \text{mS}$$

(b) For loads within the circuit's drive capability, Eq. (4-17) applies:

$$I_L = 2\ \text{V} \times -1\ \text{mS} = -2\ \text{mA}$$

(c) The maximum load resistance is determined by relating I_L and $V_{o(max)}$ as follows:

$$R_{L(max)} = \frac{V_{o(max)}}{I_L}$$

$$= \frac{12\ V}{2\ mA}$$

$$= 6\ k\Omega$$

If $R_L > 6\ k\Omega$, then the op amp will saturate.

(d) The maximum value of V_{in} that may be used is the voltage that causes either: (1) $I_L > I_{o(max)}$ or (2) $V_o > V_{o(max)}$, whichever occurs first. In this case, because $R_L = 0\ \Omega$, condition 1 will prevail. (The proof of this is left for the student to perform.) Hence

$$V_{in(max)} = \pm I_{o(max)} R_1$$

$$= 10\ mA \times 1\ k\Omega$$

$$= \pm 10\ V$$

Floating load current sources like those in Fig. 4-18 perform quite well; however, more often than not, the load must be referred to ground. In these cases, the VCIS of Fig. 4-19 would be used. This circuit is called the Howland current source.

For equal-valve resistors,

$$I_L = \frac{-V_{in}}{R} \tag{4-21}$$

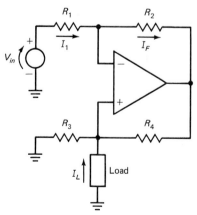

Fig. 4-19 Howland current source.

The transfer parameter g_m is defined as

$$g_m = \frac{-1}{R} \tag{4-22}$$

Consequently

$$V_o = V_{in} g_m \tag{4-23}$$

The ICVS

For low-power applications, the op amp can be used to implement a very effective ICVS, or transresistance amplifier. The main idea here is that the output voltage produced by the op amp is to be proportional to its input current. Because of this, the ICVS is commonly referred to as a current-to-voltage (I/V) converter. The basic op amp I/V converter is shown in Fig. 4-20.

The operation of the I/V converter is very straightforward. In Fig. 4-20, a current source provides the input to the amplifier. Inspection of the circuit reveals

$$-I_F = I_1 = I_{in}$$

Basic op amp analysis yields the following result:

$$V_o = I_F R_F$$

Therefore

$$V_o = -I_{in} R_F \tag{4-24}$$

The value of R_F defines the transresistance of the amplifier. That is,

$$r_m = R_F \tag{4-25}$$

The bias current compensation resistor R_B is set equal to R_F in order to minimize output offset voltage.

Two commonly encountered devices that may be modeled as current sources are the photodiode and the phototransistor. Figure 4-21 illustrates one possible method of amplifying the output of a photodiode (I_S) and, in the process, converting the signal into a voltage. The photodiode may be approximately modeled as a current source, in which I_S is proportional to the intensity of the light impinging on the PN junction. Circuits such as that shown in Fig. 4-21 (albeit more complex) are used in fiber optic data communication systems.

4-3 VOLTAGE-TO-CURRENT AND CURRENT-TO-VOLTAGE CONVERSION

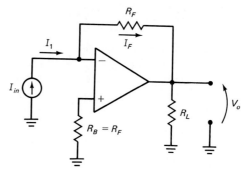

Fig. 4-20 Op amp current-to-voltage converter.

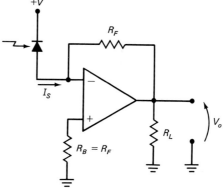

Fig. 4-21 ICVS photodiode light sensor.

The ICIS (Current Amplifier)

As mentioned earlier, current amplifiers are not often implemented using op amps. For the sake of completeness, one possible current amplifier is shown in Fig. 4-22. In this circuit, $I_L = I_{in}$. This circuit is analogous to the voltage follower, for which $V_o = V_{in}$.

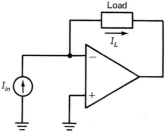

Fig. 4-22 Current-controlled current source.

4-4 ♦ A VOLTAGE AMPLIFIER VARIATION

Based on the op amp circuits that have been presented up to this point, if very high voltage gain is required, there are two alternatives: (1) Design a single-stage gain block with very large R_F/R_1 ratio, or (2) set up several gain stages and divide the gain among the various stages. Most of the time, the second alternative is preferable, because of the practical limitations that apply to the values of R_F and R_1. Also, because thermally generated noise is proportional to resistance (and temperature), a high R_F value coupled with very high voltage (noise) gain can cause severe noise problems.

It is possible to obtain very high voltage gain from a single closed-loop op amp, without the need to resort to extremely high or low resistor values or extremely high resistor ratios. The circuit of Fig. 4-23 is used to meet these requirements. The derivation of the voltage gain expression for this curcuit is as follows.

Applying the virtual ground concept at node A,

$$I_1 = \frac{V_{in}}{R_1}$$

and

$$I_2 = -I_1 = \frac{-V_{in}}{R_1}$$

Since the left side of R_2 is at virtual ground and the lower terminal of R_3 is at physical ground, and both are connected at node B, R_2 and R_3 are effectively in parallel with each other. This means that R_2 and R_3 will have equal voltage drops, and we may write

$$I_2 R_2 = I_3 R_3$$

Now we may define I_3 as

$$I_3 = \frac{I_2 R_2}{R_3}$$

And since $I_2 = -I_1 = -V_{in}/R_1$, we may write

$$I_3 = \frac{-V_{in} R_2}{R_1 R_3}$$

Applying KCL to node B,

$$I_4 = I_2 + I_3$$

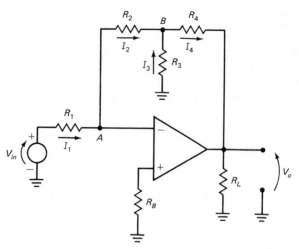

Fig. 4-23 Op amp with modified feedback loop.

The output voltage is now given by

$$V_o = I_4 R_4 + I_2 R_2 = I_3 R_3 + I_4 R_4$$

Because $I_2 = -I_1$, we may write

$$V_o = I_4 R_4 - I_1 R_2$$

Now, substituting identities for I_1 and I_4 in terms of V_{in}, we obtain

$$\begin{aligned} V_o &= R_4 (I_3 + I_2) - \frac{V_{in} R_2}{R_1} \\ &= R_4 \left(\frac{-V_{in} R_2}{R_1 R_3} - \frac{V_{in}}{R_1} \right) - \frac{V_{in} R_2}{R_1} \\ &= \frac{-V_{in} R_2 R_4}{R_1 R_3} - \frac{V_{in} R_4}{R_1} - \frac{V_{in} R_2}{R_1} \end{aligned} \qquad (4\text{-}26)$$

Dividing by V_{in} yields

$$A_v = \frac{V_o}{V_{in}} = -\frac{R_2 R_4}{R_1 R_3} - \frac{R_4}{R_1} - \frac{R_2}{R_1}$$

Alternatively, we may write

$$A_v = -\left(\frac{R_2 R_4}{R_1 R_3} + \frac{R_4}{R_1} + \frac{R_2}{R_1} \right)$$

This may be further simplified to

$$A_v = -\left(\frac{R_2+R_4}{R_1} + \frac{R_2R_4}{R_1R_3}\right) \quad (4\text{-}27)$$

EXAMPLE 4-8

The circuit in Fig. 4-23 has the following component values and input voltage: $R_2 = R_4 = R_L = 10\ \text{k}\Omega$, $R_1 = R_3 = 1\ \text{k}\Omega$, $V_{in} = 50\ \text{mV}$. Determine: (a) A_v, (b) V_o, (c) I_o.

(a) The voltage gain is found by direct application of Eq. (4-27):

$$A_v = -120$$

(b) $V_o = -120 \times 50\ \text{mV}$
$\quad\ \ = -6\ \text{V}$

(c) The output current may be found in a number of different ways. Here, we shall use

$$I_o = \frac{V_o}{R_L} + \frac{V_o}{R_1} - \frac{V_{in}}{R_1} - \frac{V_{in}R_2}{R_1R_3}$$

$\quad = -0.6\ \text{mA} - 6\ \text{mA} - 0.05\ \text{mA} - 0.5\ \text{mA}$

$\quad = -7.15\ \text{mA}$

Chapter Review

The op amp summer is an extremely versatile circuit that is capable of linearly mixing or adding multiple input signals. Because the inverting input of the op amp is at virtual ground, multiple sources connected to a summing amplifier will exhibit no interaction, under normal conditions. Inputs may also be applied to the noninverting input of a summing amp.

 The differential amplifier can be considered to be a special case of the summing amp in which the output is proportional to the difference between two input signals. The differential amp is used to form the instrumentation amplifier, which is used primarily in the processing of bridge circuit outputs. In such amplifiers, provisions for gain adjustment and input guarding are normally provided.

 In general, the op amp may be used to implement the four basic amplifier types: VCVS, VCIS, ICVS, and ICIS. The use of negative feedback is essential to the implementation of any of these four op amp-based configurations. Each amplifier type has certain characteristics that make it suitable for a given application.

Questions

4-1. In general, are instrumentation amplifiers designed for optimum ac or dc performance?

4-2. State two terms that are used to describe the inverting input of an operational amplifier.

4-3. What phenomenon is primarily responsible for the inverting summing amplifier's elimination of input source interaction?

4-4. In general, what effect will increasing the number of inputs have on the performance of a summing amplifier?

4-5. Explain why high CMRR is important in instrumentation amplifier applications.

4-6. In general, what is the main factor that determines the CMRR of an instrumentation amplifier?

4-7. What gain parameter is associated with the VCIS? What units are used to represent this parameter?

4-8. What technique is used to reduce the effects of leakage currents between instrumentation amplifier inputs?

4-9. List the four general types of amplifiers that may be implemented using the op amp.

4-10. Why would a single-ended-input–single-ended-output amplifier be inappropriate for the processing of the output of an active bridge?

Problems

4-1. The circuit shown in Fig. 4-3 has the following specifications: $R_1 = R_2 = 18$ kΩ, $R_F = 68$ kΩ, $R_L = 2.2$ kΩ, $V_1 = -500$ mV, and $V_2 = -500$ mV. Determine: **(a)** I_F, **(b)** V_o, **(c)** I_L, **(d)** the op amp output current I_o, **(e)** R_B for minimum offset.

4-2. Given $R_F = 25$ kΩ, design a five-input averager with equally weighted inputs.

4-3. Refer to Fig. 4-4. Given that all input voltages are equal to 3 V, determine the value of R_F such that $V_o = -10$ V.

4-4. Design a four-input summer that will realize the expression

$$V_o = -\left(\frac{V_1 + V_2 + 2V_3 + 3V_4}{4}\right)$$

4-5. Refer to Fig. 4-5. Assume that $R_1 = 10$ kΩ, $R_2 = 18$ kΩ, $R_A = 27$ kΩ, $R_B = 6.8$ kΩ, $R_F = 38.6$ kΩ, $R_L = 1$ kΩ, $V_1 = 518$ mV, $V_2 = 1.4$ V, and $V_3 = 5$ V. Determine: **(a)** I_1, **(b)** I_2, **(c)** I_F, **(d)** I_o, **(e)** V_o.

4-6. The circuit in Fig. 4-9 has the following component values: $R_1 = R_1' = 1.5$ kΩ, $R_F = R_F' = 54$ kΩ, $V_1 = -4.00$ V, and $V_2 = -3.92$ V. Determine V_o for each of the following CMRR values. **(a)** CMRR = ∞, **(b)** CMRR = 90 dB, **(c)** CMRR = 50 dB.

4-7. Refer to Fig. 4-23. Given $R_2 = R_4 = 22$ kΩ and $R_3 = 4.7$ kΩ, determine: **(a)** R_1 such that $A_v = 100$, **(b)** R_B for minimum offset.

4-8. Refer to Fig. 4-24. To what value must R_5 be adjusted such that $V_o = 0$ V? (Assume CMRR = ∞.)

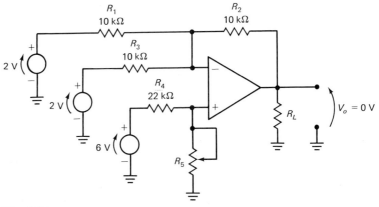

Fig. 4-24

4-9. Refer to Fig. 4-15. Given $R_2 = R_2' = R_1 = R_1' = 10$ kΩ and $R_F = R_F' = 20$ kΩ, determine the value of R_G such that $A_d = 100$.

4-10. Design a floating load V/I converter such that $g_m = 5$ mS.

4-11. Refer to Fig. 4-19. **(a)** Determine R such that $I_L = 2$ mA when $V_{in} = 5$ V. **(b)** Write the expression for i_L if $v_{in} = 8 \cos 100t$ V.

4-12. See Fig. 4-25. Assume that the bridge is in balance when $T = 100°$ C. That is, $R_1 = R_2 = R_3 = R_4 = 20.000$ kΩ. The thermistors are matched,

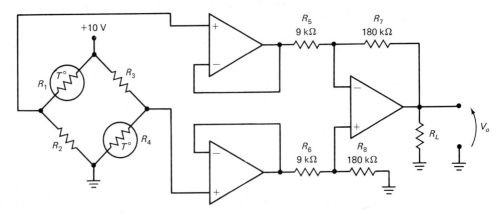

Fig. 4-25

and their resistance is related to temperature (over the range of interest) by

$R = 0.14T^2 + 521.94$

(T is absolute temperature in kelvins.) Determine V_o for the following temperatures: **(a)** $T = 110°$ C, **(b)** $T = 90°$ C. (Note: absolute temperature K = °C + 273°.)

4-13. Design a summing amp for which $V_o = -(V_1 + V_2 + V_3 + V_4)$.

4-14. Assume that the op amp used in Prob. 4-13 has GBW = 10 MHz. Determine the bandwidth of the circuit.

4-15. Determine the minimum and maximum bandwidth for the circuit in Fig. 4-26. The op amp is a 741.

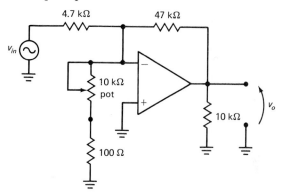

Fig. 4-26

4-16. A certain differential amplifier has $A_d = 112$ and CMRR = 55 dB. Determine the common-mode output voltage V_{oCM} given $v_{iCM} = 2 \sin 2\pi 100t$ V.

4-17. A certain differential amplifier has the following inputs and outputs: $V_{id} = 50$ mV, $V_{od} = 6.00$ V, $V_{iCM} = 5.00$ V, and $V_{oCM} = 500$ mV. Determine the CMRR (in decibels) of the circuit.

OP AMP APPLICATIONS: PART B

This chapter begins with a more generalized investigation of op amp circuit operation in the familiar inverting and noninverting configurations, using complex impedances in the feedback loop. This provides a foundation for understanding additional important op amp applications, such as active filters. Also covered in this chapter are op amp-based integrators and differentiators. These circuits find wide use in signal processing and process control applications. Nonlinear op amp circuits are also presented. Specifically, logarithmic and antilogarithmic amplifiers, analog multipliers, and precision rectifiers are discussed.

5-1 ♦ THE OP AMP WITH COMPLEX IMPEDANCES

Operational amplifier-based circuits are not limited to purely resistive feedback and input (gain-setting) elements, or to purely resistive loads. The occurrence of purely resistive elements in an op amp circuit is actually just a special case of the complex impedances that may be used. When dealing with impedances (or admittances), the more general symbology of Fig. 5-1 is often used. The rectangles represent the various impedances that appear in the circuit. In this section, we shall consider the behavior of op amp circuits that contain elements such as those of Fig. 5-1, under steady-state sinusoidal input signal conditions.

The Inverting Configuration

Recall that the voltage gain of the inverting amplifier in the purely resistive case is given by

$$A_v = -\frac{R_F}{R_1}$$

where the minus sign indicates phase inversion. For ac signals, the voltage gain of the circuit effectively caused a 180° phase shift at the output. Viewing

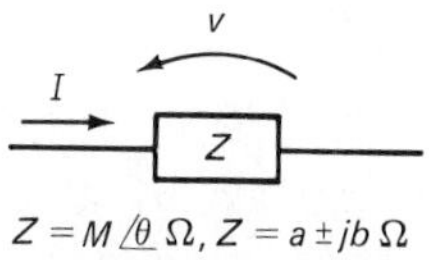

5-1 THE OP AMP WITH COMPLEX IMPEDANCES

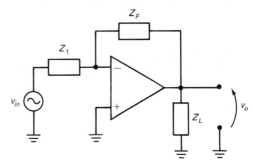

Fig. 5-1 General symbol for an impedance.

Fig. 5-2 General impedance symbology applied to op amp circuit.

this as a complex gain, we could write

$$A_v = M\,\underline{/\theta} \tag{5-1}$$

where M is the magnitude of the voltage gain and θ is the phase angle of the voltage gain. In the purely resistive case, $M = R_F/R_1$ and the phase angle, $\theta = 180°$, is independent of frequency. For the purposes of this discussion, we shall assume operation within the midband of the op amp. This allows the op amp to be idealized.

An inverting op amp with generalized impedances is shown in Fig. 5-2. The complex voltage gain for this circuit is given by

$$A_v = -\frac{Z_F}{Z_1} \tag{5-2}$$

where again the negative sign represents the 180° phase shift that is inherent to the inverting configuration. It is important to realize that because either the numerator, the denominator, or both in Eq. (5-2) may have a nonzero phase angle, the overall phase of the voltage gain may easily differ from 180°. This is demonstrated in the following example.

EXAMPLE 5-1

The circuit in Fig. 5-2 contains the following impedances and input voltage: $Z_F = 10 + j6.8$ kΩ, $Z_1 = 1 + j0$ kΩ, and $v_{in} = 600 \sin \omega t$ mV. Determine (a) the complex voltage gain, and (b) V_o. (c) Sketch the input and output waveforms, showing their relative phase.

Solution

The impedances are given in rectangular form, but Eq. (5-2) will be easier to work with if the impedances are converted into polar form. Coordinate conversions are performed as follows.

The general rectangular form is represented by

$$Z = a \pm jb$$

The polar form is represented by

$$Z = M \underline{/\theta}$$

Given the rectangular form, the polar form is defined as

$$M = \sqrt{a^2 + b^2} \quad \text{and} \quad \theta = \tan^{-1}\frac{b}{a}$$

Given the polar form, the rectangular form is given by

$$a = M \cos \theta + jM \sin \theta$$

Now, converting the impedances to polar form, we obtain

$$Z_1 = 1\underline{/0°} \text{ k}\Omega \quad \text{and} \quad Z_F = 12.1\underline{/34.2°} \text{ k}\Omega$$

The voltage gain is given by Eq. (5-2):

$$A_v = -\frac{Z_F}{Z_1}$$

$$= -\frac{12.1\underline{/34.2°} \text{ k}\Omega}{1\underline{/0°} \text{ k}\Omega}$$

Because the negative sign represents a 180° phase shift, we may write

$$A_v = \frac{12.1\underline{/180° + 34.2°} \text{ k}\Omega}{1\underline{/0°} \text{ k}\Omega}$$

$$= 12.1\underline{/214.2°}$$

Alternatively, we may write

$$A_v = 12.1\underline{/-145.8°}$$

Generally, it is best to use the phase angle that is lowest in absolute magnitude (in this case, −145.8°). The output voltage is the product of A_v

and V_{in}. Because V_{in} is a sine wave of fixed frequency (frequency has already been accounted for, as impedances are frequently dependent), it may be represented in shorthand form as

$$V_{in} = 600 \underline{/0°} \text{ mV}$$

Therefore, the output voltage is

$$V_o = A_v V_{in}$$
$$= 12.1 \underline{/-146°} \times 600 \underline{/0°} \text{ mV}$$
$$= 7.26 \underline{/-146°} \text{ V}$$

The input and output voltage waveforms are shown in Fig. 5-3. Notice that the output waveform is an inverted version of the input with an additional 34.2° phase shift.

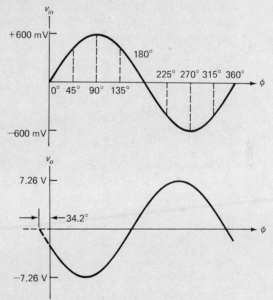

Fig. 5-3 Input and output voltages for Example 5-1.

In the preceding example, the actual types of components used in the circuit were not specified. However, it is easy to deduce that Z_1 was a 1-kΩ resistor, and apparently Z_F consisted of a 10-kΩ resistor in series with an inductor. The value of the inductor in henries cannot be determined because insufficient information was given (the exact frequency was not specified). Normally, the exact component types and their values will be known. With this information, the voltage gain and output for a given input frequency may be determined. This is demonstrated in the following example.

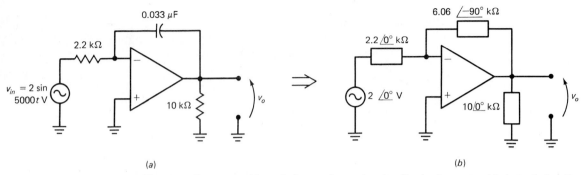

Fig. 5-4 Op amp with resistive and reactive feedback elements. (a) Actual circuit; (b) circuit with impedances in polar form.

EXAMPLE 5-2

Refer to Fig. 5-4a and b. Determine A_v and $\mathbf{V}_o$ in shorthand and full trigonometric form, and sketch the input and output waveforms, for the component values given.

Solution

Again, it is easiest to work with the impedances in polar form. For $\mathbf{Z}_1$ this is

$$\mathbf{Z}_1 = R = 2.2 \underline{/0°} \text{ k}\Omega$$

The feedback loop is closed via a 0.033-μF capacitor. At the frequency of interest, ω = 5000 rad/s (795.8 Hz), the impedance of the capacitor (considered to be purely reactive) is found to be

$$\mathbf{Z}_F = X_c = 6.06 \underline{/-90°} \text{ k}\Omega$$

Now, since the frequency of v_{in} has been accounted for in the determination of the impedances, we may use the following shorthand form to simplify our notation:

$$\mathbf{V}_{in} = 2 \underline{/0°} \text{ V}$$

The voltage gain at the frequency of interest is

$$A_v = -\frac{6.06 \underline{/-90°} \text{ k}\Omega}{2.2 \underline{/0°} \text{ k}\Omega}$$

$$= \frac{6.06 \underline{/-90° + 180°} \text{ k}\Omega}{2.2 \underline{/0°} \text{ k}\Omega}$$

$$= 2.75 \underline{/90°}$$

The output voltage is given by

$$\mathbf{V}_o = 2.75 \underline{/90°} \times 2\underline{/0°} \text{ V}$$
$$= 5.5 \underline{/90°} \text{ V}$$

In complete trigonometric form, this is

$$v_o = 5.5 \sin(5000t + 90°) \text{ V}$$

The observant reader should recognize that the output voltage is a cosine waveform, which may be expressed as

$$v_o = 5.5 \cos 5000t \text{ V}$$

The input and output waveforms are shown in Fig. 5-5.

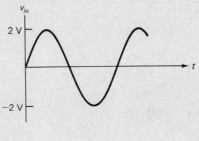

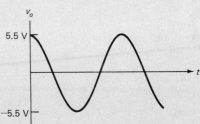

Fig. 5-5 Input and output voltages for Example 5-2.

It is interesting to note that for the circuit in Fig. 5-4, for sinusoidal signals the output will always lead the input by 90°, regardless of the frequency. This is a very important point, and the reader is encouraged to verify the validity of this statement, perhaps by changing the frequency of v_{in} in Fig. 5-3 and reworking the example. Also, if the positions of the resistor and the capacitor are switched, the output will always lag the input by 90° regardless of the signal frequency. However, these 90° phase relationships will hold only as long as Z_1 and Z_F are pure (or nearly pure) resistances and reactances and the op amp is in the inverting configuration.

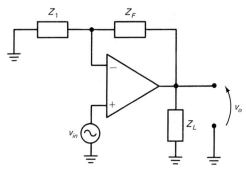

Fig. 5-6 Noninverting op amp with feedback.

The Noninverting Configuration

The generalized noninverting op amp configuration is shown in Fig. 5-6. The previously derived voltage gain expressions again hold for this circuit, and in generalized form they are

$$A_v = 1 + \frac{Z_F}{Z_1} \tag{5-3a}$$

or

$$A_v = \frac{Z_1 + Z_F}{Z_1} \tag{5-3b}$$

Usually, the second form is easier to work with; however, either expression may be used. The following example illustrates the application of the second form.

EXAMPLE 5-3

The circuit in Fig. 5-6 contains the following: $Z_1 = 2.2 + j0$ kΩ, $Z_F = 0 - j6.06$ kΩ, $v_{in} = 2 \sin 5000t$ V ($\mathbf{V}_{in} = 2\underline{/0°}$ V). Determine **(a)** A_v and **(b)** $\mathbf{V}_o$; **(c)** sketch the input and output waveforms.

Solution

In applying Eq. (5-3b), it is easiest to begin with Z_1 and Z_F in rectangular form. This yields

$$A_v = \frac{(2.2 + j0) + (0 - j6.06)}{2.2 + j0} \text{ k}\Omega$$

$$= \frac{2.2 - j6.06 \text{ k}\Omega}{2.2 + j0 \text{ k}\Omega}$$

Converting the numerator and denominator into polar form and performing the division yields

$$A_v = \frac{6.45\ \underline{/-70°}\ \text{k}\Omega}{2.2\ \underline{/0°}\ \text{k}\Omega}$$

$$= 2.93\ \underline{/-70°}$$

The output voltage may now be determined. In shorthand form, we obtain

$$\mathbf{V}_o = A_v \mathbf{V}_{in}$$
$$= 2.93\ \underline{/-70°} \times 2\underline{/0°}\ \text{V}$$
$$= 5.86\ \underline{/-70°}\ \text{V}$$

The full trigonometric form is

$$v_o = 5.86 \sin(5000t - 70°)\ \text{V}$$

The input and output waveforms are shown in Fig. 5-7.

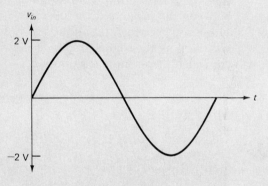

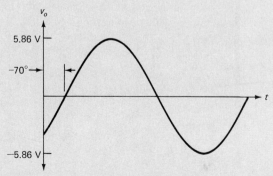

Fig. 5-7 Input and output voltages for Example 5-3.

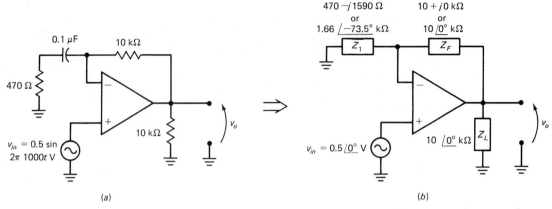

Fig. 5-8 Noninverting op amp. (*a*) Actual circuit; (*b*) circuit with external components expressed as generalized impedances in polar and rectangular form.

The preceding example used the same component values and input signal as Example 5-2 (Z_F is a pure reactance, in this case capacitive, and Z_1 is purely resistive). In fact, the only difference between Examples 5-3 and 5-2 was the swapping of input terminals. The important thing to note here is that the noninverting configuration did not result in a 90° phase relation between the input and the output. It can be demonstrated that the relative phase shift between the input and output of the circuit used in Example 5-3 will vary as the signal frequency is changed. This is not the case for the equivalent inverting configuration.

EXAMPLE 5-4

Determine A_v and V_o (in shorthand form) for the circuit in Fig. 5-8*a*.

Solution

In this circuit, Z_F is purely resistive, but Z_1 is complex, consisting of a resistor in series with a capacitor. Dealing with Z_1 first, we find the reactance of the capacitor to be

$$X_C = \frac{1}{\omega C}$$

$$= 1.59 \underline{/-90°} \text{ k}\Omega$$

This leads to

$$Z_1 = 470 - j1590 \text{ }\Omega$$

and

$$Z_F = 10 + j0 \text{ k}\Omega$$

Now applying Eq. (5-3b), we obtain

$$A_v = \frac{(470 - j1590) + (10{,}000 + j0)}{470 - j1590}\,\Omega$$

$$= \frac{10.47 - j1.59 \text{ k}\Omega}{0.47 - j1.59 \text{ k}\Omega}$$

Converting into polar form and dividing yields

$$A_v = \frac{10.6\underline{/-8.6°}\text{ k}\Omega}{1.7\underline{/-73.5°}\text{ k}\Omega}$$

$$= 6.2\underline{/64.9°}$$

The expression for V_o is

$$\mathbf{V}_o = A_v \mathbf{V}_{in}$$

$$= 6.2\underline{/64.9°} \times 0.5\underline{/0°}\text{ V}$$

$$= 3.1\underline{/64.9°}$$

The analysis of these types of circuits is not much more difficult than analysis of the op amp with resistive circuit elements, except that the bookkeeping is more tedious and it is often hard to visualize the phase relationships between input and output signals. In the next section, we shall see how circuits like those covered here are applied to realize specific mathematically related operations.

5-2 ♦ DIFFERENTIATORS AND INTEGRATORS

Two of the most fundamentally important mathematical operations are differentiation and integration. The operational amplifier can be used to perform either of these operations. In fact, the term *operational amplifier* was originally derived from the use of electronic circuitry to perform various mathematical operations. A calculus background is required for a complete understanding of differentiation and integration. However, this discussion avoids as much of this higher math as possible; where necessary, intuitive explanations of the operations covered are presented.

The Differentiator

A differentiator is a circuit whose output is proportional to the derivative of its input. Basically, the derivative of a function is the instantaneous slope or rate of change of that function. What we in electronics call a signal, mathe-

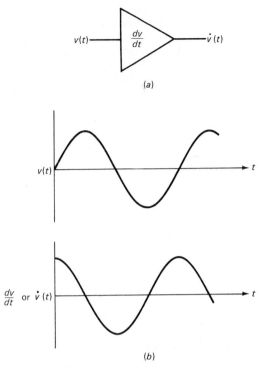

Fig. 5-9 Differentiator. (*a*) Symbol; (*b*) input-output phase relationships for sinusoidal inputs.

maticians call a function. In electronic applications, the input to a differentiator (or integrator) is a signal, which will generally be a function of time. Therefore, the output of the differentiator is proportional to the rate of change of the input signal, with respect to time. Most often, we will be dealing with signals that are described by the trigonometric functions sine and cosine, although in practice other simpler and more complex functions (signals) may also have to be dealt with.

Differentiator Fundamentals

The block diagram symbol for a differentiator and the mathematical symbol for the operation of differentiation are shown in Fig. 5-9*a*. The output of a differentiator (or the result of taking the derivative of the input) is termed the *derivative* of the input. Here, the input is $v(t)$, instantaneous voltage as a function of time. The derivative is commonly symbolized by placing a dot over the dependent variable. The input-output relationships for a sinusoidal input to the differentiator are shown in Fig. 5-9*b*. Notice that when the input undergoes its maximum rate of change (at positive- and negative-going zero crossings), the derivative reaches its maximum positive and negative values. Also, when the rate of change of the input is zero (at the positive and negative peaks), the output of the differentiator is also zero. Disregarding amplitudes for the time being, it can be seen that the output (derivative) leads the input

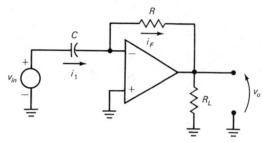

Fig. 5-10 An op amp differentiator.

by 90°. Thus an input of the form $V \sin \omega t$ will produce an output of the form $\omega V \cos \omega t$. The phase is frequency-independent.

The basic op amp differentiator is shown in Fig. 5-10. This is an inverting circuit, and its output will be inverted with respect to that of a true differentiator. We may determine the output of this circuit for sinusoidal inputs, using the methods presented in the last section [applying Eq. (5-2)]. In effect, a sinusoidal signal may be differentiated without directly resorting to the use of calculus. However, to reduce our mathematical workload even further, we will take advantage of the fact that the output of the op amp differentiator will always lag the input by 90° (inversion of the true derivative). This tells us that the phase of the output is $-90°$ with respect to the input. The gain of the circuit is given by

$$A_v = -\omega RC \,\underline{/\!-90°} \tag{5-4}$$

By selecting proper values for R and C, we can make the gain of the differentiator at the frequency of interest as high or low as required.

Practical Differentiators

There are a few problems associated with op amp differentiators. The main problem is noise sensitivity. The gain of the ideal differentiator is zero at dc, and it increases with frequency at a rate of 20 dB/decade. This means that high-frequency noise will tend to be amplified greatly, possibly affecting the output. In order to reduce the gain to high-frequency noise, a resistor is placed in series with the input capacitor, as shown in Fig. 5-11. The problem

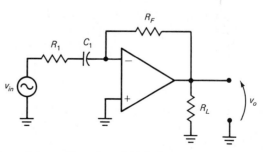

Fig. 5-11 Differentiator with reduced high-frequency response.

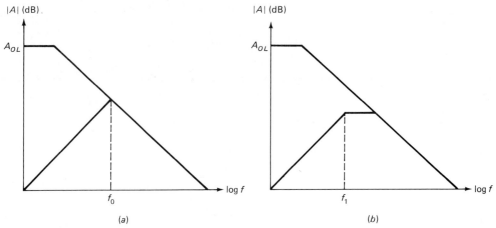

Fig. 5-12 Differentiator Bode plots. (*a*) Unmodified circuit response; (*b*) modified circuit response.

with adding this resistance is that it may severely alter the phase characteristics of the circuit. That is, the output will no longer lag the input by 90°.

The gain characteristic of the unmodified differentiator of Fig. 5-10 is shown superimposed on a typical op amp open-loop Bode plot in Fig. 5-12*a*. The differentiator will operate correctly up to the point where the two curves intersect (f_0). Beyond this frequency, the circuit will no longer behave as a differentiator. Placing a resistance in series with the capacitor causes the differentiator gain to level off at f_1, as shown in Fig. 5-12*b*. This frequency is given by

$$f_1 = \frac{1}{2\pi R_1 C} \tag{5-5}$$

In practice, R_1 may be chosen such that $10R_1 < R_F$. This introduces some phase error, but also reduces high-frequency gain and noise.

EXAMPLE 5-5

Determine v_o for Fig. 5-10, given $R = 2.2$ kΩ, $C = 0.1$ μF, and **(a)** $v_{in} = 2 \sin 1000t$ V, **(b)** $v_{in} = 2 \sin 4000t$ V.

Solution

(a) $A_v\big|_{\omega = 1 \text{ krad/s}} = -\omega RC$

$\qquad\qquad\qquad\quad = -0.22$

$\qquad\therefore v_o = -0.44 \cos 1000t$ V

(b) $A_v\big|_{\omega = 4 \text{ krad/s}} = -\omega RC$

$\qquad\qquad\qquad\quad = -0.88$

$\qquad\qquad \therefore v_o = -1.76 \cos 4000t \text{ V}$

Differentiation of Nonsinusoidal Inputs

When nonsinusoidal signals are applied to the differentiator, the use of calculus is generally required for the evaluation of the resulting output signal. There are, however, several commonly encountered signals whose output responses can be determined quite easily. An understanding of the action of the capacitor makes most of these output responses intuitively plausible. Figure 5-13 illustrates the true derivatives and the derivatives that are produced by the op amp differentiator in response to several common signals. A brief discussion of each input and its associated output is presented next. Since the sinusoidal input was discussed previously, it shall be skipped here. Refer to Figs. 5-10 and 5-13 during these descriptions.

THE CONSTANT VOLTAGE INPUT A constant (voltage) has zero rate of change; therefore, the derivative of a constant is zero. In Fig. 5-10, $I_F = -I_1$, and $V_o = I_F R_F$. In order to obtain current flow through a capacitor, the voltage across the capacitor must be time-varying. A constant voltage produces zero capacitor current; therefore

$$V_o = 0 \text{ V} \qquad v_{in} = \text{const}, k \qquad (5\text{-}6)$$

THE STEP FUNCTION A step function ideally makes an instantaneous transition from one level (usually zero) to some other level. At the instant of the step, the rate of change is infinite; therefore the ideal op amp differentiator should produce a negative-going spike (impulse) of infinite amplitude with zero width. This would occur because the capacitor acts as a short circuit during the transition between levels. In practice, the pulse is limited to slightly less than the supply voltage, and the pulse width depends on the slew rate of the op amp and the load on the output of the amplifier.

THE SQUARE WAVE A square wave may be considered to be essentially a series of positive and negative step functions. This input will produce a series of negative- and positive-going impulses at the output of the differentiator.

THE RAMP A ramp input increases (positively or negatively) linearly with time. Since the rate of change of this signal (the derivative) is constant, a constant current which is proportional to the rate of change of the input

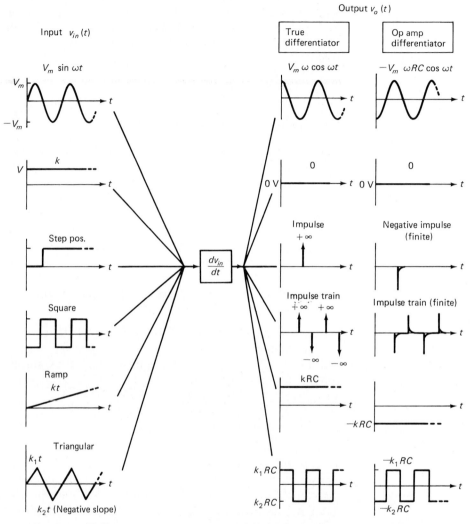

Fig. 5-13 Differentiator input-output relations for various common inputs.

voltage will flow through the input capacitor. This in turn produces a constant I_F, and hence V_o is constant. The output expression for this input is

$$V_o = -RCk \quad \text{linear ramp input} \tag{5-7}$$

where k is the slope of the function in volts per second.

THE TRIANGULAR WAVEFORM A triangular waveform can be thought of as a series of positive- and negative-going ramps. The output of the differentiator will alternate between voltage levels that are proportional to the slope of the waveform during positive and negative excursions. The output of the differentiator is given by

$$V_o = -RCk_n \quad \text{triangular input} \tag{5-8}$$

where k_n is the slope of the input (in volts per second). If the input is asymmetrical ($|k_1| \neq |k_2|$), then the magnitude of V_o will differ for positive- and negative-going portions of the input.

EXAMPLE 5-6

The differentiator in Fig. 5-10 has the following component values: $R = 10$ kΩ, $C = 0.47$ μF. The input to the circuit is a triangle wave (symmetrical) of 1 V_{P-P} at 1 kHz. Sketch the input and output signals, labeling their peak amplitudes and periods.

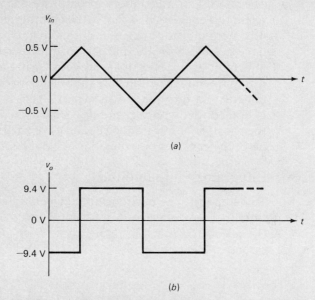

Fig. 5-14 Voltage waveforms for Example 5-6.

Solution

Since the triangle wave is symmetrical, the slopes of the positive- and negative-going portions will be equal in magnitude ($|k_1| = |k_2|$), as shown in Fig. 5-14a. This also means that the output will be symmetrical about 0 V. For linear functions, the slope k is given by $k = \Delta V/\Delta t$. Now, since the period of the signal is 1 ms ($T = 1/f$) and the signal goes from -0.5 to $+0.5$ V (and vice versa) in 0.5 ms ($T/2$), the slope of the signal is

$$k = \frac{1 \text{ V}}{0.5 \text{ ms}}$$

or

$$k = 2000 \text{ V/s}$$

> The peak amplitudes are given by Eq. (5-8) [or Eq. (5-7)] as
>
> $$\pm V_{o(pk)} = -RCk$$
> $$= (10 \text{ k}\Omega)(0.47 \text{ }\mu\text{F})(\pm 2000)$$
> $$= \pm 9.4 \text{ V}$$
>
> This waveform is shown in Fig. 5-14b.

The Integrator

The process of integration is complementary to that of differentiation. The relation is analogous to that between multiplication and division. In general, it is more difficult to gain an intuitive grasp of integration than of differentiation, and so for the purpose of this discussion, it is best to think of an integrator as a device that "undoes" what a differentiator does. The block diagram symbol for an integrator and the mathematical notation for the integral of a function are shown in Fig. 5-15. The function being integrated is termed the *integrand*, and *dt* is called the *differential*. The C term is called the constant of integration. Because we shall be dealing with *definite integrals* here, the constant of integration will be disregarded.

Integrator Fundamentals

To help clarify the relationship between differentiation and integration, consider that the derivative of a function yields the instantaneous rate of change (slope) of that function. Now, recall from Example 5-6 that for the special case of a linear ramp,

$$\dot{v}(t) = \frac{dv}{dt} = \frac{\Delta V}{\Delta t} \quad \text{for linear functions}$$

Here, the exact value of the derivative may be obtained by normal division (this is not the case for nonlinear functions). Conversely, integration produces the equivalent of the continuous sum of the values of the function at infinitely many infinitesimally small increments of *t*. For example, in Fig. 5-16, the integral of $v(t)$ from t_1 to t_2 is the area under the curve in volt-seconds. When the limits of the integration are defined, as in Fig. 5-16, we are evaluating what is called a *definite integral*. If we can mathematically describe the function, we can also mathematically evaluate the integral of that function.

Certain functions (or signals) may be too ill-defined or complex to integrate mathematically. In such cases, an op amp integrator can be used to evaluate the integral of the function electronically.[*]

---

[*] Both integrators and differentiators (and summers) are combined and used to perform mathematical operations in analog computer circuits and in signal processing applications. In general, high-speed digital computers have replaced analog computers, but integrators and differentiators are still widely used.

5-2 DIFFERENTIATORS AND INTEGRATORS

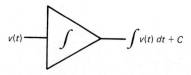

Fig. 5-15 Symbol for an integrator and the notation for the mathematical operation performed.

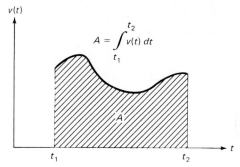

Fig. 5-16 Evaluation of the integral of the function yields the area under the curve of that function.

The basic op amp integrator is shown in Fig. 5-17. Like the op amp differentiator, the op amp integrator produces an output that is opposite in sign to the true integral of the input. For sinusoidal inputs, the output of the integrator may be determined using the techniques developed earlier in this chapter. However, just as the differentiator produced an output with a constant phase relationship relative to the input, the same is true of the integrator. A true integrator will produce an output that lags the sinusoidal input by 90°. The output of the op amp integrator will lead the input by 90°, as a result of the use of the inverting configuration. Ideally, the output of the integrator will maintain the 90° phase lead, regardless of frequency.

The gain magnitude of the integrator for sinusoidal inputs is given by

$$|A_v| = \frac{1}{\omega RC} \tag{5-9}$$

Notice that in Eq. (5-9) the denominator approaches zero as ω approaches zero. This implies infinite gain at zero frequency (dc). It is shown later that this is not quite the case, as Eq. (5-9) only applies to sinusoidal inputs.

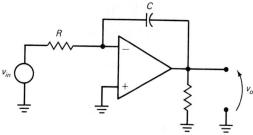

Fig. 5-17 Basic op amp integrator.

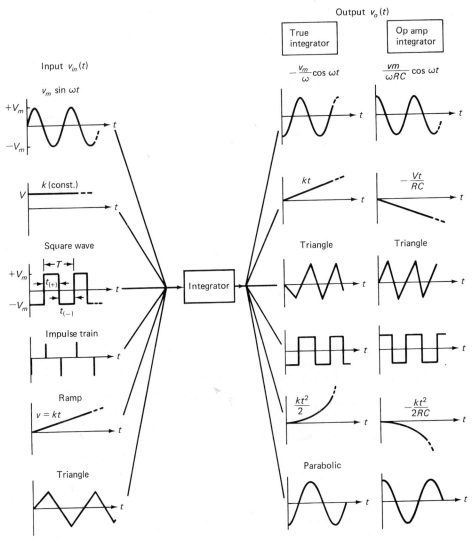

Fig. 5-18 Integrator input-output relations for commonly occurring signals.

Integration of Nonsinusoidal Inputs

Not all inputs are sinusoidal; therefore it is important to be familiar with the other commonly encountered signals. A representative sampling of typical waveforms is presented in Fig. 5-18. The complementary relationship between the integrator and the differentiator is easily seen by comparing Figs. 5-18 and 5-13.

The following is a list of descriptions of the various input-output relationships presented in Fig. 5-18. Again, because the sinusoidal case is easily analyzed using methods discussed earlier, this input shall be neglected here.

Also, the output of the integrator is assumed to be zero at the instant the input signal is applied. The reasons for this assumption (and an explanation of why it is not always the case) are explained in the following subsections.

THE CONSTANT VOLTAGE INPUT A constant voltage applied to the integrator will result in a linear ramp output. Observe that as time progresses, the area under the input function curve increases linearly. From the circuit operation perspective, the generation of this output can be understood by applying the virtual ground concept. Since $I_1 = V_{in}/R$, the output of the op amp must force an equal but opposite feedback current I_F to flow through the capacitor in order to maintain virtual ground at the inverting input terminal. Recall that current will flow through a capacitor only if the voltage across the capacitor is time-varying. In this case, in order to cause $-I_F = I_1$, the output of the op amp must increase (positively or negatively) linearly with time. Thus a linear ramp is produced. The ramp voltage at a given time after the application of the input is given by

$$v_o = -\frac{V_{in}t}{RC} \tag{5-10}$$

assuming that $v_o = 0$ V at the start of integration.

THE SQUARE WAVE Given a square-wave input, the output of the op amp integrator will alternately ramp up and down, with polarity opposite to that of the applied input voltage. A square wave ($t_{(+)} = t_{(-)}$) that is symmetrical about 0 V will produce a symmetrical triangle wave with peak amplitudes given by the application of Eq. (5-10).

For more general rectangular waveforms in which $t_{(+)} \neq t_{(-)}$, the output of the op amp integrator will ramp in positive and negative directions for differing lengths of time, in accordance with the input voltage levels present. For example, an input with $t_{(+)} = 10$ ms and $t_{(-)} = 2$ ms will cause the output to ramp five times longer in the negative direction than in the positive direction. This is illustrated in Fig. 5-19. Eventually, the op amp will saturate under such conditions.

To determine the output of the integrator at a given time, Eq. (5-10) is applied in a successive manner. To illustrate this point, let us assume that Fig. 5-19 represents the input and output voltages for Fig. 5-17 with $R = 10$ kΩ and $C = 1$ μF. Assuming that the output is initially zero ($V_0 = 0$ V at t_0), at t_1 the output is

$$V_1 = -\frac{V_m t_{(+)}}{RC}$$

$$= -\frac{2 \text{ V} \times 10 \text{ ms}}{10 \text{ kΩ} \times 1 \text{ μF}}$$

$$= -2 \text{ V}$$

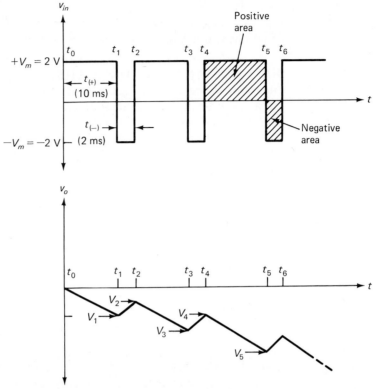

Fig. 5-19 Example of response of integrator with $C = 1\ \mu F$ and $R = 10\ k\Omega$ to a bipolar rectangular waveform.

Note that $t_{(+)} = t - t_0$. The output at t_2 is found by adding V_1:

$$V_2 = V_1 + \left(-\frac{V_m t_{(-)}}{RC}\right)$$

$$= -2\ V - \frac{-2\ V \times 2\ ms}{10\ k\Omega \times 1\ \mu F}$$

$$= -2\ V + 0.4\ V$$

$$= -1.6\ V$$

The output voltages present at the remaining points in time are found in a similar manner. In terms of calculus, we are taking advantage of the following relationship:

$$\int_{t_0}^{t_n} v(t)\,dt = \int_{t_0}^{t_1} v(t)\,dt + \int_{t_1}^{t_2} v(t)\,dt + \cdots + \int_{t_{n-1}}^{t_n} v(t)\,dt \tag{5-11}$$

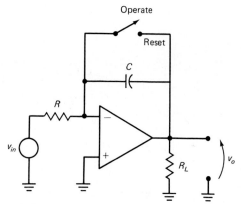

Fig. 5-20 Reset switch added to force integrator initial conditions to zero.

Thus, integration is a linear operation, where the output is the sum (continuous) of an arbitrary number of subintegrals starting at $t=0$ and ending at the upper limit of the integration period.

It is sometimes necessary to discharge the integrator's capacitor prior to the application of a signal. A reset switch may be placed across the capacitor for this purpose, as shown in Fig. 5-20.

THE IMPULSE TRAIN INPUT Application of an alternating-polarity impulse train will result in a rectangular output waveform. The exact amplitude of a given output level depends on the area of the impulses, and is beyond the scope of this text. Impulses are often referred to as delta functions $\delta(t)$.

THE RAMP INPUT Application of a linear ramp input will result in a parabolic output voltage. The output voltage at a given instant in time is given by

$$v_o = -\frac{kt^2}{2RC} \tag{5-12}$$

where k is the rate of change of v_{in} in volts per second.

THE TRIANGULAR INPUT The output of the integrator in response to a triangular input is found in a manner similar to that for a ramp input. This output bears a close resemblance to a sinusoidal waveform, although it is not sinusoidal, but rather parabolic in shape.

EXAMPLE 5-7

Determine the output voltages produced by the cascaded integrators in Fig. 5-21 at $t = 0.5$ s. Assume that the integrators are reset to 0 V at the start of the operation.

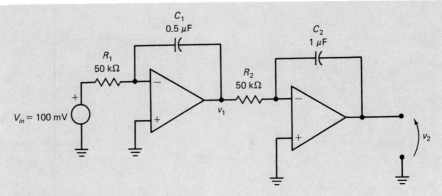

Fig. 5-21 Circuit for Example 5-7.

Solution

The input is a dc voltage of 100 mV; therefore, the first integrator will produce a negative-going ramp, given by Eq. (5-10):

$$v_1 = -\frac{V_{in}t}{RC}$$

$$= -\frac{0.1t \text{ V·s}}{0.025 \text{ s}}$$

In general form, the expression for v_1 is

$$v_1 = -4t \text{ V}$$

At the time of interest, $t = 0.5$ s, the output is

$$V_1\big|_{t=0.5\text{ s}} = \frac{-0.1 \text{ V} \times 0.5 \text{ s}}{0.025 \text{ s}}$$

$$= -2 \text{ V}$$

The output of the second integrator is given by Eq. (5-12), because the input is a linear ramp:

$$v_2 = -\frac{kt^2}{2RC}$$

$$= -\frac{-4t^2}{2RC}$$

$$= \frac{4t^2 \text{ V}}{0.1 \text{ s}}$$

At the time of interest ($t = 0.5$ s),

$$v_2\big|_{t=0.5\,s} = \frac{4V \times 0.5^2}{0.1}$$

$$= \frac{1}{0.1}$$

$$= 10 \text{ V}$$

The input and output voltage waveforms for this example are shown in Fig. 5-22.

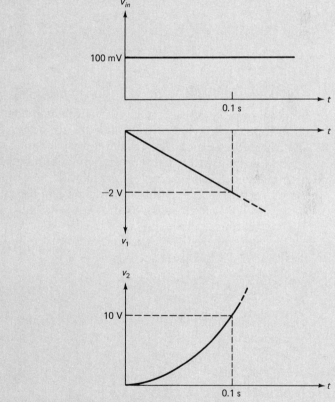

Fig. 5-22 Voltage waveforms for Example 5-7.

As demonstrated in the previous examples, it is not necessary to fully understand the mathematical details behind integrators and differentiators in order to make use of them or understand their functions in many systems.

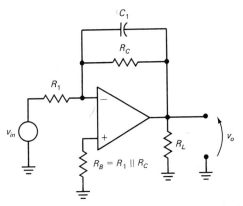

Fig. 5-23 Adding R_C reduces dc gain of the integrator, reducing drift.

It is important, however, that the responses of these building blocks to commonly encountered signals be understood.

Practical Integrators

Because the integrator effectively accumulates voltage over time, the presence of an input offset voltage will cause the capacitor to charge up, producing an error in the output. The smaller the capacitor, the more quickly the offset error builds up with time. Three obvious approaches to help alleviate this problem are the use of a larger capacitor, the use of low-offset op amps, and the use of a bias compensation resistor R_B on the noninverting terminal. R_B is made equal to R_1 in value, as shown in Fig. 5-23. Another commonly used method that helps eliminate offset errors is the placement of a resistor R_C in parallel with the feedback capacitor. This is also shown in Fig. 5-23. Typically, the value of R_C is chosen such that

$$R_C > 10R_1$$

This ensures that the response of the integrator is not dominated by the relationship $v_o = -R_C/R_1$ over the useful range of signal frequencies.

5-3 ◆ NONLINEAR OP AMP CIRCUITS

Most typical applications require the op amp and its ancillary components to act in a linear manner. For passive devices such as resistors and capacitors, this means that the *I-V* relationship for a given device should be described by a linear equation (Ohm's law). For the op amp, linear operation means that its input and output voltages are related by a *constant* of proportionality. That is, A_v should be constant, and independent of V_{in}. We have made these assumptions all along, but there are certain applications in which it is desirable to make an op amp behave in a nonlinear manner. Such applications often require the use of logarithmic and antilogarithmic amplifiers, which are the topics of this section.

The Logarithmic Amplifier

A logarithmic amplifier is an amplifier for which the output voltage is proportional to the logarithm of the input voltage. In order to generate such a relationship between input and output voltages, a device or devices that behave nonlinearly (logarithmically) must be used to control the gain of the op amp. One such device is the semiconductor diode. The forward transfer characteristics of silicon diodes are closely described by Shockley's equation:

$$I_F \cong I_s e^{V_F/\eta V_T} \tag{5-13}$$

where I_s is the diode saturation (leakage) current, e is the base of the natural logarithms ($e \cong 2.71828$), V_F is the forward voltage drop across the diode, and V_T is the thermal equivalent voltage for the diode (at room temperature, 20° C, $V_T \cong 26$ mV). The index η (Greek *eta*) varies from about 2 for currents of about the same magnitude as I_S to approximately 1 for higher values of I_F.

Basic Log Amp Operation

Closing the feedback loop of an inverting op amp with a diode will result in an amplifier whose output voltage is proportional to the logarithm of its input voltage. Such an amplifier is shown in Fig. 5-24.

The analysis of the logarithmic amplifier is relatively straightforward. Once again, the concept of virtual ground is central to the circuit's operation.

Because the inverting terminal is at virtual ground, we may write

$$I_1 = \frac{V_{in}}{R_1} \tag{5-14}$$

and

$$I_F = -I_1 \tag{5-15}$$

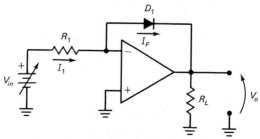

Fig. 5-24 Simple logarithmic amplifier.

which produces

$$I_F = \frac{-V_{in}}{R_1} \tag{5-16}$$

The magnitude of the output voltage is equal to the voltage drop across the diode, which is obtained by solving Eq. (5-13) for V_F. This yields

$$V_o = -V_F = -\eta V_T \ln \frac{I_F}{I_S} \tag{5-17}$$

Substituting Eq. (5-16) into (5-17) and performing some algebra yields

$$V_o = -\eta V_T \ln \frac{V_{in}}{R_1 I_S} \tag{5-18}$$

In order to apply Eq. (5-18) and obtain reasonably useful results, it is necessary to know the diode saturation current at the temperature of operation and the value of η. At higher current levels ($I_F > 1$ mA), most diodes begin to behave somewhat linearly. This occurs because ohmic resistances (contact resistance and bulk resistance) are of the same order of magnitude as the diode forward resistance. Recall that the instantaneous forward resistance (dynamic resistance) of a diode is given by the familiar equation

$$r_D = \frac{26 \text{ mV}}{I_F}$$

Because r_D decreases rapidly at higher currents, logarithmic amplifiers are generally restricted to operation with $I_F < 1$ mA.

EXAMPLE 5-8

Refer to Fig. 5-24. Given $R_1 = 10$ kΩ and $T = 20°$ C, and using a silicon diode with $I_S = 25$ nA, determine V_o for the following input voltages: $V_{in} = 0.01$ V, 0.1 V, 1.0 V. Assume that $\eta = 1$.

Solution

Applying Eq. (5-18) and entering all constants,

$$V_o = -0.026 \text{ V} \times \ln\left(\frac{V_{in}}{2.5 \times 10^{-4}}\right)$$

For $V_{in} = 0.01$ V,

$$V_o = -0.026 \times \ln 40$$
$$= -0.096 \text{ V}$$

The remaining output voltages are determined in a similar manner, producing

$V_{in} = 0.10$ V $V_o = -0.156$ V

$V_{in} = 1.00$ V $V_o = -0.216$ V

Plots of $|V_o|$ versus V_{in} for Example 5-8 on linear and semilogarithmic graphs are shown in Fig. 5-25. On the linear graph, it is obvious that the voltage gain of the amplifier is very high for low voltages ($V_{in} < 0.01$ V) and very low for high input voltages ($V_{in} > 0.10$ V). The circuit produces a straight-line plot on the semilogarithmic graph, emphasizing the logarithmic nature of the amplifier's transfer characteristic. The transfer characteristics of logarithmic amplifiers are usually expressed in terms of the slope of the V_o versus V_{in} plot in millivolts per decade. For the circuit used in Example 5-8, the slope is found as follows, for the 10 to 1 change in V_{in} from 0.10 V to 1.00 V:

$m = 0.563$ V $- 0.443$ V

$= 60$ mV/decade

Ideally, any diode will make the op amp exhibit this 60 mV/decade characteristic. In practice, η may actually be closer to 2 than 1, and it also varies with I_F. Variation of η is minimized by restricting the diode current to a 2- or 3-decade range. In Example 5-8, if we had used $\eta = 2$, the slope of the transfer characteristic would have been 120 mV/decade. This is illustrated by the upper curve in Fig. 5-25b. A real diode will yield a slope somewhere between 60 and 120 mV/decade, depending on the magnitude of I_F.

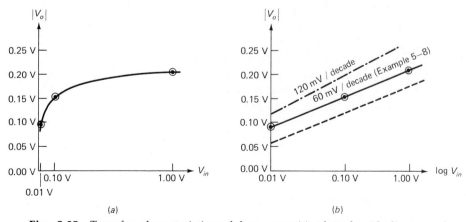

Fig. 5-25 Transfer characteristics of log amp: (a) plotted with linear scales; (b) plotted with logarithmic abscissa.

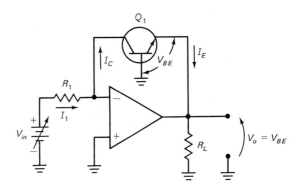

Fig. 5-26 Transdiode log amp.

While η affects the slope of the transfer curve, I_S determines the y intercept. For example, the lower curve in Fig. 5-25b would be produced for Example 5-8 if I_S were lower than 25 nA. Notice that the slope of this curve is 60 mV/decade, indicating η = 1.

The circuit in Fig. 5-24 can be used only with positive input voltages. Operation with a negative input voltage can be achieved if the direction of the diode is reversed. This will result in a positive output voltage for a given input.

Additional Log Amp Variations

Often, a transistor will be used as the logging element in a logarithmic amplifier. Figure 5-26 shows this alternative, with the transistor connected in what is called the transdiode configuration. The operation of this circuit is basically the same as that of the circuit in Fig. 5-24. The equation describing the behavior of the transistor is

$$I_C \cong I_{ES} e^{V_{BE}/V_T}$$

where I_{ES} is the emitter saturation current and V_{BE} is the drop across the base-emitter junction. Transistor logging elements generally allow operation of the log amp over wider current ranges than do equivalent diode-based circuits (greater dynamic range).

PN junction saturation current is highly temperature-dependent. As a general rule of thumb, I_S (and I_{ES}) doubles for every 10° C increase in temperature. Recall also that V_T varies with temperature. As a consequence of this, the logging element must be held at a constant temperature for accurate operation. There are commercially available monolithic transistor arrays with built-in heating circuitry. The heating circuitry maintains a constant temperature, resulting in constant I_{ES} and V_T.

The Antilogarithmic Amplifier

The output of an antilog amp is proportional to the antilog of the input voltage. With a diode as the logging element, the basic antilog amp is shown

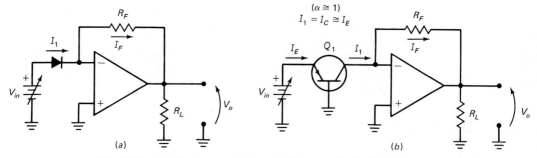

Fig. 5-27 Antilog amp circuits. (*a*) Using a diode; (*b*) using a transdiode.

in Fig. 5-27. Using Eq. (5-13) and applying basic op amp theory, the following output voltage expression is obtained:

$$V_o = -R_F I_S \, e^{V_{in}/V_T} \tag{5-19a}$$

An equivalent antilog amp using the transdiode logging element is shown in Fig. 5-27*b*. Assuming that for Q_1, $\alpha = 1$, the output of this circuit is given by the following equation:

$$V_o = -R_F I_{ES} \, e^{V_{in}/V_T} \tag{5-19b}$$

As with the log amp, it is necessary to know the saturation currents and to tightly control junction temperature for meaningful use of these equations.

EXAMPLE 5-9

The circuit in Fig. 5-27*a* has the following specifications: $R_F = 200 \ \Omega$, $I_S = 100$ pA, and $T = 20°C$. Determine V_o for the following inputs: $V_{in} = 0.40$ V, 0.46 V, 0.52 V. Assume that $\eta = 1$.

Solution

Using Eq. (5-19*a*) and $V_{in} = 0.40$ V, we obtain

$$V_o = -(200)(100 \times 10^{-12})(e^{0.40/0.026})$$
$$= -(2 \times 10^{-8})(4.802 \times 10^6)$$
$$= -0.096 \text{ V}$$

Similarly, for the remaining voltages we get

$$V_o = -0.965 \text{ V}$$

$$V_o = -9.803 \text{ V}$$

The preceding example illustrates the radical change that is possible in V_o, given a relatively small change in V_{in}. In order to show the relationship between the antilog amp and the log amp, the increments in V_{in} in Example 5-9 were chosen to be 60 mV. Each 60-mV increase in V_{in} resulted in a tenfold increase in V_o. This is the opposite of the log amp, for which a tenfold increase in V_{in} resulted in a 60-mV increase in V_o.

Logarithmic Amplifier Applications

Logarithmic amplifiers are used in several areas. One area of application uses log and antilog amps to form analog multipliers. Both analog multipliers and logarithmic amplifiers are also used in analog signal processing applications. These applications are presented briefly in this section.

Analog Multipliers

Log and antilog amps may be used to perform the operations of multiplication and division electronically. The operation of these amplifiers in such appli-

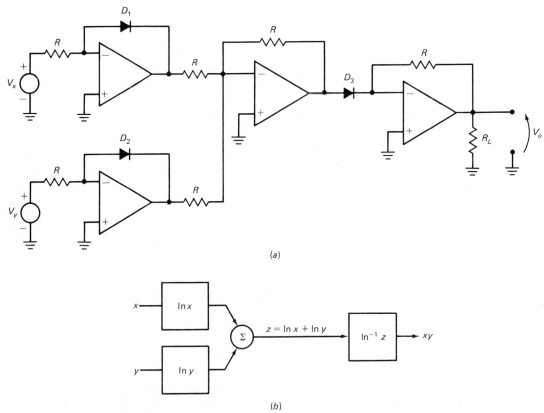

Fig. 5-28 Using log and antilog amps to implement a multiplier. (*a*) The circuit; (*b*) block diagram representation.

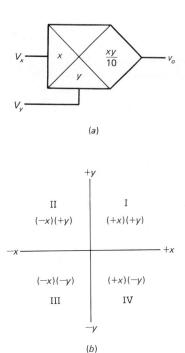

Fig. 5-29 Analog multiplier. (*a*) Symbol; (*b*) the four quadrants of operation.

cations is based on the following mathematical relationships:

$$\ln xy = \ln x + \ln y$$

$$\ln \frac{x}{y} = \ln x - \ln y$$

The circuit in Fig. 5-28*a* produces an output that is proportional to the product of two positive inputs, V_x and V_y. A block diagram representation of this circuit is shown in Fig. 5-28*b*. This is referred to as a *one-quadrant multiplier*. The term *one-quadrant* refers to the fact that the inputs must both be of the same polarity (positive in the case of Fig. 5-28). The general symbol for an analog multiplier and the multiplier quadrants are illustrated in Fig. 5-29. Generally, commercially available analog multiplier inputs are limited to ±10 V, and the output is scaled by 0.1. This makes the full-scale output of the multiplier ±10 V, a convenient range to work with.

If the multiplier of Fig. 5-29*a* was a *two-quadrant multiplier*, one of its inputs (say the *x* input) would be restricted to positive voltages only, while the other input could be driven either positively or negatively. In such a case, operation of the circuit would be restricted to quadrants I and IV in Fig. 5-29*b*. *Four-quadrant multipliers* may have any combination of polarities on their inputs, allowing operation in all four quadrants.

The circuit in Fig. 5-30*a* produces an output that is proportional to the square of the input voltage, while the circuit in Fig. 5-30*b* produces an output that is proportional to the square root of the input.

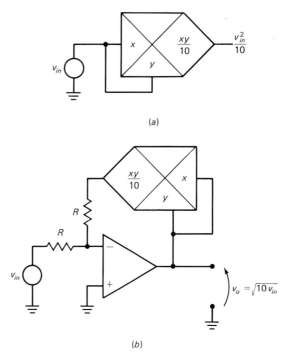

Fig. 5-30 Implementation of mathematical operations using the analog multiplier. (*a*) Squaring circuit; (*b*) square root circuit.

EXAMPLE 5-10

The input to the four-quadrant multiplier in Fig. 5-30*a* is given by $v_{in} = 2 \sin 500t$ V. Determine the expression for v_o and make a sketch of this waveform.

Solution

Since the input is a sine wave, we must resort to the use of trigonometry to determine the output. The output of the squarer is of the form

$(x \sin \theta)(y \sin \phi)$

The following trigonometric identity is applied in this situation:

$(x \sin \theta)(y \sin \phi) = \dfrac{xy}{2}[\cos(\theta - \phi) - \cos(\theta + \phi)]$

Now, since the x and y inputs are the same, we may assign the following identities:

$x = y = 2$ V $\theta = \phi = 500t$ rad

Substituting into the trigonometric identity yields

$$v_o = \frac{4\text{ V}}{20}(\cos 0 - \cos 1000t)\text{ V}$$

$$= \frac{4\text{ V}}{20} - \frac{4\text{ V}}{20}\cos 1000t$$

$$= 0.2 - 0.2\cos 1000t\text{ V}$$

The input and output voltage waveforms are shown in Fig. 5-31. Notice that squaring a sine function creates a dc offset and doubles the frequency of the input signal. These can be quite useful properties.

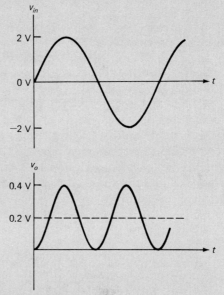

Fig. 5-31 Waveforms for Example 5-10.

Signal Processing

Many transducers produce output voltages that vary nonlinearly with the physical quantity being measured (recall the thermistor used in Prob. 4-12). Often, it is desirable to linearize the outputs of such devices. Logarithmic amplifiers and analog multipliers may be used for this purpose. Suppose that the output voltage from a light-sensitive transducer follows the solid inverse exponential curve shown in Fig. 5-32. If this voltage was to be monitored on a meter, the scale would be compressed at the high end, like that of an analog ohmmeter. This signal could be linearized by processing it through an exponential amplifier prior to sending it to the meter. The dashed curve in Fig. 5-32 represents the I/O (input-output) characteristics of this amplifier.

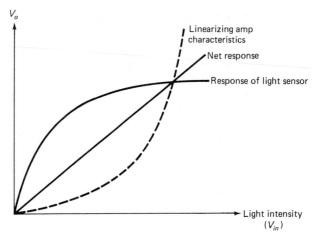

Fig. 5-32 Linearization of a signal by use of a circuit with complementary transfer characteristics.

The net result of this processing is the constant-slope curve in the middle of the graph.

Some applications require just the opposite type of signal processing. For example, a device called a pressure transmitter produces an output voltage (or current) that is proportional to the difference in pressure between the two sides of a strain gage sensor. The basic ideas behind the pressure transmitter are shown in Fig. 5-33. A venturi is used to create a pressure differential across the strain gage, and the output of the transmitter is proportional to this pressure differential. This tells us something about the ratios of the diameters of the pipe and the venturi, but suppose that we wanted to know the rate of flow through the pipe. It turns out that the fluid flow rate through

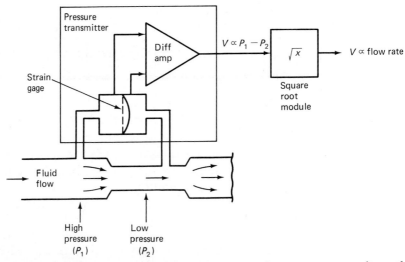

Fig. 5-33 Use of a square root amplifier to produce an output voltage that is directly proportional to flow rate.

the pipe is proportional to the square root of the pressure differential, detected by the strain gage. Now, if the output of the transmitter is processed through a square root amplifier, we will obtain an output that is directly proportional to flow rate. This is generally a more useful quantity to monitor, and commercial pressure transmitters often include built-in square root circuitry.

5-4 ◆ PRECISION RECTIFIERS

Op amps can be used to form nearly ideal rectifiers. The idea behind this application is to use negative feedback to effectively make the op amp behave like a rectifier with near-zero barrier potential and with a linear I/O characteristic. The transconductance curves for a typical silicon diode and an ideal diode are shown in Fig. 5-34.

Precision Half-Wave Rectifier

A half-wave precision rectifier is shown in Fig. 5-35. The operation of this circuit is explained in two parts. The solid arrows represent current flow for

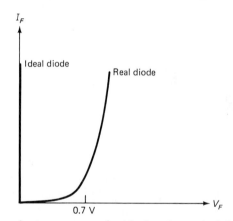

Fig. 5-34 Transconductance curves for ideal and practical diodes.

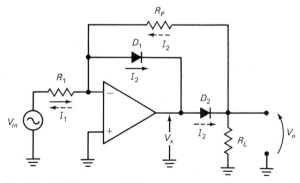

Fig. 5-35 Precision half-wave rectifier.

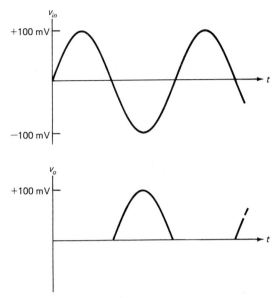

Fig. 5-36 Typical I/O voltage waveforms for a precision HW rectifier.

positive half-cycles of v_{in}, and the dashed arrows represent current flow during negative half-cycles.

Assuming that the signal source is going positive, the output of the op amp begins to go negative, forward-biasing D_1. Since D_1 is forward-biased, the output of the op amp (V_x) will reach a maximum level of about -0.7 V, regardless of how far positive v_{in} goes. This is insufficient to appreciably forward-bias D_2, and V_o will remain at 0 V.

On negative-going half-cycles (the dashed arrows), D_1 is reverse-biased and D_2 is forward-biased. Negative feedback effectively reduces the barrier potential of D_2 to 0.7 V/A_{OL}, which is very near zero for most purposes. The gain of the circuit to the negative-going portions of v_{in} is given by the usual inverting op amp equation

$$A_v = \frac{-R_F}{R_1}$$

The application of negative feedback also reduces the nonlinearity of the diode transconductance curve. Typical input and output voltage waveforms for the precision half-wave rectifier are shown in Fig. 5-36.

Precision Full-Wave Rectifier

The precision full-wave rectifier is shown in Fig. 5-37. The operation of this circuit is easiest to understand if it is analyzed for each half-cycle of the input signal. Current flow for positive half-cycles are shown as solid arrows, while negative half-cycle currents are shown as dashed arrows.

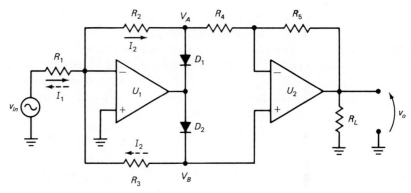

Fig. 5-37 Precision full-wave rectifier.

The positive half-cycle causes D_1 to become forward-biased, while reverse-biasing D_2. Under these conditions, $V_B = 0$ V and

$$v_A = -v_{in}\frac{R_2}{R_1} \quad \text{for positive half-cycles of } v_{in} \tag{5-20}$$

The output of U_2 is now given by

$$v_o = -v_A\frac{R_5}{R_4} \tag{5-21}$$

Substituting Eq. (5-20) into (5-21) produces

$$v_o = v_{in}\frac{R_2 R_5}{R_1 R_4} \tag{5-22}$$

On negative-going half-cycles, the output of U_1 is driven positive, forward-biasing D_2 and reverse-biasing D_1. Under these conditions, $V_A = 0$ V and

$$v_B = -v_{in}\frac{R_3}{R_1} \quad \text{for negative half-cycles of } v_{in} \tag{5-23}$$

Op amp U_2 is operating in the noninverting configuration now, and its output is given by

$$v_o = v_B\left(1 + \frac{R_5}{R_4}\right) \tag{5-24}$$

Substituting Eq. (5-23) into (5-24) produces

$$v_o = -v_{in}\left(\frac{R_3}{R_1} + \frac{R_3 R_5}{R_1 R_4}\right) \tag{5-25}$$

If $R_3 = R_1/2$, both half-cycles of the input will receive equal gain, which is generally the desired situation.

Precision rectifiers are useful when the signal to be rectified is very low in amplitude and where good linearity is required. The frequency and power-handling limitations of the op amps used in the circuit generally limit the use of precision rectifiers to low-power applications from a few hundred kilohertz on down.

The precision full-wave rectifier is often referred to as an *absolute magnitude circuit*. This term is generally applied when the rectifier is used to produce the absolute magnitude of a function in an analog computer circuit, or in a variation of the four-quadrant multiplier called the quarter-square multiplier.

Chapter Review

Op amps have proven themselves to be one of the most versatile of all electronic building blocks. Op amps may be used with resistive and reactive components inside and outside of the feedback loop. In such circuits, with sinusoidal inputs, the general op amp analysis methods are applicable, with phasor algebra applied as needed.

Op amp integrators and differentiators are very widely used to perform continuous, real-time operations on analog signals. The differentiator produces an output that is negatively proportional to the rate of change of the input voltage. The output of the integrator is proportional to the area under the voltage versus time curve of the input signal. Both of these circuits are also used in wave-shaping applications as well.

Logarithmic and antilogarithmic amplifiers are nonlinear circuits that rely on the exponential nature of the PN junction for their operational characteristics. Logarithmic amplifiers tend to be rather temperature-sensitive, and special precautions are necessary for accurate operation of these circuits. Log and antilog amps form the basic foundation for analog multiplier circuits. An analog multiplier is a circuit that provides an output voltage that is proportional to the product of its input voltages. Analog multipliers, logarithmic amps, and antilog amps find use in various signal processing applications.

Precision half- and full-wave rectifiers are designed using op amps and semiconductor diodes. Such circuits have nearly ideal rectifier characteristics. Precision rectifiers are generally limited to low-power, low-frequency applications.

Questions

5-1. Why is the noninverting op amp configuration unsuitable for use in integrator and differentiator applications?

5-2. In the noninverting configuration, if a reactive device is to be used in the feedback loop of the op amp, and a resistor is connected from the inverting terminal to ground, what type of reactive component is required to produce a leading phase at the output of the circuit?

5-3. How do the outputs of op amp-based integrators and differentiators differ from their mathematically derived counterparts?

5-4. What phase shift (relative to v_{in}) will be produced at the output of an op amp differentiator? At the output of an op amp integrator?

5-5. In general, what type of waveform is produced at the output of an integrator when a square-wave input is applied?

5-6. Why are input offset voltages a serious problem in the realization of op amp integrators?

5-7. What effect does variation in η for the logging element have on the V_o versus V_{in} transfer characteristic of a logarithmic amplifier?

5-8. In the case of an antilog amplifier, does A_v effectively increase or decrease as V_{in} is increased in magnitude?

5-9. A certain logarithmic amplifier produces an output voltage that changes at a rate of 120 mV/decade with V_{in}. What can be said about the equation describing the transconductance of the logging element used in this circuit?

5-10. What type of analog multiplier can be used to produce the product of input voltages of any polarity?

5-11. State the main advantages of op amp-based precision rectifiers over PN junction types. What are the advantages of passive PN junction rectifiers over op amp active rectifiers?

Problems

5-1. Refer to Fig. 5-2. Given v_{in} = 250 sin $2\pi 1000t$ mV, Z_1 = 200 + j600 Ω, and Z_F = 2.7 + j3.3 kΩ, write the expression for v_o.

5-2. The circuit in Fig. 5-2 has Z_1 formed by a 0.22-μF capacitor while Z_F is formed by a 1-kΩ resistor in series with a 100-mH inductor. Write the expression for v_o given v_{in} = 500 sin 8000t mV.

5-3. Repeat Prob. 5-2 for v_{in} = 500 sin 10,000t mV.

5-4. Refer to Fig. 5-6. Given Z_1 = 800 $\underline{/20°}$ Ω, Z_F = 18 $\underline{/0°}$ kΩ, and $\mathbf{V}_{in}$ = 170 $\underline{/90°}$ mV, write the shorthand expression for v_o.

5-5. Refer to Fig. 5-6. Z_1 is formed by a 2.2-kΩ resistor and Z_F is formed by a 0.68-μF capacitor. Given v_{in} = 1 sin 200t V, write the expression for v_o and sketch the v_{in} and v_o waveforms.

5-6. Given the conditions stated in Prob. 5-4, write the shorthand expression for the current flowing through the feedback element (that is, the current I_F through Z_F).

5-7. The differentiator in Fig. 5-10 has $C = 0.01\ \mu\text{F}$ and $r = 22\ \text{k}\Omega$. The input voltage is $v_{in} = -10{,}000t$ V. Write the expression for v_o.

5-8. Sketch the output voltage waveform for the circuit in Fig. 5-38. Label the peak and mean voltages.

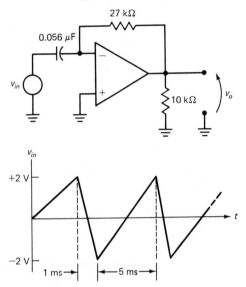

Fig. 5-38

5-9. Sketch the output voltage waveform for the circuit in Fig. 5-39. Label the peak and mean voltages.

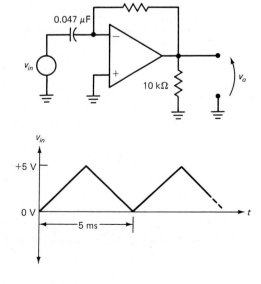

Fig. 5-39

5-10. Write the expression for v_o for Fig. 5-39, given $v_{in} = 3 \cos 6283t$ V.

5-11. Sketch the output voltage waveform for the circuit in Fig. 5-40, given the input shown. Assume that the integrator is initially reset to $V_o = 0$ V. Label all pertinent voltages.

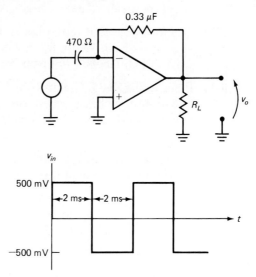

Fig. 5-40

5-12. Refer to Fig. 5-40. Sketch the output voltage waveform that should be produced for the first three cycles of the input voltage shown in Fig. 5-41. Label all output voltage peaks.

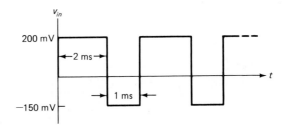

Fig. 5-41

5-13. Refer to Fig. 5-21. Write the expressions for v_1 and v_2 given $v_{in} = 10 \sin 2\pi 100t$ V. Assume that both integrators are initially reset.

5-14. Refer to Fig. 5-21. Given $v_{in} = 500t$ V, determine the value of v_1 at $t = 10$ ms. Assume that the integrators are initially reset.

5-15. Refer to Fig. 5-24. Given $R_1 = 2.2$ kΩ, $I_S = 50$ pA, $T = 20°C$, and $\eta = 1$, determine V_o for $V_{in} = 4$ V and $V_{in} = 8$ V.

5-16. Refer to Fig. 5-24. Given $R_1 = 10$ kΩ, $I_S = 20$ pA, $T = 20°C$, and $\eta = 1$, determine the value of V_{in} that will produce $V_o = -0.500$ V.

5-17. Refer to Fig. 5-27a. Given $R_F = 1$ kΩ, $I_S = 1$ nA, $T = 20°C$, and $\eta = 1$, determine V_o for $V_{in} = 0.35$ V and $V_{in} = 0.40$ V.

5-18. Given the conditions of Prob. 5-17, determine the values of V_{in} that will produce $V_o = 1.00$ V and 12.00 V.

5-19. The four-quadrant multiplier in Fig. 5-29 has the following input voltages: $V_x = -5$ V, $v_y = 5 \sin \omega t$ V. Sketch the output voltage waveform. Label the peak values of the waveform.

5-20. The four-quadrant multiplier in Fig. 5-29 has the following input voltages: $v_x = 2 \sin 100t$ V, $v_y = 2 \cos 100t$ V. Sketch the input and output waveforms, labeling the peak values of each.

5-21. Refer to Fig. 5-35. Given $R_F = 10$ kΩ and $R_1 = 1$ kΩ, sketch the input and output voltage waveforms for $v_{in} = 1 \sin \omega t$ V. Label all peak values and determine the average value of V_o.

5-22. Refer to Fig. 5-37. Given $R_1 = R_2 = R_4 = 1$ kΩ, determine the values required for R_5 and R_3 such that the gain magnitude of the circuit is 2 for both half-cycles of the input.

5-23. Refer to Fig. 5-37. Given $R_5 = R_4 = 10$ kΩ and $R_1 = 1$ kΩ, determine the required values for R_2 and R_3 such that negative-going inputs are amplified by a factor of 2 and positive-going inputs are amplified by a factor of 4. Make a sketch of the transfer characteristics (V_o versus V_{in}) of the circuit.

CHAPTER 6
ACTIVE FILTERS

Operational amplifiers find wide application in the design of active filters. There are many ways to characterize a filter, but in general one of the two following descriptions will apply. A filter is a circuit that is designed to pass frequencies within a specific range, while rejecting all frequencies that fall outside this range. Another class of filters is designed to produce an output that is delayed in time or shifted in phase with respect to the filter's input.

Filters may be constructed using only passive components (resistors, capacitors, and inductors), and, appropriately enough, such filters are called *passive filters*. An *active filter* is a filter whose characteristics are augmented through the use of one or more amplifiers. Active filters are generally constructed using op amps, resistors, and capacitors only. This inductorless design approach is one of the advantages of the active filter over the passive filter. Also, in many cases, the use of active filters allows many filter parameters to be adjusted continuously and at will. This is often not possible or practical in traditional passive filter designs. In this chapter, we shall investigate some of the more common active filters and their applications.

6-1 ♦ FILTER FUNDAMENTALS

There are five basic types of filters: low-pass (LP), high-pass (HP), bandpass (BP), bandstop (notch or band-reject), and all-pass (or time-delay) filters. Typical frequency response curves for the LP, HP, BP, and bandstop filter types are shown in Fig. 6-1. The vertical axes (ordinates) of these graphs are scaled in terms of decibel gain magnitude H as a function of input signal frequency ω in rad/s. $|H(j\omega)|$ is basically just an alternative notation used to denote the frequency-dependent voltage gain of a filter. The complex filter response is given by

$$H(j\omega) = |H(j\omega)| \; \underline{/\theta(j\omega)}$$

189

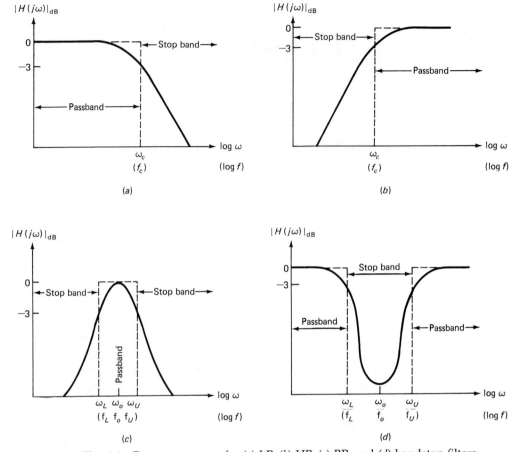

Fig. 6-1 Response curves for (*a*) LP, (*b*) HP, (*c*) BP, and (*d*) bandstop filters.

When signal frequencies are expressed in hertz, it is common practice to express filter response as $|H(jf)|$. The horizontal axes (abcissas) of the graphs are logarithmically scaled in terms of frequency. It is standard practice to use such semilogarithmic graphs to plot filter response curves.

The fifth class of filters, the all-pass or time-delay filters, will be discussed later in the chapter.

Filter Terminology

The range of frequencies that a filter will allow to pass, either amplified or relatively unattenuated, is termed the filter's *passband*. All other frequencies are considered to fall into the filter's *stop band(s)*. In most cases, the frequency at which the gain of the filter drops by 3.01 dB from that of the passband determines where the stop band begins. This frequency is called the *corner frequency* f_c. Normally, it is sufficiently accurate to assume that the response of the filter is down by 3 dB at the corner frequency.

The reason that the -3-dB point is chosen is because a 3-dB decrease in voltage gain translates to a reduction of 50 percent in the power being delivered to the load that is being driven by the filter. Because of this, f_c is often called the *half-power point*. Decibel voltage gain is actually intended to be a logarithmic representation of power gain. Power gain is related to decibel voltage gain by the following equation:

$$A_P = 10 \log \frac{P_o}{P_{in}} \tag{6-1}$$

where $P_o = \dfrac{V_o^2}{Z_L}$

$P_{in} = \dfrac{V_{in}^2}{Z_{in}}$

Substituting,

$$A_P = 10 \log \frac{V_o^2/Z_L}{V_{in}^2/Z_{in}}$$

which rearranges to form

$$A_P = 10 \log \frac{V_o^2 Z_{in}}{V_{in}^2 Z_L} \tag{6-2}$$

Now, if we assume that $Z_L = Z_{in}$, Eq. (6-2) may be written as

$$A_P = 10 \log \frac{V_o^2}{V_{in}^2}$$

$$= 10 \log \left(\frac{V_o}{V_{in}}\right)^2 \tag{6-3}$$

Making use of the logarithmic relationship $\log x^2 = 2 \log x$ yields

$$A_p = 20 \log \frac{V_o}{V_{in}} \tag{6-4a}$$

$$= 20 \log A_v \tag{6-4b}$$

Equations (6-4a) and (6-4b) tell us that when the input impedance of the filter equals the impedance of the load being driven by the filter, the power gain is dependent on the voltage gain of the circuit only. In active filter work, input impedance and load impedance are always considered to be equal. Also, even though Eq. (6-4) describes the power gain of the filter, because we are working with voltage ratios, the gain is expressed as a voltage gain, in decibels. That is,

$$|H(j\omega)|_{dB} = 20 \log \frac{V_o}{V_{in}} \qquad (6\text{-}5a)$$

or

$$|H(j\omega)|_{dB} = 20 \log A_v \qquad (6\text{-}5b)$$

Notice in Fig. 6-1a and b that once the frequency is well into the stop band, the rate of increase of attenuation is constant. Recall from previous discussions of the op amp Bode plot that rolloff is expressed in decibels per decade. The ultimate rolloff rate of a given filter is determined by the order of that particular filter. As discussed previously, a first-order section will have a rolloff of -20 dB/decade. A second-order filter will have an ultimate rolloff rate of -40 dB/decade. The reason that we are using the term ultimate rolloff is because some filters may begin to roll off at a rate greater than that which is implied by the filter order initially. In general, the ultimate rolloff of a filter may be determined using the following equation:

$$\text{Rolloff} = -20n \quad \text{dB/decade} \qquad (6\text{-}6a)$$

where n is the order of the filter. Rolloff may also be expressed in terms of decibels per octave. An octave is a twofold increase or decrease in frequency. The mathematical relationship here is

$$\text{Rolloff} = -6n \quad \text{dB/octave} \qquad (6\text{-}6b)$$

where, again, n is the order of the filter.

In between the relatively flat portion of the passband and the region of constant rolloff in the stop band is a region that is referred to as the *transition region*. This is illustrated in Fig. 6-2, using a low-pass response curve. Given two filters of the same order, if one has a greater initial increase in attenuation in the transition region, that filter will have a greater attenuation at any given frequency in the stop band. An important parameter that has a great effect on the shape of an LP or HP filter's response in the passband, stop band, and transition region is termed the *damping coefficient* α. The damping coefficient will generally range from 0 to 2. Filters with lower damping tend to exhibit peaking in the passband (and sometimes in the stop band as well) and more rapid and radically varying transition-region response attenuation. Filters with higher damping coefficients tend to pass through the transition region more smoothly, and they also tend not to exhibit peaking in the passband and stop band. Damping will be quantified in later sections of the text.

As shown in Fig. 6-1, HP and LP filters each have a single corner frequency. The BP and bandstop filters each have two corner frequencies,

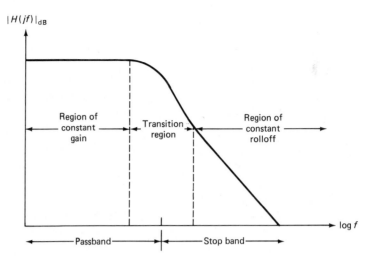

Fig. 6-2 LP response curve characteristics.

labeled f_L and f_U, plus a third frequency denoted as f_o, the center frequency. Because a logarithmic frequency scale is used, f_o appears centered between f_L and f_U. If the upper and lower corner frequencies are known, the center frequency may be found using the following equation:

$$f_o = \sqrt{f_L f_U} \tag{6-7}$$

The center frequency is the *geometric mean* of f_L and f_U. The bandwidth of a BP or bandstop filter is defined as

$$BW = f_U - f_L \tag{6-8}$$

Another important parameter that is associated with bandpass and bandstop filters is termed Q. This is defined by

$$Q = \frac{f_o}{BW} \tag{6-9}$$

A BP filter with a high Q will pass a relatively narrow range of frequencies, while a BP filter with a lower Q will pass a wider range of frequencies. Like HP and LP filters, bandpass filters will exhibit a constant ultimate rolloff rate, which is determined by the order of the particular BP filter.

Basic Filter Theory Review

The simplest filters are the first-order low- and high-pass *RC* sections shown in Fig. 6-3a and b, respectively. Circuits such as these will always exhibit a passband gain of slightly less than unity. Assuming negligible loading of the

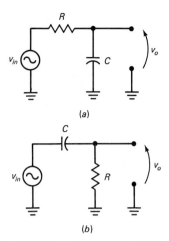

Fig. 6-3 First-order RC filters: (a) low-pass and (b) high-pass.

filter, the amplitude response (voltage gain) of the low-pass section of Fig. 6-3a is given by

$$H(j\omega) = \frac{-jX_C}{R - jX_C} = \frac{1}{1 + j\omega RC} \qquad (6\text{-}10)$$

This is a complex number, which can be resolved into gain magnitude and phase using the relationships

$$H(j\omega) = \frac{X_c}{\sqrt{R^2 + X_c^2}} \; \underline{/-\tan^{-1}\frac{R}{X_c}} \qquad (6\text{-}11)$$

The corner frequency f_c for a first-order LP (or HP) RC section is found by equating the magnitudes of R and X_C and solving for frequency as follows:

$$R = X_C$$

$$= \frac{1}{2\pi f C}$$

$$\frac{1}{f_c} = 2\pi RC$$

$$f_c = \frac{1}{2\pi RC} \qquad (6\text{-}12)$$

The gain (in decibels) and the phase response of the first-order LP filter may also be expressed in terms of input frequency f and the corner frequency f_c using

$$H(jf)|_{dB} = 20 \log \frac{1}{\sqrt{1 + (f/f_c)^2}} \; \underline{/-\tan^{-1}\frac{f}{f_c}} \qquad (6\text{-}13)$$

Similarly, the amplitude and phase response of the equivalent HP section are given by

$$H(jf)|_{dB} = 20 \log \frac{1}{\sqrt{1 + (f_c/f)^2}} \quad \angle \tan^{-1}\frac{f_c}{f} \qquad (6\text{-}14)$$

EXAMPLE 6-1

Refer to Fig. 6-3a. Given $C = 0.01$ μF, determine the value of R such that $f_c = 1$ kHz. Determine the numerical and decibel gain magnitude, and the phase of v_o relative to v_{in} for $v_{in} = 5 \sin 2\pi 10^3 t$ V.

Solution

Since C and f_c are given, we begin by solving Eq. (6-12) for R. This yields

$$R = \frac{1}{2\pi f C}$$

$$= 15.92 \text{ k}\Omega$$

The numerical gain magnitude and the phase shift of the filter at $f = 10$ kHz ($\omega = 2\pi f$) is found using a slightly modified version of Eq. (6-13).

$$H(jf) = \frac{1}{\sqrt{1 + (f/f_c)^2}} \quad \angle -\tan^{-1}\frac{f}{f_c}$$

$$= 0.1 \angle -84.3°$$

In decibels, the gain magnitude is

$$|H(jf)|_{db} = 20 \log 0.1$$

$$= -20 \text{ dB}$$

This is to be expected, as the first-order section has a rolloff rate of -20 dB/decade, and the input signal frequency is one decade above the corner frequency f_c.

In the preceding example it was assumed that the filter was unloaded, or that the load resistance being driven was high enough to be ignored. This is more the exception than the rule. Figure 6-4a illustrates the effect of loading on filter response. Loading tends to make the filter's response very droopy, which is quite undesirable. In order to prevent such loading, filter sections may be isolated using high-input-impedance buffers, such as the noninvert-

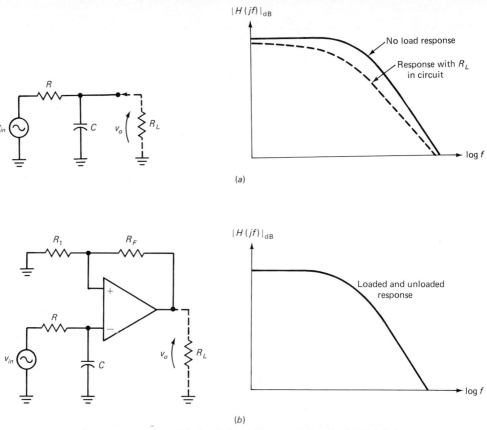

(a)

(b)

Fig. 6-4 Effects of loading on RC LP filter and buffered RC LP filter.

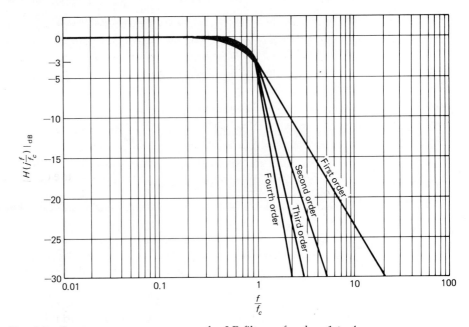

Fig. 6-5 Frequency response curves for LP filters of orders 1 to 4.

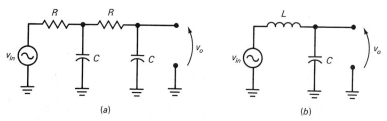

Fig. 6-6 Second-order LP filters: (a) RC; (b) LC.

ing op amp in Fig. 6-4b. The op amp effectively isolates the load from the filter, allowing minimal deviation from the desired response.

The amplitude/phase response of the filter in Fig. 6-4b is easily found by slightly modifying Eq. (6-13):

$$H(jf)|_{dB} = 20 \log \frac{A}{\sqrt{1 + (f/f_c)^2}} \quad \bigg/ -\tan^{-1}\left(\frac{f}{f_c}\right) \qquad (6\text{-}15)$$

where A is the closed-loop gain of the op amp.

Higher-order filters may be realized by cascading basic RC sections. A normalized plot of responses for RC filters of orders 1 through 4 is shown in Fig. 6-5. This plot is based on the use of filters with negligible loading. Notice that the higher the order of the filter, the more closely its response resembles that of the ideal brick wall filter.

A second-order LP filter may be designed using two cascaded RC sections, or by using an LC section, as shown in Fig. 6-6a and b, respectively.

Under the best of conditions (no interstage or output loading), the second-order RC filter will yield a response curve like that shown in Fig. 6-5. Unavoidable loading effects will result in a somewhat more droopy response than the ideal. The LC filter is not restricted to one single response shape, however. The corner frequency of the LC filter in Fig. 6-6b is given by the familiar equation

$$f_c = \frac{1}{2\pi\sqrt{LC}}$$

Notice that the product of L and C determines the corner (resonant) frequency of the circuit. A given LC product, however, can be achieved using infinitely many different inductor and capacitor combinations. This gives the LC filter much more flexibility in terms of response shape. Using a high C to L ratio results in a low damping coefficient and peaking in the response curve. Using a lower C to L ratio (and maintaining the same LC product) results in a higher damping coefficient and a smoother response characteristic. The effects of the damping coefficient on a second-order LP filter are illustrated in Fig. 6-7. If the filter has $\alpha = 1.414$, the response is as flat as possible in the passband. This is called *critical damping*, and it is the same response shape that is obtained from a second-order RC filter with negligible loading. Lower

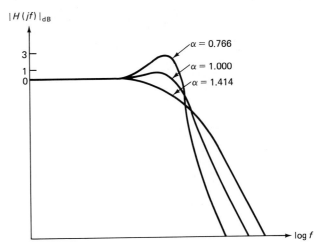

Fig. 6-7 Effects of damping on the response of a second-order LP *LC* filter.

damping coefficients result in peaking near the corner and more rapid attenuation in the transition region. In all cases, however, the ultimate rolloff will be -40 dB/decade for second-order filters.

Filters are often named after special polynomials that mathematically describe their response curves. For example, the Butterworth polynomial is associated with the maximally flat response attained with $\alpha = 1.414$; therefore, such filters are often called Butterworth filters. Filters with peaked response in the passband are called Chebyshev filters, because their response curves are described by Chebyshev polynomials. The amount of peaking that occurs is used to further classify Chebyshev filters. For example, a filter producing the response curve with the highest peak in Fig. 6-7 is called a 3-dB Chebyshev, because the amplitude of the peak is 3 dB. A particular damping factor is associated with a given filter response shape. In general, filters with damping coefficients of less than 1.414 are called *underdamped*, while those with damping coefficients greater than 1.414 are said to be *overdamped*. This is why filters with $\alpha = 1.414$ are said to be critically damped.

The damping factor of a filter also affects the location of the corner frequency. Butterworth (critically damped) filters have no deviation in f_c from the expected value. Underdamped filters experience an effective increase in f_c, while overdamped filters experience a decrease in f_c, relative to an equivalent Butterworth design. Several of the major filter types and their characteristics are presented below. The damping factors (α) given apply only to second-order filters. The frequency scaling factors k_f indicate the relative increase or decrease from f_c of an equivalent filter with $\alpha = 1.414$. For example, it will be shown that the corner frequency of a second-order Butterworth active filter is given by

$$f_c = \frac{1}{2\pi\sqrt{C_1 C_2 R_1 R_2}}$$

If the damping of the filter is changed, the new corner frequency would be given by

$$f_c = \frac{k_f}{2\pi\sqrt{C_1 C_2 R_1 R_2}}$$

Bessel Filter

This filter provides nearly linear phase shift as a function of frequency. It has droopy passband response with gradual rolloff and very low overshoot for transient inputs (no ringing).

$\alpha = 1.732$

$k_f = 0.785$

Butterworth Filter

Allowing the flattest possible passband, this filter is generally the most popular in use.

$\alpha = 1.414$

$k_f = 1$

Chebyshev Filters

These filters allow peaking in passband, with more rapid transition-region attenuation. The higher the peaking, the more nonlinear the phase response becomes, and the more rapid the transition-region attenuation becomes. Chebyshev filters tend to overshoot and ring in response to transients.

1-dB Chebyshev: $\alpha = 1.045$

                          $k_f = 1.159$

2-dB Chebyshev: $\alpha = 0.895$

                          $k_f = 1.174$

3-dB Chebyshev: $\alpha = 0.767$

                          $k_f = 1.189$

There are many other response shape options available. A few of them are Thompson response, elliptical response, and Paynter response. These are also very useful, but the response types just presented will meet most filter needs.

Second-order filters are very frequently encountered in many applications. The normalized phase responses for the second-order LP and HP filters described previously are shown in Fig. 6-8. The decibel gain magnitude equations for second-order LP and HP filters are presented below, in terms of the damping coefficient.

Second-Order LP

$$|H(jf)|_{dB} = 20 \log \frac{A}{\sqrt{(f/f_c)^4 + (\alpha^2 - 2)(f/f_c)^2 + 1}} \qquad (6\text{-}16a)$$

For the Butterworth response, $\alpha = 1.414$, and Eq. (6-16a) reduces to

$$|H(jf)|_{dB} = 20 \log \frac{A}{\sqrt{(f/f_c)^{2n} + 1}} \qquad (6\text{-}16b)$$

where n is the order of the Butterworth filter.

Second-Order HP

$$|H(jf)|_{dB} = 20 \log \frac{A}{\sqrt{(f_c/f)^4 + (\alpha^2 - 2)(f_c/f)^2 + 1}} \qquad (6\text{-}17a)$$

Again, for Butterworth response, $\alpha = 1.414$ and Eq. (6-17a) reduces to

$$|H(jf)|_{dB} = 20 \log \frac{A}{\sqrt{(f_c/f)^{2n} + 1}} \qquad (6\text{-}17b)$$

where, again, n is the order of the Butterworth filter.

It is shown in the next section that the numerators of Eqs. (6-16) and (6-17) represent the closed-loop gain of the op amp around which the active filter is designed.

6-2 ◆ ACTIVE LP AND HP FILTERS

Recall that it is not possible to produce a passive *RC* filter with a damping coefficient lower than 1.414. Using passive filter techniques, one must resort to inductor-capacitor designs in such cases. At low frequencies, the inductors required to produce many response shapes tend to be excessively large, heavy, and expensive. Also, inductors generally tend to pick up electromagnetic interference quite readily. All of these undesirable inductor characteristics have helped to increase the popularity of active filter circuits.

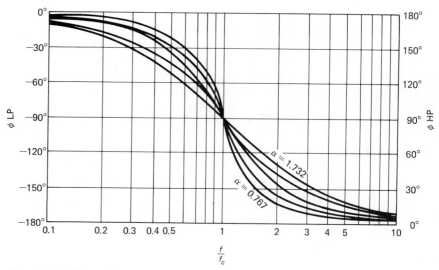

Fig. 6-8 Effects of damping on the phase response of a second-order low-pass LC filter.

Sallen-Key LP and HP Filters

The Sallen-Key realizations of active HP and LP filters are extremely popular. There are two versions of HP and LP Sallen-Key filters, and both use the op amp in the noninverting configuration as a VCVS.

Unity Gain Sallen-Key VCVS

The most basic active filter is the unity gain, second-order Sallen-Key VCVS. As a notational convenience, we shall refer to any member of this filter class as a unity gain VCVS. Low-pass and high-pass versions of the unity gain VCVS are shown in Fig. 6-9a and b, respectively.

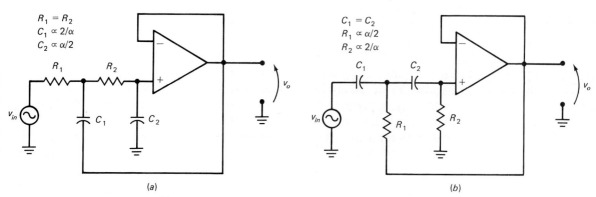

Fig. 6-9 Sallen-Key unity gain VCVS second-order filters: (a) LP; (b) HP.

The corner frequencies for both HP and LP unity gain VCVSs with Butterworth response are given by

$$f_c = \frac{1}{2\pi\sqrt{C_1 C_2 R_1 R_2}} \tag{6-18}$$

As was discussed earlier, if the damping coefficient is some value other than 1.414, as in the case of a Chebyshev or Bessel response, the appropriate frequency scaling factor must be included in the f_c expression. The decibel gain magnitude of the LP unity gain VCVS is determined using Eq. (6-16), while that of the HP version is found using Eq. (6-17). In both cases, the numerator, A, has a value of 1.

In order to set the damping coefficient of an LP unity gain VCVS to a desired value, and produce a cutoff frequency of 1 rad/s, we set $R_1 = R_2 = 1\ \Omega$, while C_1 and C_2 are chosen such that

$$C_1 = \frac{2}{\alpha} \quad \text{farads} \tag{6-19}$$

$$C_2 = \frac{\alpha}{2} \quad \text{farads} \tag{6-20}$$

This procedure is called normalization. While the very low corner frequency and low impedance levels of the normalized filter are probably not very useful from a practical standpoint, the normalized filter does provide a convenient design starting point. We shall make use of techniques called frequency and impedance scaling to produce a useful design. The following rules apply.

1. To scale impedance while maintaining a constant corner frequency, multiply all resistors by the scale factor and divide all capacitors by that same scale factor. The impedance scaling factor k_z is the ratio of the desired impedance Z_{new} to the existing impedance Z_{old}. In equation form, this is

$$k_z = \frac{Z_{new}}{Z_{old}} \tag{6-21}$$

2. To scale frequency while maintaining a given impedance, divide all capacitors by the frequency scaling factor k_f.

    The corner frequency may also be scaled by multiplying all resistances by the scaling factor, while leaving the capacitors at a given value. The scaling factor is the ratio of the desired corner frequency f_{new} to the existing corner frequency f_{old}:

$$k_f = \frac{f_{new}}{f_{old}} \tag{6-22}$$

The following example demonstrates the use of the scaling rules in the design of an LP unity gain VCVS.

EXAMPLE 6-2

Using the normalized unity gain VCVS LP filter as a starting point, design a second-order LP unity gain VCVS for f_c = 1 kHz and Butterworth response. The impedance level of the circuit is to be 10 kΩ.

Solution

Refer to Fig. 6-9a. The required damping coefficient is 1.414. This yields C_1 = 1.414 F and C_2 = 0.707 F. Now, applying rule 1, the impedance scaling factor is

$$k_z = \frac{10 \text{ k}\Omega}{1 \Omega}$$

$$= 10,000$$

Multiplying the resistors and dividing the capacitors by 10,000 produces

$$R_1 = R_2 = 10 \text{ k}\Omega$$

and

$$C_1 = 141.4 \text{ μF} \qquad C_2 = 70.7 \text{ μF}$$

To scale the corner frequency, we begin by converting the normalized corner frequency from radians per second into hertz:

$$f_c = \frac{\omega}{2\pi} \qquad (\omega = 1 \text{ rad/s})$$

$$= 0.1592 \text{ Hz}$$

Rule 2 is applied, with the scaling factor determined as follows:

$$k_f = \frac{1 \text{ kHz}}{0.1592 \text{ Hz}}$$

$$= 6281$$

Dividing the capacitor value produces

$$C_1 = 0.023 \text{ μF} \qquad \text{and} \qquad C_2 = 0.01 \text{ μF}$$

This circuit is shown in Fig. 6-10.

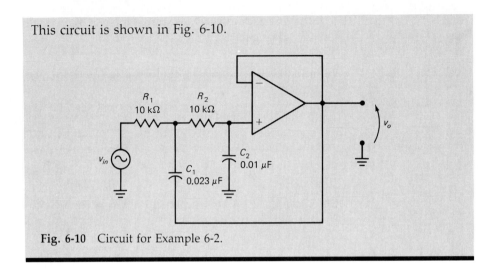

Fig. 6-10 Circuit for Example 6-2.

Now that we have an LP Butterworth circuit scaled up to $f_c = 1$ kHz, using 10-kΩ resistors, it is possible to apply the scaling procedure to obtain any other impedance or f_c values desired.

In order to obtain a useful form of the HP unity gain VCVS, a slightly different approach must be taken. As before, we set the corner frequency to 1 rad/s, but now the capacitors are made equal in value, at 1 farad, while R_1 and R_2 are chosen such that

$$R_1 = \frac{\alpha}{2} \tag{6-23}$$

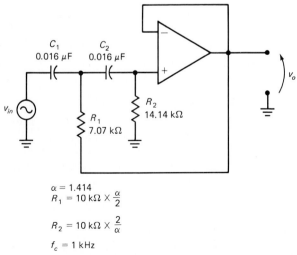

$\alpha = 1.414$
$R_1 = 10\ \text{k}\Omega \times \dfrac{\alpha}{2}$

$R_2 = 10\ \text{k}\Omega \times \dfrac{2}{\alpha}$

$f_c = 1$ kHz

Fig. 6-11 High-pass unity gain VCVS.

and

$$R_2 = \frac{2}{\alpha} \tag{6-24}$$

Using the same scaling rules as before, it is possible to obtain a Butterworth HP filter with $f_c = 1$ kHz at an impedance level of 10 kΩ, as shown in Fig. 6-11.

EXAMPLE 6-3

Using scaling techniques, determine the component values required for Fig. 6-11 to give $f_c = 500$ Hz. Assume that Butterworth response is desired.

Solution

To scale the frequency, we apply rule 2 and simply divide the capacitor values by 0.5 (500 Hz / 1 kHz = 0.5). This yields $C_1 = C_2 = 0.032$ μF. The resistor values remain the same. This circuit is shown in Fig. 6-12.

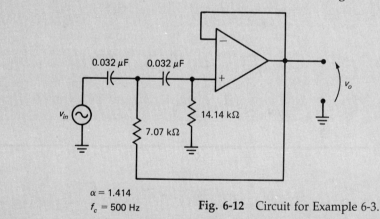

Fig. 6-12 Circuit for Example 6-3.

Since the op amp is used as a unity gain noninverting voltage follower, the unity gain VCVS will provide the highest upper corner frequency possible with a given op amp. Recall that the op amp's small-signal performance is limited by its closed-loop break frequency. In the noninverting mode with $A_v = 1$, the op amp will maintain relatively constant gain up to the unity gain crossing frequency. This is the major advantage of the unity gain VCVS.

Equal-Component Sallen-Key VCVS

Although they do get maximum bandwidth from the op amp, unity gain VCVS filters can be somewhat difficult to design and analyze. Also, because

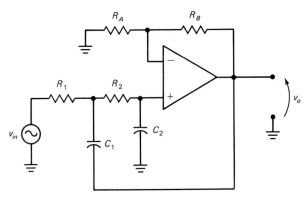

Fig. 6-13 Low-pass equal-component Sallen-Key VCVS.

strict component ratios must be maintained, it is rather difficult to vary the parameters of the filter independently. The equal-component Sallen-Key VCVS filters provide quite effective solutions to these problems.

As the name implies, equal-component Sallen-Key VCVS filters (herein referred to as just equal-component VCVSs) are designed using equal-valued frequency-determining components. That is, $R_1 = R_2$ and $C_1 = C_2$. An LP equal-component VCVS is shown in Fig. 6-13. Notice that the gain of the circuit is determined by R_A and R_B, which will generally not be equal in value. The design of the equal-component VCVS requires the gain of the op amp to be set at some value that will produce the desired damping coefficient. More is said about this later. Assuming Butterworth response, the corner frequency of an equal-component VCVS filter is given by Eq. (6-18), which is repeated here for convenience:

$$f_c = \frac{1}{2\pi\sqrt{R_1 R_2 C_1 C_2}} \tag{6-18}$$

Because the frequency-determining resistor and capacitor pairs are equal, Eq. (6-18) may be simplified to

$$f_c = \frac{1}{2\pi RC} \quad R = R_1 = R_2, \ C = C_1 = C_2 \tag{6-25}$$

Notice that this is the same expression for f_c as that of the simple RC section.

One of the convenient features of the equal-component VCVS is that an LP filter may be converted into an HP filter with the same corner frequency simply by swapping the positions of resistors R_1 and R_2 with capacitors C_1 and C_2 in the circuit. This simple transformation is not possible with the unity gain VCVS.

Because we have forced $R_1 = R_2$ and $C_1 = C_2$, it is necessary to adjust the gain of the op amp to various fixed values in order to produce a given damping coefficient (response shape). The relationship between the op amp's voltage gain and the damping coefficient is given by

$$A_v = 3 - \alpha \qquad (6\text{-}26)$$

The following useful relationship between R_A, R_B, and the damping coefficient is developed by equating the gain expression for the noninverting op amp with Eq. (6-26):

$$R_B = R_A (2 - \alpha) \qquad (6\text{-}27)$$

EXAMPLE 6-4

Design an equal-component LP filter with a 1-dB Chebyshev response and $f_c = 1200$ Hz. Assume that $C_1 = C_2 = 0.022$ μF and $R_A = 18$ kΩ.

Solution

Since this is to be a 1-dB Chebyshev filter, in order to obtain $f_c = 1200$ Hz, Eq. (6-25) must be multiplied by 1.159. That is,

$$f_c = \frac{1.159}{2\pi RC} \qquad C = 0.022 \text{ μF}$$

Solving for R, we obtain

$$R_1 = R_2 = R = 7.0 \text{ kΩ}$$

The damping coefficient of the 1-dB Chebyshev is 1.045. Using Eq. (6-27), the value for R_B is determined:

$$R_B = R_A (2 - \alpha)$$
$$= 17.2 \text{ kΩ}$$

The analysis of a second-order equal-component VCVS requires a reverse application of the design procedure. Specifically, the analysis steps are:

1. Determine the passband gain of the filter and calculate the damping coefficient. The response type is determined by comparing the calculated damping coefficient with those listed for the common filter responses.
2. Apply the appropriate frequency scaling factor to Eq. (6-25), and calculate the corner frequency of the filter.

The application of these two steps is demonstrated in the following example.

EXAMPLE 6-5

Determine the response shape and corner frequency of the HP filter shown in Fig. 6-14.

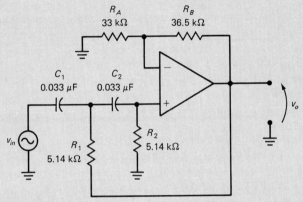

Fig. 6-14 Circuit for Example 6-5.

Solution

The passband gain of the filter is found using the noninverting op amp gain equation, producing

$A_v = 2.106$

The damping coefficient is found by solving Eq. (6-27) for α, giving

$\alpha = 0.894$

Comparing this damping coefficient with those presented earlier shows that Fig. 6-14 will very closely approximate a 2-dB Chebyshev response. Now the corner frequency of the filter is found by multiplying Eq. (6-25) by the scaling factor 1.174. This produces

$f_c = 1.1$ kHz

Higher-Order LP and HP Filters

Active filters with orders of greater than two are obtained by cascading first- and second-order sections as required. The overall order of a filter that is designed in this manner is equal to the sum of the orders of the individual sections that are used. For example, a third-order filter may be constructed by cascading a first-order section and a second-order section. A fourth-order filter may be constructed by cascading two second-order sections, and so on. This concept is illustrated in Fig. 6-15.

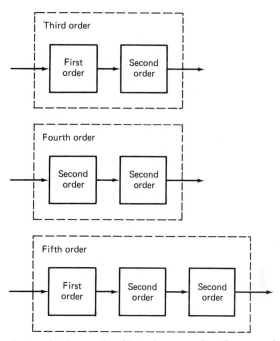

Fig. 6-15 Producing higher-order filters by cascading lower-order sections.

Although it is easy to produce a given order filter by cascading first- and second-order sections, obtaining a particular response shape is not quite as simple. For example, cascading two active second-order Butterworth sections will produce a fourth-order filter, but the overall response will not be that of a Butterworth filter (maximally flat passband). In order to produce a given response, the various sections used to produce the filter must be designed with specific damping coefficients and corner-frequency scaling factors taken into account. The mathematics behind the determination of these parameters is rather complex, and therefore, this text presents only the results that are required to produce the common response shapes for third- and fourth-order filters. Table 6-1 gives the frequency scaling factors k_f and damping coefficients required to produce third-order filters by cascading first- and second-

Table 6-1 REQUIRED DAMPING COEFFICIENTS AND FREQUENCY SCALING FACTORS FOR THIRD-ORDER RESPONSES

Overall response shape	First-order section	Second-order section	
	k_f	α	k_f
Bessel	1.328	1.477	1.454
Butterworth	1	1.000	1
1-dB Chebyshev	0.452	0.496	0.911
2-dB Chebyshev	0.322	0.402	0.913
3-dB Chebyshev	0.299	0.326	0.916

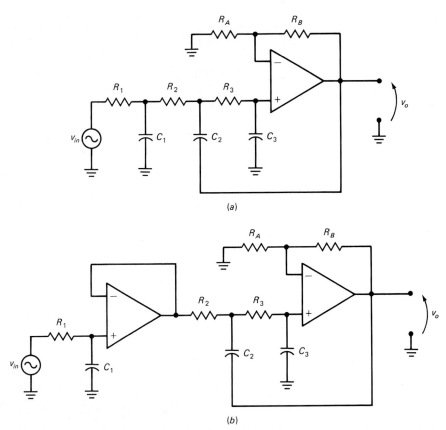

Fig. 6-16 Third-order active LP filters. (*a*) Minimum op amp implementation; (*b*) op amp isolation of first section.

order sections. Notice that the first-order sections has no damping coefficient given in the table. This is so because the damping of a simple RC section is fixed at a value slightly greater than 1.414.

When dealing with higher-order filters, all second-order sections used here will be of the equal-component VCVS type, because they are easier to analyze and design, and because LP–HP conversions are performed simply by swapping frequency-determining resistors and capacitors. Two variations of the third-order active LP filter are shown in Fig. 6-16. For notational convenience, the gain-setting resistors of the second-order section have been labeled R_A and R_B. In Fig. 6-16*a*, the first-order RC section is connected directly to the second-order section. This is the simplest approach, but the second section will tend to load down the first section, producing an overall response that has slightly greater damping than desired. Isolating the first section, as in Fig. 6-16*b*, eliminates the loading effects of the second section, at the expense, of course, of adding an extra op amp.

In order to use the circuit of Fig. 6-16*a* effectively, the impedance level of the first section should be much lower than that of the second section. As a general rule, scaling the impedance of the first-order section to around 1/10 of the impedance level of the second section will result in good performance. The disadvantage to this approach is that the signal source will be required to drive a rather low-impedance point, which may be undesirable.

EXAMPLE 6-6

Refer to Fig. 6-16b. Given $C_1 = C_2 = C_3 = 0.01\ \mu\text{F}$ and $R_A = 10\ \text{k}\Omega$, determine the required values for the remaining components such that a Bessel response is produced with $f_c = 5$ kHz.

Solution

Referring to Table 6-1, we find that k_f for the first-order section is 1.328. Applying this factor to the corner frequency equation for the simple RC section and solving for R, we obtain

$$R_1 = \frac{1.328}{2\pi f_c C}$$

$$= 4.23\ \text{k}\Omega$$

The second-order section requires $k_f = 1.454$ and $\alpha = 1.477$. Application of k_f to Eq. (6-25) yields

$$R_2 = R_3 = \frac{1.454}{2\pi f_c C}$$

$$= 4.63\ \text{k}\Omega$$

Now, applying Eq. (6-27), the value of R_B is found to be

$$R_B = R_A (2 - \alpha)$$

$$= 5.23\ \text{k}\Omega$$

Fourth-order filters are designed by cascading two second-order filter sections, as shown in Fig. 6-17. The necessary damping coefficients and frequency scaling factors for fourth-order filters are given in Table 6-2.

Chebyshev filters of order greater than two will exhibit multiple peaks, or ripples in the passband. The higher the order of the filter, the more ripples occur in the passband. The passband response of 3-dB Chebyshev filters of orders two through four are shown in Fig. 6-18. From this diagram it can be seen that the corner frequency of the Chebyshev filter is defined as the frequency at which the ripple channel ends.

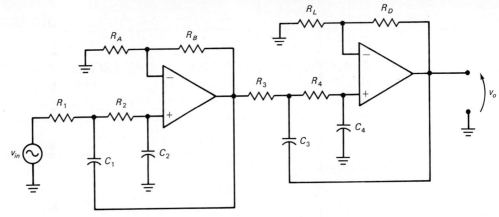

Fig. 6-17 Realization of a fourth-order active filter.

Table 6-2 DAMPING COEFFICIENTS AND FREQUENCY SCALING FACTORS FOR FOURTH-ORDER RESPONSES

Overall response shape	First section		Second section	
	α	k_f	α	k_f
Bessel	1.916	1.436	1.241	1.610
Butterworth	1.848	1.000	0.765	1.000
1-dB Chebyshev	1.275	0.502	0.281	0.943
2-dB Chebyshev	1.088	0.466	0.244	0.946
3-dB Chebyshev	0.929	0.433	0.179	0.950

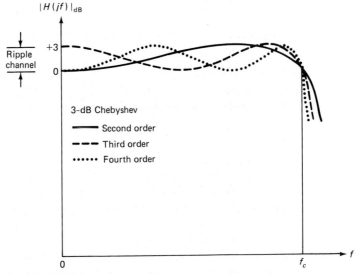

Fig. 6-18 Passband response for 3-dB LP Chebyshev filters of orders 2 to 4.

EXAMPLE 6-7

Refer to Fig. 6-17. Given $C_1 = C_2 = 0.022$ μF, $C_3 = C_4 = 0.01$ μF, and $R_A = R_C = 10$ kΩ, determine the required values for the remaining components such that the filter has a Butterworth response with $f_c = 2.5$ kHz.

Solution

Starting with the first section, and referring to Table 6-2, we see that for Butterworth response, $\alpha = 1.848$ and $k_f = 1.000$. Applying Eqs. (6-25) and (6-27), we obtain

$$R_1 = R_2 = \frac{1}{2\pi f_c C} \quad C = 0.022 \text{ μF}$$

$$= 2.89 \text{ kΩ}$$

$$R_B = R_A (2 - \alpha)$$

$$= 1.52 \text{ kΩ}$$

The second section is designed in a similar manner, with $k_f = 1.000$ and $\alpha = 0.765$. This produces

$$R_3 = R_4 = \frac{1}{2\pi f_c C} \quad C = 0.01 \text{ μF}$$

$$= 6.37 \text{ kΩ}$$

$$R_D = R_C (2 - \alpha)$$

$$= 12.35 \text{ kΩ}$$

Filters of the fourth order generally have rapid enough gain rolloff in the stop band for most applications. Higher-order filters are also implemented using active filter design techniques, although the damping coefficients and frequency scaling factors for these designs are not presented here.

6-3 ♦ BANDPASS FILTERS

In this section, several of the more widely used bandpass active filters are discussed. Bandpass (BP) (and bandstop) filters are rather easily designed using active filter techniques. Active bandpass and bandstop filters have several advantages over their passive counterparts. One obvious advantage is inductorless design. Other major advantages are ease of tuning and independent parameter adjustment (Q, f_o, and bandwidth), electronic control of parameters, and the option of adjustable passband gain.

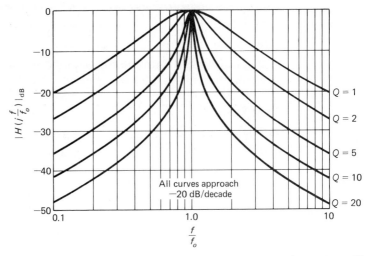

Fig. 6-19 Second-order one-pole BP normalized response for various Q's.

The major performance parameter associated with bandpass and bandstop filters is Q. Q is the reciprocal of the filter damping coefficient. The reason that BP and bandstop filters are described in terms of Q rather than α is because, in general, the damping coefficient of BP or bandstop filters will be quite small, making α somewhat inconvenient to work with. The Q of a BP or bandstop filter is a measure of the sharpness of the response around the filter's center frequency, f_o. Typical BP filter amplitude response curves for Q's of 1 to 20 are shown in Fig. 6-19. Notice that regardless of the Q of the filter, the slope of the curve ultimately approaches a constant value, which in this case is -20 dB/decade. The ultimate rolloff of a bandpass filter is determined by the order of that particular filter. The minimum BP filter order is 2.

The response curves in Fig. 6-19 are those for a *second-order, single-pole* BP filter. Generally, BP filters are of even order, with equal ultimate rolloff rates on either side of f_o. That is, on a semilog graph, the response curve will be symmetrical about the center frequency. A convenient way of visualizing the relationship between the order of a BP filter and its amplitude response curve is to assume that on each side of f_o, the ultimate rolloff rate will be that of a HP or LP filter of one-half the order of the BP filter. Thus, a fourth-order BP filter will ultimately roll off at -40 dB/decade in the stop bands. The mathematical explanation of the poles of a filter is beyond the scope of this book, but in general a second-order BP will have one pole, a fourth-order BP will have two poles, and so on.

Wideband BP Filters

A bandpass filter is generally considered to have a wide passband if the Q of the filter is 1 or less. Thus, the term *wideband* is used in a relative sense here. Such low-Q filters are generally designed by using cascaded LP and

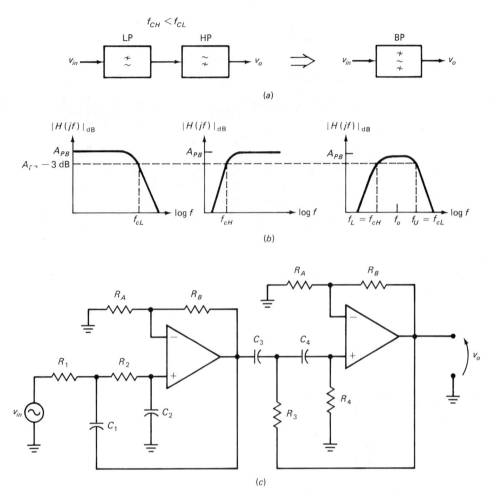

Fig. 6-20 BP filter using cascaded LP and HP sections. (*a*) Block diagram; (*b*) response curves; (*c*) circuit realization.

HP filters. The block diagram representation for such a filter is shown in Fig. 6-20*a*. In this design, the LP section corner frequency is higher than that of the HP section. The net response is that of a bandpass filter whose order is given by the sum of the orders of the HP and LP sections. This is illustrated in Fig. 6-20*b*. A fourth-order, two-pole realization of this circuit is shown in Fig. 6-20*c*.

EXAMPLE 6-8

The circuit in Fig. 6-20 is designed such that the low-pass section has $f_c = 10$ kHz, while the HP section has $f_c = 500$ Hz. Determine the bandwidth, the center frequency f_o, and the filter's Q.

Solution

The bandwidth is found using Eq. (6-8):

$$BW = f_U - f_L$$
$$= 10 \text{ kHz} - 500 \text{ Hz}$$
$$= 9.5 \text{ kHz}$$

The center frequency is determined using Eq. (6-7):

$$f_o = \sqrt{f_L f_U}$$
$$= \sqrt{500 \times 10{,}000}$$
$$= 2236 \text{ Hz}$$

The Q is found using Eq. (6-9):

$$Q = \frac{f_o}{BW}$$
$$= 0.235$$

Multiple-Feedback Bandpass Filters

Most applications that call for the use of a BP filter require the Q to be much higher than unity. For Q's in the range from 2 to approximately 20, a circuit called the multiple-feedback bandpass (MFBP) filter is used. The MFBP is a one-op amp circuit with a second-order, single-pole amplitude response characteristic. One variation of the MFBP is shown in Fig. 6-21. The center frequency of this filter is given by

$$f_o = \frac{1}{2\pi\sqrt{R_1 R_2 C_1 C_2}} \tag{6-28}$$

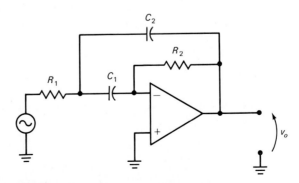

Fig. 6-21 Multiple-feedback BP active filter.

To ease component selection and to reduce the number of variables, the capacitors are set equal to each other. That is,

$$C_1 = C_2 = C$$

For design purposes, the values of the resistors, based on the desired filter characteristics, are determined by the application of the following equations:

$$R_1 = \frac{Q}{2\pi f_o A_v C} \tag{6-29}$$

$$R_2 = \frac{A_v}{2\pi f_o Q C} \tag{6-30}$$

The passband voltage gain is given by

$$A_v = -Q\sqrt{\frac{R_2}{R_1}} \tag{6-31}$$

It is possible to continuously vary the center frequency of the BP filter in Fig. 6-21 without changing the gain or the Q by simultaneously varying R_1 and R_2 and keeping the ratio R_2/R_1 constant. However, this is generally not practical. A variation of the multiple-feedback BP that allows for easy adjustment of f_o is shown in Fig. 6-22.

The primary design equations for Fig. 6-22 are presented below. As before, to ease design and analysis, we make $C_1 = C_2 = C$.

$$f_o = \frac{\sqrt{(1/R_2 C^2)(1/R_1 + 1/R_3)}}{2\pi} \tag{6-32}$$

$$R_1 = \frac{Q}{2\pi f_o A_v C} \tag{6-33}$$

$$R_2 = \frac{Q}{\pi f_o C} \tag{6-34}$$

$$R_3 = \frac{Q}{2\pi f_o C(2Q^2 - A_v)} \tag{6-35}$$

$$A_v = -\frac{R_2}{2R_1} \tag{6-36}$$

One of the tradeoffs in obtaining easy adjustment of f_o is a restriction between the allowable passband gain and the Q of the filter. Examination of Eq. (6-35) leads to the following inequality, which must be met in order to obtain a finite positive value for R_3:

$$A_v < 2Q^2 \tag{6-37}$$

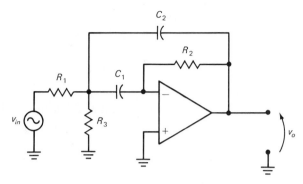

Fig. 6-22 Modified multiple-feedback BP filter.

EXAMPLE 6-9

Design a three-resistor MFBP filter such that $f_o = 1$ kHz, $A_v = 1$, and $Q = 2$. Assume $C_1 = C_2 = 0.022$ μF. Use standard resistors in the final component value specifications. Determine the percent change in f_o and A_v caused by the use of standard resistors in the design.

Solution

The inequality of Eq. (6-37) is met; therefore, we proceed in the design process by applying the appropriate design equations, yielding

$R_1 = 14.5$ kΩ (use 15 kΩ)

$R_2 = 28.9$ kΩ (use 27 kΩ)

$R_3 = 2.07$ kΩ (use 2.2 kΩ)

Applying Eqs. (6-32) and (6-36) and using the standard resistors selected, we find

$f_o = 1005$ Hz (+0.5% change)

$A_v = -0.9$ (−10% change)

The center frequency of the three-resistor MFBP filter is changed by selecting a new value for R_3 according to the following equation:

$$R_3' = R_3 \left(\frac{f_{\text{old}}}{f_{\text{new}}}\right)^2 \tag{6-38}$$

where R_3' is the new value of R_3 and f_{new} is the desired center frequency.

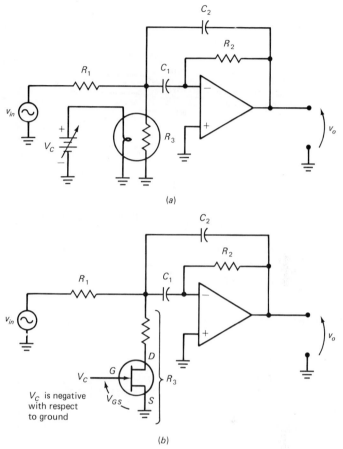

Fig. 6-23 Electronically adjustable center frequency using (a) a photocoupler and (b) a JFET.

It is possible to control the center frequency of the three-resistor MFBP electronically by replacing R_3 with a voltage- or current-variable resistor. One device that could be used to replace R_3 is a photocoupler. This modification is shown in Fig. 6-23a. A photocoupler is a light-dependent resistor (LDR) encapsulated with a light source. The resistance of the LDR decreases as the lamp current (and intensity) increases. Varying the lamp current will vary the center frequency of the filter. Photocouplers respond relatively slowly and tend to have a very nonlinear response to the change in light-source current. These characteristics can cause problems in some applications.

A JFET may also be used as a voltage-controlled resistor, as shown in Fig. 6-23b. A negative control voltage applied to the gate will drive the JFET toward pinchoff, increasing the drain-to-source resistance (r_{DS}). In order to use the JFET effectively, V_{DS}, and hence the input voltage, must be held to a maximum of about 500 mV$_{P-P}$. For voltages within these limits, the JFET will act essentially like a linear resistance whose value is dependent on V_{GS}.

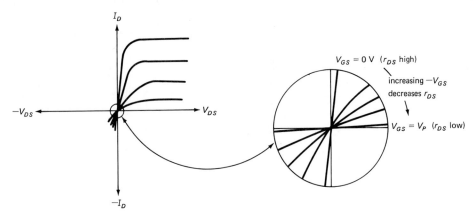

Fig. 6-24 N-channel JFET drain curves expanded about the origin.

The drain curves for a typical N-channel JFET are shown in Fig. 6-24. It is seen that for low values of V_{DS}, the FET is operating in the ohmic region, which is fairly linear.

Some BP Filter Applications

One possible application for an electronically controlled BP filter is in the design of a simple spectrum analyzer. A spectrum analyzer displays the amplitudes of the various frequency components that comprise a signal. Figure 6-25 shows one possible approach that could be taken to the design of a spectrum analyzer. Basically, as the sweep oscillator varies f_o of the BP filter, it also drives the horizontal input of an oscilloscope. The output of the BP filter is amplified and rectified, and applied to the vertical input of the scope. The frequency components that exist in the input signal are filtered out at different times during a sweep, causing peaks to appear on the scope display. The horizontal scale of the scope represents frequency, while the vertical scale represents voltage. Notice that this is not a real-time instrument. That is, changes in the frequency content of the input signal are not shown at all points in time.

The circuit in Fig. 6-25 would be of limited practical value because of the frequency limitations of most op amps and the wide f_o sweep range required. Also, various nonlinear relationships between the sections of the circuit would have to be compensated for. Nevertheless, this principle is used as the basis for many spectrum analyzer designs.

Most readers are probably familiar with a piece of audio equipment called a graphic equalizer. Graphic equalizers consist of a bank of variable-gain BP filters that are used to boost or attenuate signal components at several fixed frequencies. The block diagram for a six-band graphic equalizer is shown in Fig. 6-26. In this circuit, the BP filters are set to various center frequencies (f_{o1}, f_{o2}, f_{o3}, etc.) within the audio-frequency range. The potentiometers on the outputs of the filters allow each frequency band to be attenuated or amplified. The BP outputs are summed, producing a signal that is tailored

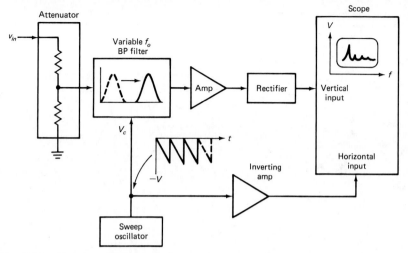

Fig. 6-25 Conceptual diagram for a spectrum analyzer.

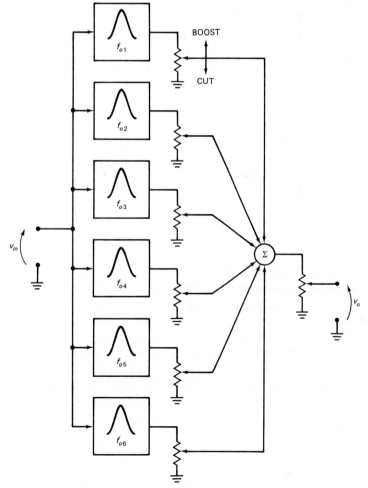

Fig. 6-26 Block diagram for a six-band graphic equalizer.

221

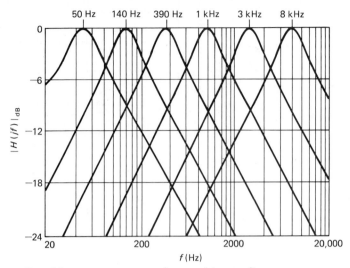

Fig. 6-27 Possible response curves for graphic equalizer.

to suit the operator's taste, or to compensate for room acoustics, or to overcome deficiencies in other parts of the system. The poteniometer on the output acts as a master signal level control.

The audio spectrum ranges from 20 Hz to 20 kHz. This is a four-decade range. To cover such a wide range of frequencies with only six BP filters, the filters would normally be chosen to cover the spectrum in a manner similar to the plot shown in Fig. 6-27. In this example, the filters have Q's of about 2, producing identically shaped response curves on the semilog scaled graph. Relative low Q is desirable, so that there are no large "holes" or gaps in the audio spectrum. If more filters were to be used, higher-Q filters could be used. Notice that the spectrum is not divided linearly, but rather in a logarithmic manner. This is typical of most graphic equalizers.

6-4 ♦ BANDSTOP FILTERS

Bandstop (notch or band-reject) filters are used to reject or attenuate undesired frequency components from signals. (Any undesired signal may be referred to as noise.) A commonly encountered noise is the 60-Hz hum that is induced by the ac power lines. Quite often, the 60-Hz noise that is induced in signal-carrying lines is of the same order of magnitude as the signal to be processed. In such cases, a 60-Hz notch filter could be used, in addition to proper shielding and grounding, to reduce this noise to a minimum. This section presents a few of the more common active bandstop filter designs.

A bandstop response can be produced by summing the outputs of HP and LP filters with overlapping amplitude response curves. For want of a better name, this type of filter will be called a *composite bandstop filter*. The main idea here is to set f_c for the LP filter at a higher frequency than for the

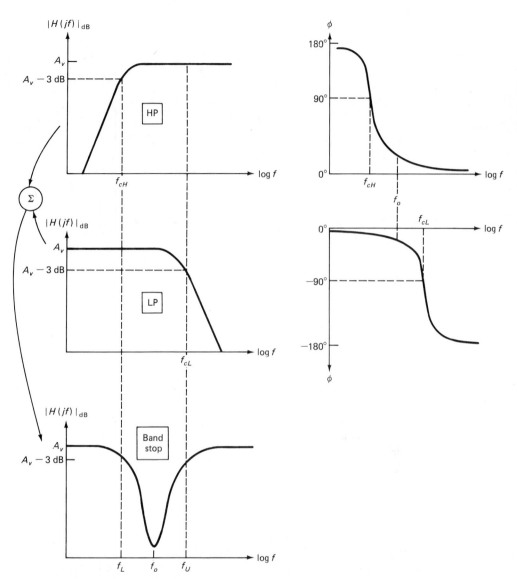

Fig. 6-28 Combining the responses of second-order HP and LP filters to produce a bandstop response.

HP filter. Figure 6-28 illustrates this concept. The amplitude response of the output of such a filter cannot be determined using the amplitude response curves of the LP and HP sections alone. The bandstop response of this circuit makes sense only when the phase response curves of the HP and LP filters are considered as well as their amplitude response curves. Notice in Fig. 6-28 that the outputs of the HP and LP filters are always out of phase with each other by nearly 180°. The critical point occurs when $f = f_o$, where the outputs of the filters are equal in amplitude and 180° out of phase with each

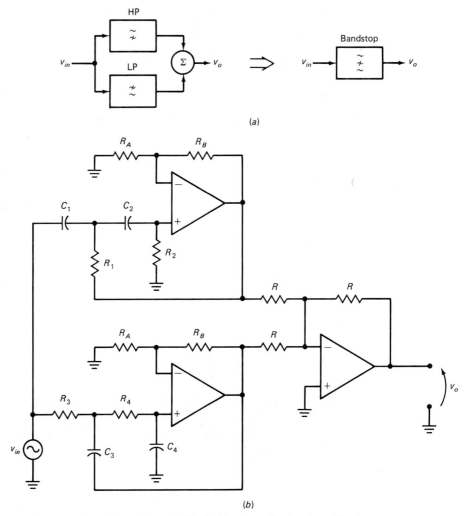

Fig. 6-29 Bandstop filter. (a) Block diagram; (b) circuit realization.

other. This results in cancellation of the signals at the output of the summer. However, because the outputs are being summed and because one filter will be producing an output voltage of much greater amplitude for frequencies on either side of f_o, two passbands are produced.

A block diagram representation for the composite bandstop filter is shown in Fig. 6-29a. An implementation using second-order equal-component VCVS filters in shown in Fig. 6-29b. To produce a predictable response, both filters should be of the same order with the same response shape (usually Butterworth response).

The Q of a bandstop filter is determined in the same manner as for a BP filter. For the highest Q, the HP and LP filters should have identical corner frequencies. The null frequency of the bandstop filter is determined using

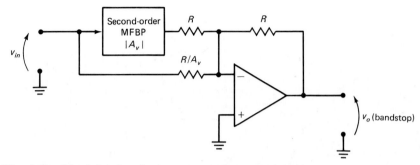

Fig. 6-30 Obtaining bandstop response from a MFBP filter.

Eq. (6-7). Ideally, a bandstop filter produces infinite attenuation at f_o. Practically, mismatches between the amplitude and phase responses of the HP and LP sections will limit the maximum rejection to around 50 dB below the passband gain. The overall gain of the bandstop filter (in decibels) at f_o is often called the *null depth*. The greater the null depth, the more effective the bandstop filter is.

A second bandstop filter that relies on cancellation of phase-shifted signals for its response characteristics is designed around the MFBP filter. A partial schematic for such a circuit is shown in Fig. 6-30. Because of the inverting gain of the MFBP filter, the output signal is 180° out of phase with the input at f_o. The phase response of a second-order MFBP filter is shown in Fig. 6-31. When the output of the MFBP is summed with the input signal, it is possible to obtain a bandstop response because of the relative phase inversion of the two signals. Because of this operation, the null frequency of the bandstop filter is the same as f_o for the MFBP filter around which it is designed. In order to realize maximum null depth, the summing amplifier

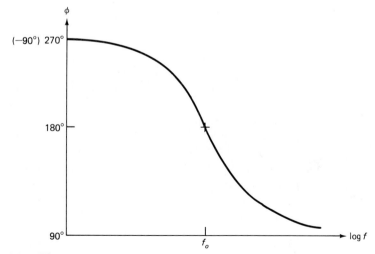

Fig. 6-31 Phase response of the second-order MFBP filter.

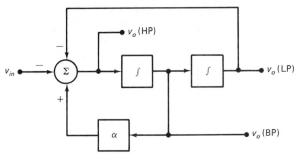

Fig. 6-32 Block diagram for the state-variable filter.

must be designed to compensate for differences between its two input signals. For example, if the MFBP filter has $A_v = 5$ and the summer has $R = 100$ kΩ, then the input resistor for the unprocessed signal should have a value of 20 kΩ for maximum null depth. Generally, a potentiometer is used to trim the summer's gain as required for best operation.

6-5 ♦ STATE-VARIABLE FILTERS

A state-variable filter is essentially an analog computer that continuously solves a second-order differential equation. There are many systems in nature that are mathematically described by such equations. For our purposes, however, we shall just consider the state-variable filter from a signal pro-

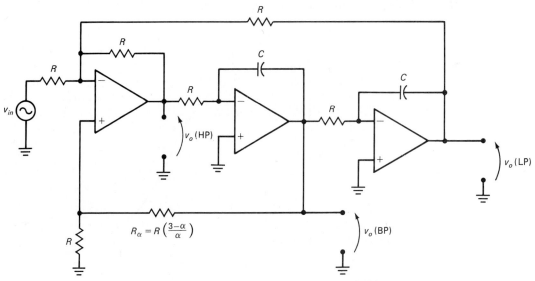

Fig. 6-33 Circuit realization for the state-variable filter.

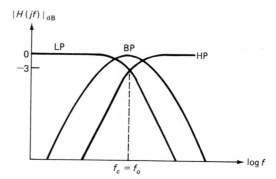

Fig. 6-34 Typical state-variable filter response curves.

cessing standpoint. The basic block diagram representation for a state-variable filter is shown in Fig. 6-32. Notice that the state-variable filter produces simultaneous HP, LP, and BP responses. For the HP and LP outputs, any practical second-order response shape can be achieved, while for the BP output, Q's of greater than 100 are easily obtained. These characteristics make the state-variable filter very flexible.

State-variable filters are constructed using three or more op amps. A typical design is shown in Fig. 6-33. Normally, both integrators use equal-value components, and, for convenience, the remaining resistors are set equal to the integrator resistors or scaled as necessary from this value. The corner frequencies of the HP and LP outputs and the center frequency of the BP output are given by the frequently recurring equation

$$f_o = \frac{1}{2\pi RC} \tag{6-39}$$

Typical response curves for the state-variable filter are shown in Fig. 6-34. The corner frequencies (and f_o) may be varied continuously, without affecting α or Q, by simultaneously varying the integrator input resistors while keeping them equal to each other. Using the implementation of Fig. 6-33, the damping coefficient (and the Q) and the passband gain are not independently adjustable. The passband gain of this circuit for the HP and LP outputs is unity. For the BP output, the gain at f_o is given by

$$A_{v(BP)} = Q = \frac{1}{\alpha} \tag{6-40}$$

For maximum flexibility, the damping coefficient and the passband gain can be made independently adjustable by adding a fourth op amp, with resistor values determined as shown in Fig. 6-35. For this circuit, the passband gain of the BP output is given by

$$A_{v(BP)} = A_v Q \tag{6-41}$$

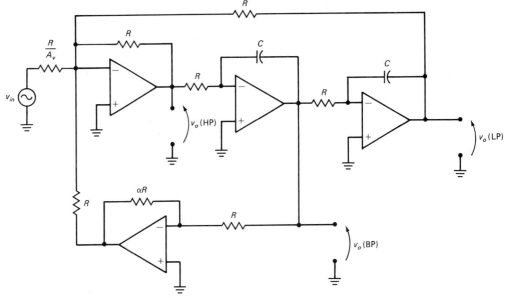

Fig. 6-35 State-variable filter with independently adjustable damping and gain.

EXAMPLE 6-10

Refer to Fig. 6-33. Given $R = 3.3$ kΩ, $R_\alpha = 3.7$ kΩ, and $C = 0.012$ µF, determine f_o and Q for the BP output, and the passband gain of the circuit, in decibels.

Solution

First, f_o is found by direct application of Eq. (6-39), producing

$f_o = 4019$ Hz

To determine the Q of the filter, the R_α relation shown in Fig. 6-33 is first solved for α, producing

$\alpha = 1.414$

Since Q is the reciprocal of α, we obtain

$Q = 0.707$

The passband gain for the bandpass output is numerically equal to the Q; therefore

$A_{v(BP)} = 20 \log 0.707$
$\phantom{A_{v(BP)}} = -3.01$ dB

6-6 ◆ THE ALL-PASS FILTER

All-pass filters are designed to provide constant gain to signals at all frequencies. That is, ideally, the all-pass filter's passband covers the entire frequency spectrum. This flat amplitude response characteristic is quite different from those of the filters discussed previously. However, a circuit that produces a frequency-dependent change in either the phase, the amplitude, or both of the signal being processed may be considered a filter. All-pass filters produce an output that is shifted in phase relative to an input signal.

An active all-pass filter and a representative plot of its input and output waveforms are shown in Fig. 6-36. This particular circuit causes the output signal to lead that of the input. For frequencies approaching zero, the phase lead approaches 180°. As frequency increases, the phase lead of the output approaches zero. The equation governing this response is

$$\phi = 2 \tan^{-1} \frac{1}{2\pi f R_1 C_1} \qquad (6\text{-}42)$$

For correct operation, the feedback and inverting input resistors must be equal to each other. The absolute value of these resistors is not critical, but for minimum offset, the parallel equivalent of these two resistors should nearly equal the value of R_1. The gain of this all-pass filter is unity, a necessary condition for normal operation. By replacing R_1 with a potentiometer (or equivalent voltage-controlled resistance), the phase angle of the output may be continuously varied.

Cascading similar all-pass sections produces an additive phase shift. Two all-pass filters cascaded will approach a maximum phase shift of 360°, three sections will approach a maximum phase shift of 540°, and so on.

A lagging phase angle may be produced by interchanging R_1 and C_1. The phase angle of the output for this configuration is given by

$$\phi = -2 \tan^{-1} (2\pi f R_1 C_1) \qquad (6\text{-}43)$$

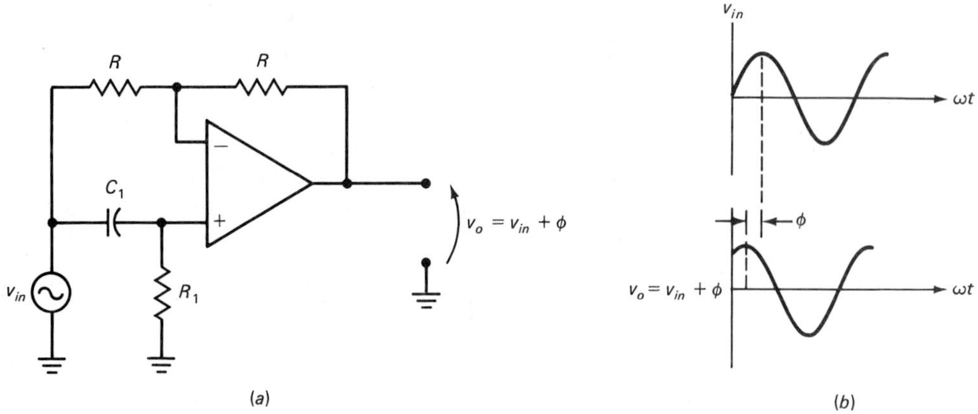

Fig. 6-36 Active all-pass filter. (a) Circuit; (b) possible I/O phase response.

EXAMPLE 6-11

Refer to Fig. 6-36. Assume that two identical circuits with $R = 10 \text{ k}\Omega$, $R_1 = 4.7 \text{ k}\Omega$, and $C_1 = 0.047 \text{ }\mu\text{F}$ are cascaded. Given $v_{in} = 2 \sin 500t$ V, determine the expression for v_o.

Solution

Because both sections are identical, Eq. (6-42) is used once to determine the phase shift of each section:

$$\phi_1 = \phi_2 = 2 \tan^{-1} \frac{1}{2\pi fRC}$$

$$= 167.4°$$

Since both sections are identical, each will contribute equally to the phase shift of the output.

$$\phi_T = \phi_1 + \phi_2$$

$$= 334.8°$$

The output voltage expression is

$$v_o = 2 \sin (500t + 334.8°) \text{ V}$$

or

$$v_o = 2 \sin (500t - 25.2°) \text{ V}$$

Chapter Review

Filters generally fall into one of five categories: low-pass, high-pass, bandpass, bandstop, and all-pass. All of these filter types can be implemented using active filter techniques. Active filters generally eliminate the need for inductors, which normally provides cost, size, and weight savings, especially at low frequencies. Most active HP and LP filters use the Sallen-Key VCVS implementation. The unity gain VCVS provides maximum op amp bandwidth, while the equal-component VCVS is more flexible in terms of design and tuning.

There are many response shape options available to the filter designer. The major response shapes are the Bessel, Butterworth, and Chebyshev responses. The damping coefficient of the filter determines which response will be produced. Butterworth response is usually the most widely encountered. Regardless of the response shape used, the ultimate rolloff of a given filter is determined by the order of that filter. Rolloff is expressed in decibels per octave or decibels per decade.

The inverse of damping is Q. This is one of the primary bandpass and bandstop filter parameters. Bandpass filters are normally of even order. The multiple-feedback active BP filter configuration is one of the more widely used, for moderate Q's. Bandstop filters may be constructed using separate HP and LP sections, or by summing the input and output of a MFBP filter. Bandstop filters are used to remove unwanted frequency components from a signal.

State-variable filters provide simultaneous HP, LP, and BP responses. Relatively high Q factors can be obtained using state-variable filters. It is also possible to independently adjust the passband gain, center frequency, and damping of the state-variable filter.

All-pass filters are used to introduce a predictable phase shift into a signal. The amplitude response of all-pass filters is constant. Phase shift in either direction is possible with these types of filters.

Questions

6-1. What filter parameter determines the shape of an HP or LP response?

6-2. What is the ultimate rolloff of a sixth-order filter, in decibels per decade and in decibels per octave?

6-3. A certain circuit has a sixth-order, three-pole BP response. What is the ultimate rolloff on either side of f_o?

6-4. For a given BP filter, if f_o is increased and BW is held constant, will the Q increase or decrease?

6-5. In general, will a Chebyshev response be produced by an underdamped, overdamped, or critically damped filter?

6-6. Refer to Fig. 6-35. If significant bias-current-induced output offset voltage existed at the output, what simple modification could be made to help alleviate this problem?

6-7. In reference to Question 6-6, which output of the filter would be most likely to reflect the existence of input offset voltage?

6-8. What term is given to the maximum attenuation obtained from a bandstop filter?

6-9. A certain BP filter has f_o = 300 Hz, f_U = 1 kHz, and f_L = 100 Hz. Would this be considered a wideband or narrowband filter?

6-10. Define the term *corner frequency* as applied to a filter that exhibits a Chebyshev response.

Problems

6-1. Refer to Fig. 6-3a. Given $R = 6.8$ kΩ, determine the value required for C to produce $f_c = 5$ kHz.

6-2. Refer to Fig. 6-3a. Given $R = 10$ kΩ, determine the value required for C to produce a phase shift of $-30°$ when $f = 1$ kHz.

6-3. Refer to Fig. 6-10. Scale the circuit such that $f_c = 300$ Hz while the impedance level remains constant.

6-4. Refer to Fig. 6-10. Scale the circuit such that the impedance level is 15 kΩ and $f_c = 5$ kHz.

6-5. Refer to Fig. 6-13. Given $R_A = 27$ kΩ and $C_1 = C_2 = 0.033$ μF, determine the required values for the remaining components to produce a 3-dB Chebyshev filter with $f_c = 1.5$ kHz.

6-6. Assume that the circuit in Fig. 6-13 has the following component values: $R_A = 47$ kΩ, $R_B = 12.6$ kΩ, $C_1 = C_2 = 0.0022$ μF, and $R_1 = R_2 = 18$ kΩ. Determine the damping coefficient, the response type, and the corner frequency of the filter.

6-7. Refer to Fig. 6-14. Determine the component values required to scale the impedance of the filter to 10 kΩ.

6-8. Refer to Eq. (6-27). Express R_A in terms of the damping coefficient and R_B.

6-9. Refer to Fig. 6-17. Given $R_A = R_C = 47$ kΩ and $R_1 = R_2 = R_3 = R_4 = 12$ kΩ, determine the required values of the remaining components to generate a fourth-order 3-dB Chebyshev response with $f_c = 1$ kHz.

6-10. Repeat Prob. 6-9, using the component values given to produce a fourth-order Butterworth filter with $f_c = 2$ kHz.

6-11. A certain BP filter has $f_o = 3.6$ kHz and $f_L = 3$ kHz. Determine f_U and Q for this filter.

6-12. The circuit in Fig. 6-20c has the following component values: $R_A = 47$ kΩ, $R_B = 27.5$ kΩ, $R_1 = R_2 = R_3 = R_4 = 22$ kΩ, $C_1 = C_2 = 0.0036$ μF, and $C_3 = C_4 = 0.036$ μF. Determine the upper and lower corner frequencies, the bandwidth, the Q, and the passband gain of the filter.

6-13. Refer to Fig. 6-20. Determine the component values required to produce a bandpass filter with $f_o = 2$ kHz and $f_L = 500$ Hz. Use second-order Butterworth filters, with $R_A = 27.5$ kΩ and $C_1 = C_2 = C_3 = C_4 = 0.01$ μF.

6-14. Refer to Fig. 6-21. Given $R_1 = R_2 = 6.77$ kΩ and $C_1 = C_2 = 0.0047$ μF, determine f_o.

6-15. Refer to Fig. 6-22. Given $C_1 = C_2 = 0.0033$ μF, determine the values required for the remaining components such that $Q = 5$, $A_v = -2$, and $f_o = 5$ kHz.

6-16. Based on the results of Prob. 6-15, if R_3 is adjustable from 100 Ω to 5 kΩ, determine the upper and lower limits of f_o.

6-17. Refer to Fig. 6-30. Assuming that the MFBP filter has a passband gain of 20 dB and $R = 10$ kΩ, what value should the remaining resistor have, for greatest null depth?

6-18. Refer to Fig. 6-33. Develop an expression that defines the value of R_α in terms of Q. That is, write the equivalent (simplified) equation for R_Q.

6-19. Refer to Fig. 6-33. Given $C = 0.047$ μF, determine the values required for the resistors that will produce $f_o = 1$ kHz with $Q = 50$.

6-20. Repeat Prob. 6-19, using the circuit in Fig. 6-35.

6-21. Design an active all-pass filter that will produce a phase shift of $-45°$ when $f = 500$ Hz. Use equal-valued resistors throughout the circuit, with $C_1 = 0.86$ μF.

6-22. Sketch a rough plot of the frequency response of the circuit in Fig. 6-37.

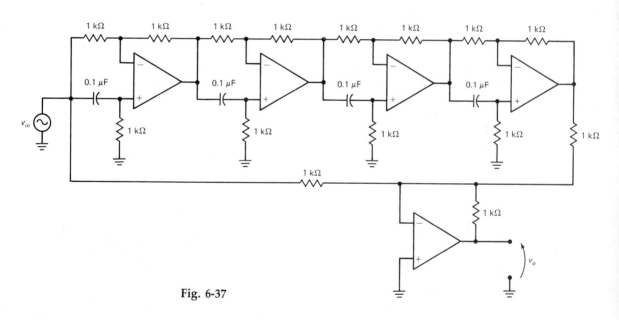

Fig. 6-37

CHAPTER 7

SPECIALIZED LINEAR ICs

The universal acceptance of operational amplifiers as fundamental linear and nonlinear circuit building blocks has inspired the production of many other specialized building blocks. Some of these devices are just op amps that have been optimized in terms of one or more characteristics. An example of such a device that was covered earlier in this book is the instrumentation amplifier. Several other devices that are closely related to the basic op amp are discussed in this chapter. Specifically, comparators fall into this category. Many other specialized devices may be rather distantly related or totally unrelated to the op amp. A few such devices are covered in this chapter, including current difference amplifiers, operational transconductance amplifiers, balanced modulators, phase-locked loops, analog switches, and track-and-hold amplifiers.

7-1 ♦ COMPARATORS

A comparator is a nonlinear circuit that switches its output to one of two levels, based on a comparison between an input voltage and a reference voltage. Depending on the application, comparators may be operated from either bipolar or single-polarity power supplies. Comparators are often used to form an interface between digital circuits and the analog world, and therefore single-polarity power-supply operation is the most common case.

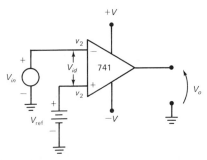

Fig. 7-1 Simple op amp comparator.

The Op Amp as a Comparator

Nearly any op amp, operating in the open-loop configuration, can be used as a comparator. Consider the circuit shown in Fig. 7-1. This circuit consists of an open-loop 741 op amp powered by a bipolar power supply. The noninverting terminal of the op amp is connected to a reference voltage of +2.5 V, while the inverting input is connected to an unknown voltage V_{in}. Recall that the output of the op amp is given by

$$V_o = A_{OL}V_{id}$$
$$= A_{OL}(V_2 - V_1)$$
$$= A_{OL}(V_{ref} - V_{in})$$

The open-loop gain of the 741 is quite high, with 200,000 being a typical value. Assuming that $V_{in} = 0$ V, it is apparent that the output of the op amp will be saturated at a level near the positive supply rail. The output voltage, $\pm V_{sat}$ (actually, $+V_{sat}$ in this case), is usually about 0.5 V lower in magnitude than the supply voltage.

Let us assume that in Fig. 7-1, V_{in} is increasing positively. The output of the comparator will remain at $+V_{sat}$ until the input voltage is within V_{ref}/A_{OL} volts of the reference voltage. Assuming that $A_{OL} = 100,000$, such an input voltage will cause the output to assume some value in between the output saturation voltages. That is, for input voltages that are within V_{ref}/A_{OL} volts of V_{ref}, the comparator will behave linearly. Based on the op amp parameters given, the input voltage range that does not result in saturation of the output is 2.5 V $\pm$ 25 µV. A plot of the op amp comparator's transfer function (input-output relationship) is shown in Fig. 7-2. In general,

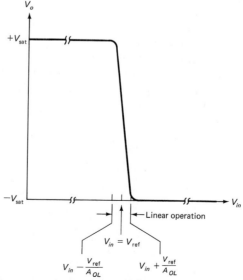

Fig. 7-2 Open-loop op amp transfer curve.

because the comparator is intended to be a two-state device, the linear region of operation is to be avoided.

EXAMPLE 7-1

The comparator of Fig. 7-3 has $V_{ref} = 2.00$ V, $+V_{sat} = 4.5$ V, $-V_{sat} = 0$ V, and $A_{OL} = 50{,}000$. Sketch the output and determine the approximate duty cycle of the output waveform if the input is $v_{in} = 4 \sin 1000t$ V.

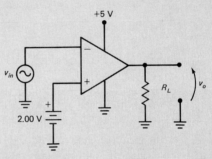

Fig. 7-3 Circuit for Example 7-1.

Solution

Because v_{in} exceeds V_{ref} during positive excursions, the output of the comparator will continuously switch between 0 V and +4.5 V (the output saturation voltages). The approximate input and output waveforms are shown in Fig. 7-4. The duty cycle of the output is defined as

$$D = \frac{T_H}{T} \quad \text{or} \quad \%D = 100\,\frac{T_H}{T}$$

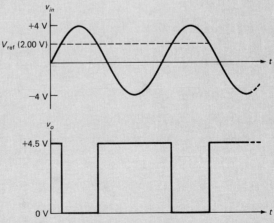

Fig. 7-4 I/O voltage waveforms for Example 7-1.

The comparator's output is high as long as v_{in} is less than 2.00 V (V_{ref}). In this case, it is most convenient to determine the time during which the output of the comparator is low, that is, the length of time during which v_{in} is greater than 2.00 V. Substituting V_{ref} for the instantaneous value of v_{in} and solving the input voltage equation for t, we obtain

$$t|_{v_{in}>2\,V} = \frac{\sin^{-1}(2/4)}{1000}\,s$$

$$= 524\,\mu s$$

(Note: Units are in radians.)

We now know that it takes 524 μs after zero crossing for v_{in} to drive the output of the comparator low. Because of the symmetry of the input, we also know that the output of the comparator will be driven high 524 μs before the next zero crossing (in the negative direction). The period of the input signal is found as follows:

$$T = \frac{2\pi}{1000}$$

$$= 6.28\,ms$$

The time for one complete half-cycle is

$$\frac{T}{2} = 3.14\,ms$$

Now, subtracting the time for which the output of the camparator is high during the positive half-cycle from $T/2$, we obtain

$$T_L = \frac{T}{2} - t|_{v_{in}>2V} - t|_{v_{in}>2\,V}$$

$$= 3.14\,ms - 524\,\mu s - 524\,\mu s$$

$$= 2.09\,ms$$

and

$$T_H = T - T_L$$

$$= 6.28\,ms - 2.09\,ms$$

$$= 4.19\,ms$$

The duty cycle of the output may now be determined:

$$D = 0.667$$

$$= 66.7\%$$

The input and reference terminals of the comparator may also be reversed. That is, the reference voltage may be applied to the inverting input, while the input voltage is applied to the noninverting input. In such a case, the output of the comparator would be driven high when v_{in} was more positive than V_{ref}.

The Zero-Crossing Detector

The zero-crossing detector is a special case of the general comparator in which the reference voltage is 0 V. A zero-crossing detector is shown in Fig. 7-5. The diodes placed in inverse-parallel across the input terminals of the comparator limit the differential input voltage V_{id} to approximately ± 0.7 V. This helps prevent possible damage to the amplifier. Because of the extremely high A_{OL}, this input voltage range is more than adequate to drive the output into saturation. Resistor R_1 serves as a current limiter, preventing the input voltage source from being damaged when either of the diodes goes into conduction. Resistor R_2 is used to help null out bias-current-induced offset voltage, and is made equal to R_1.

Zero-crossing detectors are often used to convert sinusoidal waveforms into square waves. For example, the circuit shown in Fig. 7-6 could be used to generate a 60-Hz TTL-compatible time-base, or clock, signal for a digital circuit. In other applications, zero-crossing detectors and comparators may be used to derive firing pulses for SCR power control circuits.

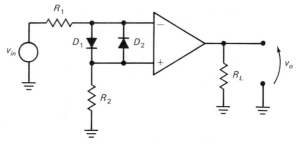

Fig. 7-5 Comparator with input clamping diodes (inverse-parallel).

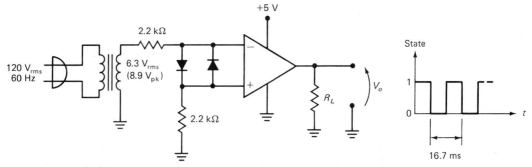

Fig. 7-6 Generating a 60-Hz time base from the ac line.

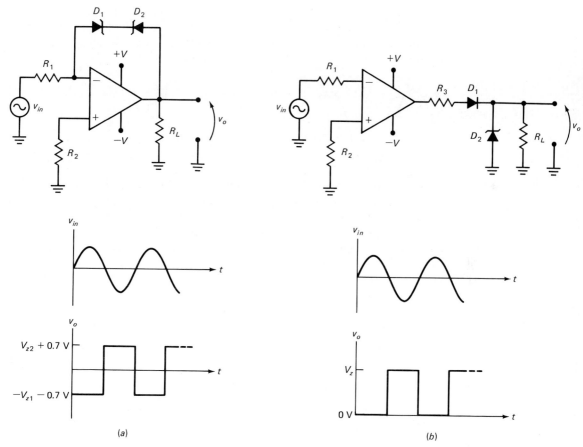

Fig. 7-7 Output limiting circuits. (a) Zener diodes placed in feedback loop; (b) zener and switching diodes placed at the output.

The output of a comparator may be limited to some value(s) less than $\pm V_{\text{sat}}$ by bounding the output with zener diodes. Two possible bounding approaches are shown in Fig. 7-7. In Fig. 7-7a, two inverse-series zeners are used to close the negative feedback loop of a zero-crossing detector. This allows a bipolar output waveform to be produced, with its peak values set by the zener breakdown voltages. In Fig. 7-7b, the output is limited to V_Z in the positive direction, with R_3 limiting the output current, protecting either the zener diode, the comparator, or both. The output will be 0 V in the negative direction, as diode D_1 will be reverse-biased.

The Comparator with Hysteresis

Comparators like those previously discussed have several drawbacks. One major problem is noise sensitivity. All signals contain some amount of noise. Suppose that the slightly noisy sinusoidal signal of Fig. 7-8 is applied to the comparator in Fig. 7-3. As may be seen in Fig. 7-8, the output of the com-

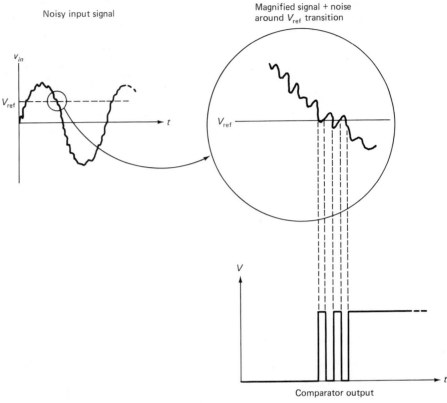

Fig. 7-8 Effects of a noisy signal on the output of a normal comparator.

parator contains several unwanted transitions that are caused by the noise. Should this circuit be used as an interface for a digital circuit such as a counter, the false transitions could be counted as, say, trigger events, resulting in erratic or ambiguous operation of the digital circuit.

The solution to the comparator's noise sensitivity problems lies in the application of positive feedback, which provides the comparator with hysteresis. By saying that the comparator has hysteresis, we mean that the upper and lower trip points or threshold voltages of the input are at different levels, with the trip point for positive-going inputs more positive than the trip point for negative-going inputs. Comparators with hysteresis are often called *Schmitt triggers*. Positive feedback is implemented as shown in Fig. 7-9. The upper trip point (V_{UT}) and lower trip point (V_{LT}) are given by

$$V_{UT} = +V_{sat} \frac{R_2}{R_1 + R_2} \tag{7-1}$$

$$V_{LT} = -V_{sat} \frac{R_2}{R_1 + R_2} \tag{7-2}$$

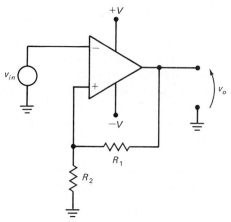

Fig. 7-9 Use of positive feedback produces hysteresis. The comparator with hysteresis is called a Schmitt trigger.

A possible transfer curve for the comparator of Fig. 7-9 is shown in Fig. 7-10a, where it is arbitrarily assumed that the output saturated positively when power was applied, and that v_{in} was initially positive-going. The transfer curve generated in Fig. 7-10a is referred to as a *hysteresis loop*. The noise immunity of the comparator with hysteresis is illustrated in Fig. 7-10b, where a rather noisy waveform is applied to the input. In this example, it would take an unusually large amount of noise to produce false triggering of the comparator. The region in between the trip points is called the *deadband*.

In addition to improved noise immunity, positive feedback also tends to speed up the switching time of the comparator. In reference to Fig. 7-9,

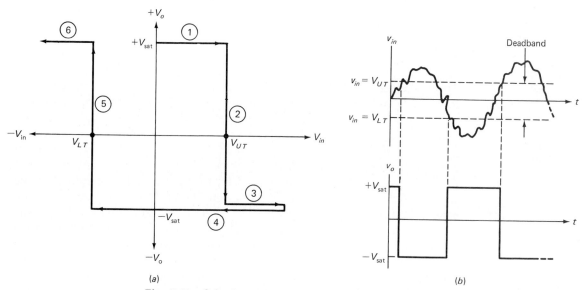

Fig. 7-10 Schmitt trigger response. (a) Transfer characteristics; (b) response to a noisy signal.

assume that initially the output is at $+V_{sat}$, and v_{in} is zero. This results in a voltage, given by Eq. (7-1), being present at the noninverting input. Now, let us assume that v_{in} begins increasing positively. Once v_{in} exceeds the noninverting input voltage, the output begins to decrease. This increases V_{id} such that the output is driven more negative, which in turn increases V_{id} further. This self-sustaining process continues until the output reaches $-V_{sat}$. A similar process occurs on transitions of the input beyond the lower trip point. Notice that if the output saturation voltages are symmetrical, the upper and lower trip points will also be symmetrical.

EXAMPLE 7-2

A certain comparator has $+V_{sat} = 4.5$ V and $-V_{sat} = -4.5$ V. Assuming that the basic circuit of Fig. 7-9 is used and $R_2 = 1$ kΩ, determine the value required for R_1 to produce $V_{UT} = 1.00$ V.

Solution

In order to determine the required value for R_1, we solve Eq. (7-1) for R_2 and substitute the given values of R_2 and $+V_{sat}$ into the resulting expression, producing

$$R_1 = 3.5 \text{ k}\Omega$$

Because of the symmetrical saturation voltages, the threshold voltages will also be symmetrical; that is, $V_{LT} = -1.00$ V.

The Window Comparator

The window comparator is used to indicate whether or not an input voltage is between the limits defined by two externally applied reference voltages. The voltage range between the two limits, or setpoints, is called a *window*. A window comparator that operates from a single-polarity supply is shown in Fig. 7-11. Because this circuit is operated from a positive supply, the upper and lower setpoints must both be greater than or equal to 0 V.

In order to obtain a better understanding of the operation of the window comparator, let us assume that $V_{UT} = +3$ V, $V_{LT} = +1$ V, and $v_{in} = 0$ V. These conditions cause the output of U_1 to drive to $+V_{sat}$, while U_2 drives to $-V_{sat}$ (about 0 V here). Under these conditions, D_1 will be forward-biased, causing V_o to assume a level of $+V_{sat} - 0.7$ V. Diode D_2 is reverse-biased, isolating the output of U_2 from R_L and preventing the outputs of U_1 and U_2 from driving each other in opposite directions. Now, if v_{in} increases to some level in between V_{LT} and V_{UT}, say 2 V, then the outputs of both U_1 and U_2 will be driven to ground potential, likewise forcing $V_o = 0$ V. When v_{in} is higher than V_{UT}, the output of U_1 will be driven to positive saturation and the output of U_2 will be driven to ground. This produces $V_o = +V_{sat}$. Hence,

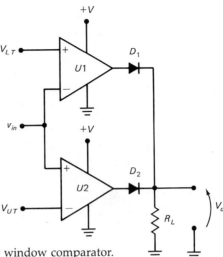

Fig. 7-11 Simple window comparator.

we can see that if v_{in} is within the "window," then $V_o = 0$ V. The output will be at $+V_{sat}$ for values of v_{in} that are outside the window.

Comparator ICs

Conventional op amps are not usually used to form comparators for use in precision applications. In these applications, specialized comparator ICs are normally used. These devices are specifically designed for rapid switching (high slew rate) and relatively high current drive capability with very good performance when operated from single-polarity power supplies. Many precision comparators have provisions for digital control of the output. One such device is the National Semiconductor LM311. An LM311 connected as a simple zero-crossing detector is shown in Fig. 7-12.

The LM311 may be operated from either a single-polarity or a bipolar power supply. The 311 has an uncommitted NPN transistor output, which may be used in either the common-emitter configuration (using pin 7 as the output terminal) or the emitter-follower configuration (if pin 1 is used as the output terminal). Strobe and balance inputs are also available on the LM311. The balance inputs are used to null out the effects of offset voltages and currents. The strobe input allows the comparison operation to be controlled by a TTL-compatible logic level, as shown in Fig. 7-12. Output currents as high as 50 mA and load voltages as high as 40 V can be handled by the LM311.

7-2 ♦ ANALOG SWITCHES AND TRACK-AND-HOLD CIRCUITS

An analog switch may be considered to be a solid-state, electronically controlled equivalent of a mechanical switch. Analog switches are used extensively in signal multiplexing, analog-to-digital conversion, and track-and-

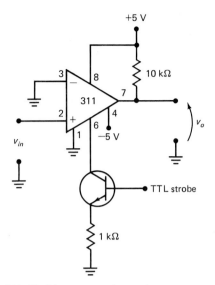

311: Pin 1 is output transistor emitter
Pin 7 is output transistor collector

Fig. 7-12 Zero-crossing detector.

hold applications. Track-and-hold (T/H) circuits (sometimes called a sample-and-hold, or S/H) are also used in digital data acquisition applications.

Analog Switches

In general, analog switches are relatively low-power, low-voltage devices. Analog switches are constructed around JFETs or MOSFETs, which are used as voltage-controlled switches. Most analog switches are designed to be controlled by the digital output produced by a TTL IC. Analog switches are available with constant on-state resistances that range from a few hundred ohms to lower than 5 Ω. In the off state, leakage currents through the analog switch terminal are typically less than 1 nA, effectively producing off-state resistances of greater than 100 MΩ.

The equivalent internal circuitry for a simple JFET SPST analog switch is shown in Fig. 7-13. The diode is used to clamp the source of the JFET to a maximum of 0.7 V. This helps prevent the FET from being turned on (when it is supposed to be off) by ac signals with positive dc offset components of nearly the same magnitude as the gate control voltage. Circuits as simple as this are generally intended to be used as current-mode switches, with the drain of the FET connected to the virtual ground point at the inverting input of an op amp. This type of application is shown in Fig. 7-14.

More sophisticated analog switches are available that can be used to control higher voltage levels. A representative device is the National Semiconductor LF11331/LF1331 Quad SPST JFET Analog Switch, shown in Fig. 7-15. This switch is designed to control analog signals that are within 5 V of V_{CC} or V_{EE}. The reference input is used to set the threshold of the control inputs.

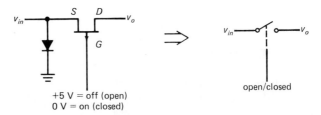

Fig. 7-13 A JFET analog switch.

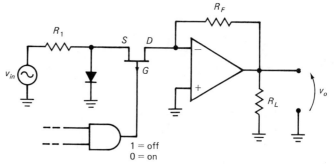

Fig. 7-14 Analog switch used in the current mode.

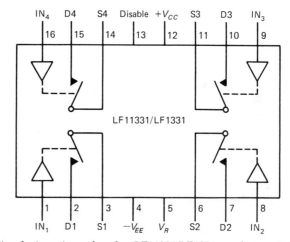

Fig. 7-15 Pin designations for the LF11331/LF1331 analog switch. (*Courtesy of National Semiconductor Corp.*)

For operation with TTL logic levels, the reference input is tied to ground. The control inputs themselves are active low. That is, a logic zero level at a given control input closes the associated analog switch. The LF11331/LF1331 is designed to operate most effectively with on-state drain currents of less than ±5 mA. The switches' off-state drain leakage currents are typically less than 1 nA at 25° C.

One of the most common applications in which analog switches are used is in the design of analog multiplexers. A multiplexer is a circuit that allows

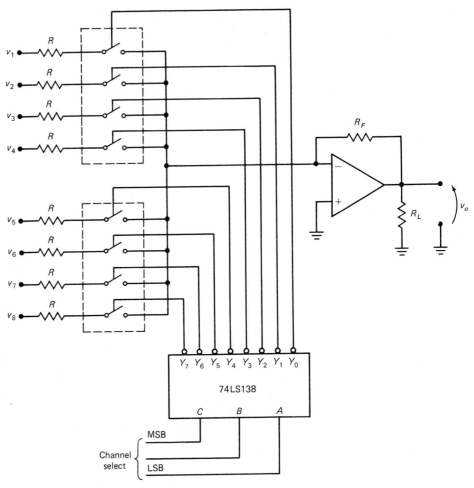

Fig. 7-16 An analog multiplexer.

one of several different signal sources to be connected to a common point. The circuit in Fig. 7-16 is an example of an 8 to 1 multiplexer. In this circuit, a 74LS138 1 of 8 decoder is used to select a given input signal source.

Track-and-Hold Circuits

Track-and-hold (T/H, or sometimes called sample-and-hold, S/H) circuits have two modes of operation, appropriately enough called the track mode and the hold mode. Generally, a digital logic level is used to select operation in either of these two modes. In the track mode, the output of the T/H mirrors the analog voltage that is applied to the input. In the hold mode, the output of the T/H will be held at the voltage level that existed at the input at the time when the hold mode was entered.

A simple T/H circuit is shown in Fig. 7-17. The heart of the circuit is the sampling capacitor C_S. The size of the sampling capacitor will normally be

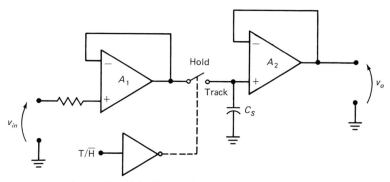

Fig. 7-17 Simple track-and-hold circuit.

some value around 0.01 µF or smaller. During the signal-tracking phase (T/$\overline{\text{H}}$ is high), op amp A_1 is connected to C_S by an analog switch. A_1 buffers the input signal to minimize source loading effects. When T/$\overline{\text{H}}$ is driven low, the analog switch opens, isolating A_1 from the sampling capacitor, which now holds the input voltage that was present at the instant the hold mode was entered. In order to maintain a nearly constant charge, C_S must be a very low-leakage capacitor, such as a polystyrene unit. Also, the input resistance of A_2 must be extremely high, which usually requires the use of an FET input op amp.

Track-and-hold circuits are usually used in conjunction with analog-to-digital converters in digital signal processing applications. The T/H allows the input voltage being digitized to be held constant while conversions are performed. Chapter 8 presents more information concerning analog-to-digital conversion. Figure 7-18 illustrates waveforms that are representative of those that might be observed during the operation of a T/H circuit. Specially designed T/H ICs are available from many sources. The data sheets supplied with these devices contain information regarding sample capacitor selection.

7-3 ♦ CURRENT DIFFERENCE AMPLIFIERS

Current difference amplifiers (CDAs) are functionally similar to standard op amps, except that they are specifically designed to operate from single-polarity power supplies. CDAs are sometimes called automotive op amps, because automobiles have a single-polarity supply (a 12-V battery), and CDAs are a natural choice for use as gain blocks in these applications. CDAs are also sometimes called Norton op amps. The schematic symbol for a CDA is shown in Fig. 7-19. The term *current difference amplifier* is derived from the basic operation of the amplifier. That is, an output voltage is generated that is proportional to the difference between the currents entering the inverting and noninverting terminals of the device. As is the case with standard op amps, CDAs have very high open-loop gain, and negative feedback is employed to reduce the gain of the circuit to some desired closed-loop value. A commonly used CDA is the National Semiconductor LM3900 Quad Norton

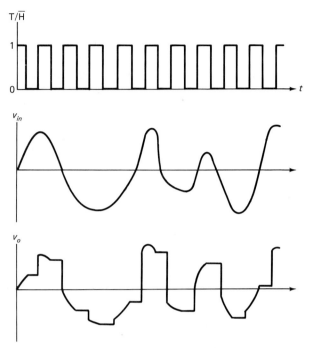

Fig. 7-18 Sampling a randomly varying voltage at a constant rate.

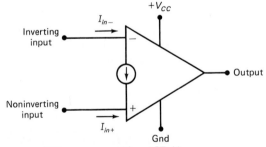

Fig. 7-19 A current difference amplifier (CDA).

amp. The LM3900 typically has A_{OL} of around 70 dB. The specs for this device will be used throughout this discussion. Assuming that the closed-loop gain of a given CDA circuit is much less than A_{OL}, the following relationship will be valid:

$$I_{in-} = I_{in+} \tag{7-3}$$

CDA Inverting Configuration

The equivalent internal circuitry of a CDA configured for operation in the inverting mode is shown in Fig. 7-20. Analysis begins at the noninverting input terminal. Diode D_1 and transistor Q_1 form a current mirror that is set

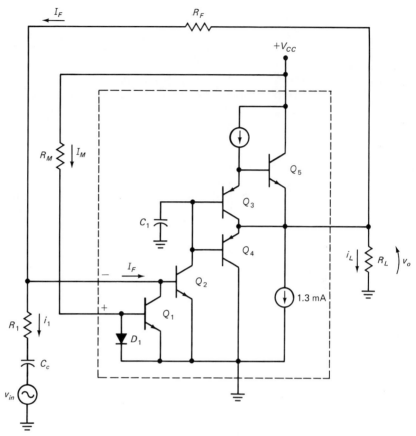

Fig. 7-20 Equivalent internal circuitry for a typical CDA.

to some user-selected value by proper selection of R_M. The mirror current is given by

$$I_M = \frac{V_{CC} - V_{BE}}{R_M} \tag{7-4}$$

For the LM3900, the maximum value of I_M is 500 μA. Typical mirror currents are in the area of 5 to 100 μA. Negative feedback supplied by R_F allows application of Eq. (7-3), yielding

$$I_F = I_M \tag{7-5}$$

This feedback current sets up a quiescent output voltage that is given by

$$V_{oQ} = I_F R_F + V_{BE} \tag{7-6}$$

The above analysis indicates that the CDA is operated as a class A amplifier. Notice that the input coupling capacitor prevents the signal source (now inactive) from affecting the quiescent feedback current. The V_{BE} term in Eq. (7-6) indicates that the inverting input terminal is at $+0.7$ V with respect to ground, which is also the potential at the noninverting input.

Assume that a sinusoidal signal is now applied to the circuit. The resultant current forced through R_1 is summed with I_F at the inverting input. The current supplied by the signal source is given by

$$i_1 = \frac{v_{in}}{R_1} \quad \text{assume } X_C = 0 \, \Omega \tag{7-7}$$

During the positive half-cycle of v_{in}, there tends to be an increase in I_{in-}. However, the effects of negative feedback cause an instantaneous reduction in I_F by a nearly equal amount, forcing both input currents to be equal, as indicated by Eq. (7-3). Thus, an ac feedback current component is created, given by

$$i_F = -i_{in} \tag{7-8}$$

The net result of this circuit action is a decrease in V_o, and phase inversion. The magnitude of the change in V_o is given by

$$\Delta V_o = \Delta I_F R_F$$

Substituting Eqs. (7-7) and (7-8) yields

$$v_o = V_{oQ} - \frac{v_{in}}{R_1} R_F \tag{7-9}$$

Analysis of the circuit operation for negative half-cycles yields similar results, indicating that I_F will increase, causing an increase in V_o.

The dc term V_{oQ} may easily be eliminated from the output voltage expression by capacitively coupling the load to the amplifier, as shown in Fig. 7-21. The resulting expression for the output voltage is

$$v_o = -\frac{v_{in}}{R_1} R_F$$

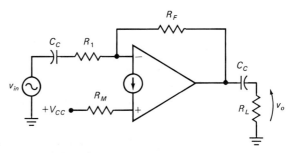

Fig. 7-21 CDA in the inverting configuration.

Dividing both sides by v_{in} results in the following voltage gain expression:

$$A_v = -\frac{R_F}{R_1} \qquad (7\text{-}10)$$

EXAMPLE 7-3

Refer to Fig. 7-21. Given $V_{CC} = +15$ V, determine the necessary values for R_M, R_F, and R_1 such that $I_M = 20$ µA, $V_{oQ} = 7.5$ V, and $A_v = -10$.

Solution

First, R_M is determined using Eq. (7-4). This yields

$$R_M = \frac{V_{CC} - V_{BE}}{I_M}$$

$$= \frac{15\text{ V} - 0.7\text{ V}}{20\text{ µA}}$$

$$= 715\text{ k}\Omega$$

The value of R_F is determined by solving Eq. (7-6) for R_F and substituting the desired value of V_{oQ} and the previously determined mirror current into the resulting expression.

$$R_F = \frac{V_{oQ} - V_{BE}}{I_F}$$

$$= \frac{7.5\text{ V} - 0.7\text{ V}}{20\text{ µA}}$$

$$= 340\text{ k}\Omega$$

Now, applying Eq. (7-10), we obtain

$$R_1 = \frac{R_F}{|A_v|}$$

$$= \frac{340\text{ k}\Omega}{10}$$

$$= 34\text{ k}\Omega$$

The inverting input of the CDA may be used as a summing junction, like the conventional op amp. Analysis of the inverting CDA with a capacitively coupled output and n inputs yields the following output voltage expression:

$$v_o = -\frac{R_F}{R_1}v_1 - \frac{R_F}{R_2}v_2 - \frac{R_F}{R_3}v_3 - \cdots - \frac{R_F}{R_n}v_n \quad (7\text{-}11)$$

The maximum output voltage that can be produced by the LM3900 CDA is approximately 1 V less than V_{CC}. With a capacitively coupled output, the maximum peak-to-peak voltage across R_L will be given by

$$v_{o(P\text{-}P)} = V_{CC} - 1 \text{ V (max)} \quad (7\text{-}12)$$

The LM3900 is guaranteed to source at least 6 mA to the load and the feedback loop. Typically, the LM3900 will source 10 mA. Because of the LM3900's output circuit design, the typical output current sink capability is -1.3 mA.

CDA Noninverting Configuration

A noninverting CDA is shown in Fig. 7-22. Unlike the standard op amp, the gain expression for this circuit is the same as that for an inverting CDA, with the exception of the polarity. That is,

$$A_v = \frac{R_F}{R_1} \quad (7\text{-}13)$$

The noninverting CDA is analyzed in the same manner as the inverting circuit; therefore, a detailed analysis is not presented here.

One interesting characteristic of the noninverting CDA that makes it more useful in some situations is the fact that the noninverting input of the CDA is a relatively low-impedance, near-ground potential node. This may be seen by examining the internal circuitry shown in Fig. 7-20. Because of these characteristics, the noninverting input is quite well suited for use as a summing junction. This is especially true when signals are summed simultaneously at the inverting and noninverting inputs of the CDA. The conventional

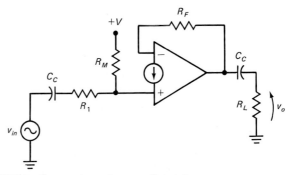

Fig. 7-22 CDA in the noninverting configuration.

op amp is not as effective in similar applications because of the extremely high input resistance of the noninverting input, plus the fact that the noninverting input will not generally be close to ground potential. Such a situation could lead to source interaction. Equivalent CDA circuits are much less likely to exhibit these possibly troublesome characteristics. Of course, the price that is paid is the requirement of capacitive coupling, which eliminates operation with dc input voltages.

7-4 ♦ OPERATIONAL TRANSCONDUCTANCE AMPLIFIERS

Another useful class of linear IC amplifiers is the transconductance amplifiers. Recall that a transconductance amplifier may also be described as being a voltage-to-current converter (V/I) or voltage-controlled current source (VCIS). It has been shown that a conventional op amp can be used as a transconductance amplifier; however, monolithic ICs are available that are designed specifically for this purpose. Such devices are called operational transconductance amplifiers, or OTAs.

The National Semiconductor LM3080 is an example of a very popular OTA. The pin diagram and equivalent internal circuitry for the LM3080 are shown in Fig. 7-23. As seen in Fig. 7-23b, the input section of the LM3080 consists of a differential pair (Q_1 and Q_2) with active collector loading provided by transistors Q_3, Q_4, Q_5, D_1, and D_3. The quiescent collector currents for the differential pair are set up by a current mirror that is formed by Q_6 and D_2. The mirror current, or amplifier bias current (I_{ABC}), is in turn determined by an external resistor.

The output of the differential pair is applied to a push-pull output stage consisting of transistors Q_7, Q_8, Q_9, D_3, and D_4. Because output-stage transistors Q_7 and Q_8 are in the common-emitter configuration, the output resistance of the amplifier is extremely high, which allows the output to approximate an ideal voltage-controlled current source.

A few of the LM3080's more useful specifications are listed below. The values given are typical, unless otherwise noted.

Slew rate	50 V/μs
Open-loop bandwidth	2 MHz
I_{out}	±350 μA (guaranteed minimum)
Output resistance	1 MΩ
Input resistance	26 kΩ
Transconductance range	10 μS – 2000 mS (linear with I_{ABC})
CMRR	110 dB
V_{IO}	0.4 mV
I_B	0.4 μA
I_{IO}	0.1 μA

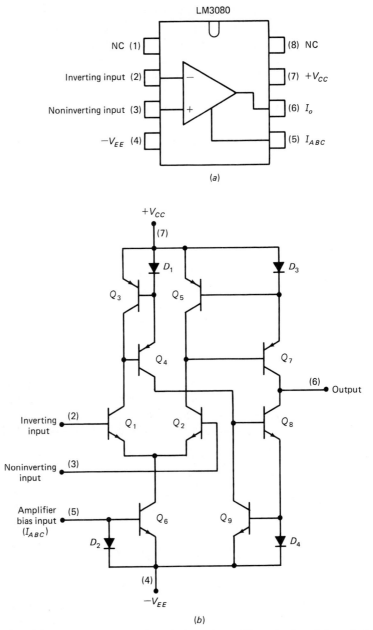

Fig. 7-23 The LM3080 OTA. (*a*) Pin designations; (*b*) equivalent internal circuitry. (*Courtesy of National Semiconductor Corp.*)

These specifications are used as a general guide in the circuit design process. Most of the values presented will vary with the amplifier's bias point, as a function of I_{ABC}, and to some extent as a function of temperature. For the purposes of this discussion, temperature will be assumed to be constant at 25°C.

All of the OTA specifications presented, except for the transconductance range, are equivalent to those that are given for standard op amps. Transconductance g_m is the fundamental gain parameter for the OTA. Transconductance has the units siemens, and is defined by the following relationship:

$$g_m = \frac{i_o}{v_{in}} \tag{7-14}$$

The relationship between g_m and amplifier bias current is defined in the next section.

Basic OTA Design Considerations

The LM3080 is usually configured for operation as either an inverting or a noninverting amplifier. The schematic diagrams for these modes of operation are shown in Fig. 7-24. In both designs, an output voltage is developed across a load resistor, through which the amplifier output current is forced.

Resistors R_1 and R_2 are set to equal values in order to reduce the effects of input offset and bias currents. The amplifier bias current is determined using the familiar current mirror equation:

$$I_{ABC} = \frac{|V_{EE}| - V_{BE}}{R_{ABC}} \tag{7-15}$$

An alternative method for connecting R_{ABC} is shown in Fig. 7-25. In this case, the bias current is given by

$$I_{ABC} = \frac{|V_{EE}| + V_{CC} - V_{BE}}{R_{ABC}} \tag{7-16}$$

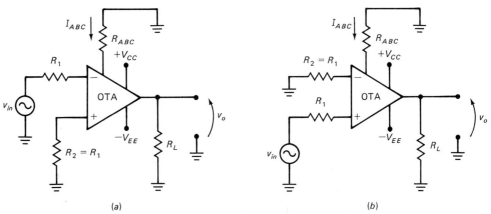

Fig. 7-24 Basic OTA amplifier configurations. (*a*) Inverting amp; (*b*) noninverting amp.

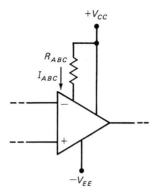

Fig. 7-25 Alternate biasing of the OTA.

The amplifier bias current should be restricted to the range defined by the inequality below:

$$0.1 \text{ μA} \leq I_{ABC} \leq 1 \text{ mA} \tag{7-17}$$

The following analysis refers specifically to the noninverting OTA configuration of Fig. 7-24b, but changing the sign associated with v_{in} makes the analysis applicable to the inverting configuration as well.

The output load current developed by the amplifier is given by

$$i_L = g_m v_{in} \tag{7-18}$$

The transconductance of the amplifier is related to I_{ABC} by the following relationship:

$$g_m = k I_{ABC} \tag{7-19}$$

where $k = 19.2 \text{ V}^{-1}$ (reciprocal volts) at 25° C

Applying Ohm's law, the output voltage developed across R_L is given by

$$v_o = i_L R_L \tag{7-20}$$

Combining Eqs. (7-18) and (7-20) yields

$$v_o = v_{in} g_m R_L \tag{7-21}$$

Equation (7-21) may now easily be solved to determine A_v for applications that require the OTA to be used as a voltage amplifier.

$$A_v = g_m R_L \tag{7-22}$$

Notice that the voltage gain is intimately related to the value of R_L.

EXAMPLE 7-4

Refer to Fig. 7-24a. This circuit is to be used as a voltage amplifier. Given $V_{CC} = |V_{EE}| = 15$ V, determine the component values required such that $I_{ABC} = 25$ µA and $A_v = -10$.

Solution

First, the value of R_{ABC} must be determined. Application of Eq. (7-15) yields

$$R_{ABC} = \frac{|V_{EE}| - V_{BE}}{I_{ABC}}$$

$$= \frac{15 \text{ V} - 0.7 \text{ V}}{25 \text{ µA}}$$

$$= 572 \text{ k}\Omega$$

The transconductance of the OTA is found using Eq. (7-19):

$$g_m = kI_{ABC}$$

$$= 19.2 \text{ V}^{-1} \times 25 \text{ µA}$$

$$= 480 \text{ µS}$$

The required value for R_L is found through application of Eq. (7-22):

$$R_L = \frac{A_v}{g_m}$$

$$= \frac{10}{480 \text{ µS}}$$

$$= 20.8 \text{ k}\Omega$$

In order to ensure that a given OTA-based amplifier will operate correctly, we must make sure that the output of the OTA can supply sufficient current to the load under all output voltages. For the preceding example, it is reasonable to expect the output voltage range to be ± 14 V. Given $R_L = 20.8$ kΩ, this means that the OTA must be able to supply a maximum of ± 672 µA to the load. The specifications presented earlier indicate that the LM3080 is only guaranteed to sink or source ± 350 µA; therefore, there is a good chance that the circuit designed in Example 7-4 will not produce the maximum available output voltage excursion. In order to ensure operation over the entire voltage range available, it would be necessary to increase the value of R_L to at least 40 kΩ. This means that R_{ABC} must be increased in order to decrease I_{ABC} and g_m, to maintain the required voltage gain of -10.

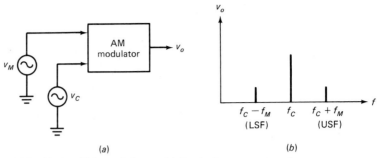

Fig. 7-26 (a) An AM modulator. (b) Typical output signal spectrum.

An OTA Modulator Application

The OTA is especially well suited for use in the generation of amplitude-modulated (AM) signals. Before this particular application is discussed, however, a brief introduction to amplitude modulation theory will be presented.

AM Fundamentals

Simply stated, amplitude modulation is a process in which the amplitude of a high-frequency signal (called the carrier) is varied in proportion to the instantaneous amplitude of a lower-frequency signal that represents some information. The device that performs this operation is called a *modulator*. Figure 7-26a illustrates the block diagram symbol for a modulator, where the modulating (information) signal is labeled v_M and the signal whose amplitude is to be modulated (the carrier) is labeled v_C. The carrier is a sinusoidal signal of some fixed frequency f_C. The modulating signal may be some complex waveform, such as speech or music. To simplify this analysis, the modulating signal will be assumed to be a constant-frequency sinusoid. The modulator combines or mixes the two input signals in a nonlinear manner (the superposition principle does not apply), producing an output that consists of frequency components at the carrier frequency f_C and components at frequencies of $f_C + f_M$ and $f_C - f_M$. A typical frequency domain representation for the output signal is shown in Fig. 7-26b. The frequency components that have been created above and below the carrier frequency are called the upper side frequency (USF) and the lower side frequency (LSF).

Figure 7-27 illustrates the time domain representations for the inputs and the output of a modulator that is supplied with a constant carrier and a modulating signal that varies in peak amplitude. The dotted outline shown around the output voltage waveform is called the *modulation envelope*.

The amount of modulation imparted to the output signal is quantified by a parameter called the *modulation index*. The modulation index is determined using the time domain representation of the output signal as shown in Fig. 7-28, using the following relationship:

$$m = \frac{V_A - V_B}{V_A + V_B} \tag{7-23a}$$

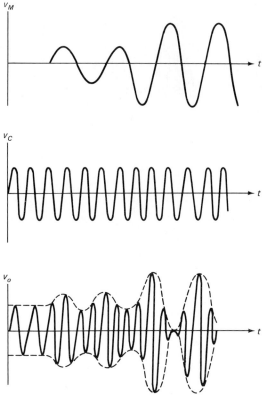

Fig. 7-27 Time domain representation of AM modulator input and output signals.

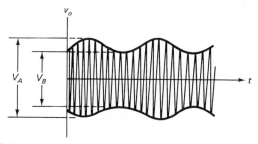

Fig. 7-28 Modulator output waveform.

or alternatively,

$$\%m = \frac{V_A - V_B}{V_A + V_B} \times 100 \qquad (7\text{-}23b)$$

Modulator time domain output waveforms and frequency domain representations for several different modulation levels are shown in Fig. 7-29. In general, modulation levels in excess of 100 percent are to be avoided, as

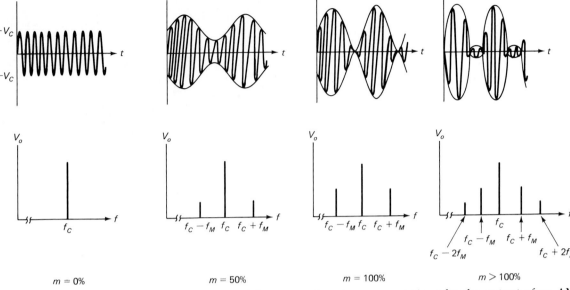

Fig. 7-29 Time and frequency domain representations for the output of an AM modulator for various levels of modulation.

undesired output frequency components will be produced, causing distortion of the modulation envelope.

OTA Modulator Analysis

A practical OTA modulator is shown in Fig. 7-30. In this discussion, the modulating and carrier voltages are defined by the following equations:

$$v_M = V_M \sin \omega_m t \quad \text{V}$$

$$v_C = V_C \sin \omega_c t \quad \text{V}$$

Basically, in this circuit the OTA is configured as a noninverting amplifier, with transconductance determined by R_{ABC}. With the modulating signal source inactive, the output of the circuit is given by direct application of Eq. (7-21). Application of v_M causes i_{abc} to be summed with I_{ABC}. Thus, the amplifier bias current and the gain of the OTA vary about the quiescent level, set by R_{ABC}. The analysis of this circuit begins with the determination of the quiescent, or no-signal, transconductance g_{mQ}.

$$g_{mQ} = kI_{ABC} \tag{7-24}$$

The modulation signal source produces a current i_{abc} that sums with I_{ABC}, producing a time-varying transconductance that is given by

$$g_m = kI_{ABC} + ki_{abc} \tag{7-25}$$

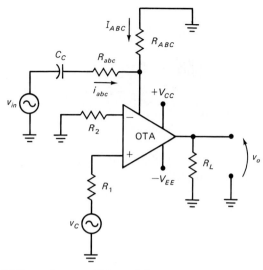

Fig. 7-30 An OTA used as an AM modulator.

Substituting Eq. (7-24) into (7-25) yields

$$g_m = g_{mQ} + ki_{abc} \tag{7-26}$$

Now, the modulating current is given by the equation

$$i_{abc} = \frac{v_M}{R_{abc}} \tag{7-27}$$

which, based on the input signals defined previously, may be expressed as

$$i_{abc} = \frac{V_M}{R_{abc}} \sin \omega_M t \tag{7-28}$$

Substituting Eq. (7-28) into (7-26) yields

$$g_m = g_{mQ} + \frac{kV_M}{R_{abc}} \sin \omega_M t \tag{7-29}$$

Substituting Eq. (7-29) into Eq. (7-21) produces the following:

$$v_o = (V_C \sin \omega_C t)\left(R_L g_{mQ} + \frac{kV_M}{R_{abc}} \sin \omega_M t\right) \tag{7-30}$$

Multiplying out Eq. (7-30) now gives us

$$v_o = R_L g_{mQ} V_C \sin \omega_C t + \left[(V_C \sin \omega_C t)\left(\frac{kV_m}{R_{abc}} \sin \omega_M t\right)\right] \tag{7-31}$$

In order to proceed further, it is now necessary to invoke the following half-angle trigonometric identity:

$$(\sin x)(\sin y) = \frac{1}{2}[\cos(x - y) - \cos(x + y)] \qquad (7\text{-}32)$$

where $x = \omega_C t$
$y = \omega_M t$

Substituting the right-hand term of Eq. (7-31) into (7-32) and multiplying by all constants produces

$$v_o = R_L g_{mQ} V_C \sin \omega_C t \\ + \left\{ \frac{1}{2} \frac{k R_L V_C V_M}{R_{abc}} [\cos(\omega_C - \omega_M)t - \cos(\omega_C + \omega_M)t] \right\} \qquad (7\text{-}33a)$$

Finally, the right-hand term of Eq. (7-33a) is expanded and the cosine terms multiplied by 1/2, producing

$$v_o = R_L g_{mQ} V_C \sin \omega_C t + \frac{k R_L V_C V_M}{2 R_{abc}} \cos(\omega_C - \omega_M)t \\ - \frac{k R_L V_C V_M}{2 R_{abc}} \cos(\omega_C + \omega_M)t \qquad (7\text{-}33b)$$

Equation (7-33b) shows that the output of the OTA will consist of three components. The first term is the amplified carrier signal. The second term represents a component of the output signal whose frequency is the difference between ω_C and ω_M. This is the lower side frequency. The third term is composed of the sum of ω_C and ω_M, which is the upper side frequency. Equation (7-33b) can be applied to a practical circuit to quantitatively determine the various parameters of the output signal's frequency components in the frequency domain.

The modulation index may also be determined analytically, using Eq. (7-33b) and the following relationship:

$$m = \frac{2 V_{SF}}{V_{OC}} \qquad (7\text{-}34)$$

where V_{SF} is the peak amplitude of either side frequency (they are symmetrical about the carrier) and V_{OC} is the peak amplitude of the output carrier frequency component. The results of the preceding analysis are used in the following example.

EXAMPLE 7-5

Refer to Fig. 7-31. Write the trigonometric expression for v_o, sketch the frequency domain representation for the output signal, and determine the modulation index.

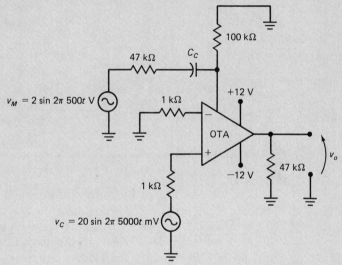

Fig. 7-31 Circuit for Example 7-5.

Solution

I_{ABC} is determined using Eq. (7-15):

$I_{ABC} = 113 \ \mu A$

The quiescent transconductance is found using Eq. (7-24):

$g_{mQ} = 2.17 \ mS$

The ac component of the amplifier bias current is now found using Eq. (7-28):

$i_{abc} = 42.6 \sin 2\pi 500 t \ \mu A$

Substituting the quantities determined above and the load resistance given in the schematic diagram into Eq. (7-33) produces the following expression for the output voltage:

$v_o = 2.04 \sin 2\pi 5000 t + 0.38 \cos 2\pi 4500 t - 0.38 \cos 2\pi 5500 t \ V$

The frequency domain representation of the output is illustrated in Fig. 7-32. The modulation index is found using Eq. (7-34):

$m = 0.37$

$= 37\%$

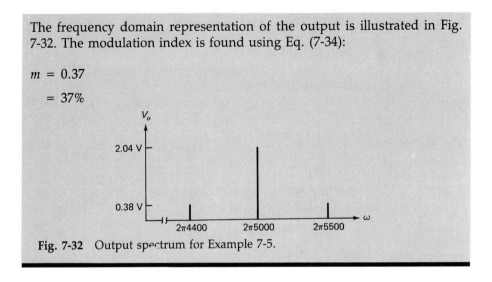

Fig. 7-32 Output spectrum for Example 7-5.

Because of the nature of the output signal spectrum produced by the amplitude modulator, such circuits are often referred to as double-sideband (DSB) modulators. The transconductance of the OTA varies with near-perfect linearity with amplifier bias current, producing a DSB output with very low distortion. That is, undesired side frequencies will be of minimal amplitude. Because of this, the LM3080 OTA is nearly ideally suited for DSB generation in low-power applications, with carrier frequencies of up to 1 MHz.

7-5 ◆ BALANCED MODULATORS

Another type of linear IC that is commonly used in radio transmitter and receiver circuitry is the *balanced modulator*. From a functional point of view, the balanced modulator may be considered to be equivalent to a four-quadrant multiplier. Thus, the output of a balanced modulator will be proportional to the product of two input signals. Balanced modulators can be used in both AM generation and demodulation applications. This section examines the operation and uses of a typical linear IC balanced modulator.

The LM1496 Balanced Modulator

An example of a readily available balanced modulator is the National Semiconductor LM1496. The equivalent internal circuitry and 14-pin DIP pinout for the LM1496 are shown in Fig. 7-33. The circuit consists of two differential pairs with cross-coupled open collectors, a biasing current source, and a modulation input section. Signals that are applied to the carrier and modulation inputs are multiplied together, and the product is scaled by the gain of the circuit. The LM1496 is designed to operate with carrier frequencies up to 100 MHz.

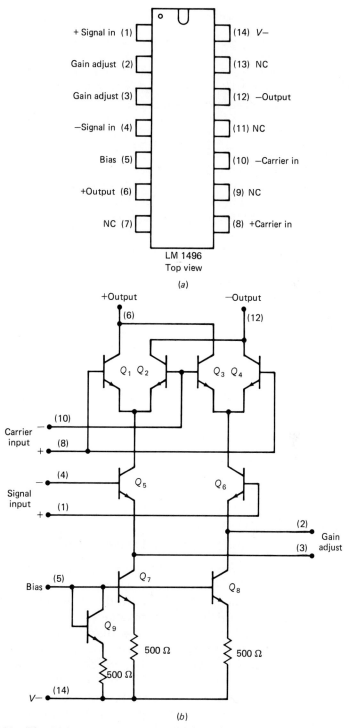

Fig. 7-33 The LM1496 balanced modulator. (*a*) Pin designations; (*b*) equivalent internal circuitry. (*Courtesy of National Semiconductor Corp.*)

Balanced Modulator Applications

Because of the inherent signal multiplying operation, it is convenient to represent the balanced modulator in simplified block diagram form as shown in Fig. 7-34. The inputs to the balanced modulator represent the carrier and modulation signals, while the factor k represents the voltage gain of the balanced modulator.

Double-Sideband Suppressed-Carrier Modulation

Balanced modulators are often used to produce what is termed *double-sideband suppressed-carrier (DSB-SC) modulation*. The term DSB-SC is derived from the fact that the original carrier-frequency component is eliminated from the output signal spectrum, leaving only the upper and lower sidebands, or side frequencies. The operation of the DSB-SC modulator can easily be understood by defining the two input signals of Fig. 7-34 as follows:

$$v_C = V_C \sin \omega_C t \text{ V} \tag{7-35}$$

$$v_M = V_M \sin \omega_M t \text{ V} \tag{7-36}$$

Given these two inputs, the output of the balanced modulator is found by applying the half-angle identity presented in Eq. (7-31), producing

$$v_o = \frac{kV_M V_C}{2} \cos(\omega_C - \omega_M)t - \frac{kV_M V_C}{2} \cos(\omega_C + \omega_M)t \text{ V} \tag{7-37}$$

The time domain representation of Eq. (7-37) for a carrier frequency of about eight times the modulating frequency is shown in Fig. 7-35a, while a typical frequency domain representation is shown in Fig. 7-35b. Both Eq. (7-37) and its graphical frequency domain interpretation clearly indicate the suppression of the original carrier frequency from the modulator's output.

Demodulation of DSB-SC Using the Balanced Modulator

A standard DSB AM signal may be demodulated simply by processing the signal through a half-wave rectifier and a low-pass RC filter. This process is illustrated in Fig. 7-36a. In this case, the negative-going portions of the AM signal are blocked by the diode, while the positive portions of the signal are passed on to a low-pass filter. The high-frequency carrier component is shunted to ground by the capacitor, leaving the original modulating signal plus a dc offset that is proportional to the amplitude of the received signal. When a DSB-SC signal is applied to the same circuit, the output is equivalent to a full-wave rectified version of the original modulating signal. Voice-modulated DSB-SC signals received on a standard AM receiver will sound garbled and unintelligible.

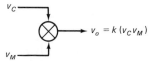

Fig. 7-34 Block diagram symbol for a balanced modulator or balanced mixer. This symbol is sometimes also used to represent a multiplier.

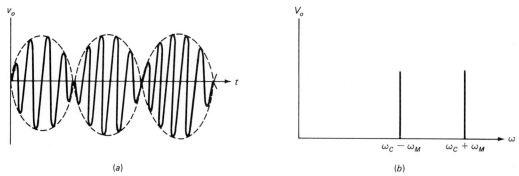

Fig. 7-35 (*a*) Time and (*b*) frequency domain representations for Eq. (7-37).

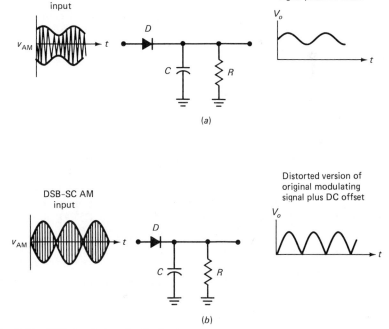

Fig. 7-36 Effects of simple diode demodulation on (*a*) a DSB AM signal and (*b*) a DSB-SC AM signal.

267

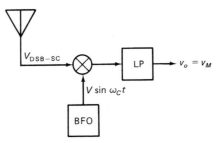

Fig. 7-37 Demodulation of a DSB-SC signal.

Demodulation of DSB-SC signals may be accomplished if the carrier is reinserted at the receiver, as shown in Fig. 7-37. The carrier frequency is produced by a beat frequency oscillator (BFO). It can be shown that the output of the balanced modulator (called a balanced mixer in this application) will consist of the original modulating signal and higher-frequency components. The high-frequency components are highly attenuated by the low-pass filter, producing an output that is a replica of the original modulating signal.

Single-Sideband Suppressed-Carrier Modulation

DSB-SC transmission is more efficient in terms of power than standard DSB AM because no power is wasted in the transmission of the carrier, which contains no information. However, both sidebands of the DSB-SC transmission contain the same information. Thus, DSB-SC transmission is no more efficient in terms of usage of the available transmission spectrum than DSB AM. That is, both standard DSB AM and DSB-SC AM require a bandwidth of $2\omega_{M(max)}$, centered about the carrier frequency. These requirements are readily apparent in Figs. 7-26 and 7-35b.

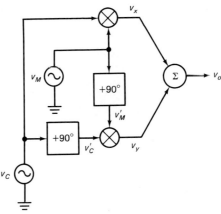

Fig. 7-38 Block diagram of an SSB-SC modulator.

Limiting the discussion to suppressed-carrier-type modulation, the elimination of one of the sidebands from a DSB-SC modulator will result in a halving of transmission bandwidth, without loss of information. Such transmission is called *single-sideband suppressed-carrier (SSB-SC) modulation*. SSB-SC modulation can be achieved in two ways. One method would require the use of a filter to remove the unwanted sideband. In practice, this technique is usually impractical because of the rapid rolloff required of the filter for good sideband suppression.

A second, more practical method of SSB-SC modulation relies on the use of balanced modulators to achieve the desired sideband suppression. The block diagram in Fig. 7-38 illustrates the construction of a SSB-SC modulator. The circuit consists of two balanced modulators, one of which is fed by 90° phase-shifted carrier and modulation signals. The 90° phase-shift sections are sometimes called *Hilbert transformers*. The output of the circuit is the sum of the voltages produced by the balanced modulators. The following analysis shows how the output signal is developed, given the following inputs:

$$v_C = V_C \sin \omega_C t \quad \text{V} \tag{7-38}$$

$$v_M = V_M \sin \omega_M t \quad \text{V} \tag{7-39}$$

The output of the upper balanced modulator is found using Eq. (7-32), producing

$$v_x = \frac{kV_M V_C}{2} \cos(\omega_C - \omega_M)t - \frac{kV_M V_C}{2} \cos(\omega_C + \omega_M)t \quad \text{V} \tag{7-40}$$

The lower balanced modulator is supplied with carrier and modulating signals that lead those applied to the upper balanced modulator by 90°. Thus, the inputs to the lower balanced modulator may be expressed in the following forms:

$$v'_C = V_C \cos \omega_C t \quad \text{V} \tag{7-41}$$

$$v'_M = V_M \cos \omega_M t \quad \text{V} \tag{7-42}$$

The following trigonometric identity may be applied to determine the form of the product of two cosines:

$$(\cos x)(\cos y) = \frac{1}{2}[\cos(x-y) + \cos(x+y)] \tag{7-43}$$

Now, using Eq. (7-43), the output of the lower balanced modulator is given by

$$v_y = \frac{kV_M V_C}{2} \cos(\omega_C - \omega_M) + \frac{kV_M V_C}{2} \cos(\omega_C + \omega_M) \quad \text{V} \tag{7-44}$$

The overall output of the circuit is found by summing Eq. (7-40) with Eq. (7-44), giving

$$v_o = \frac{kV_M V_C}{2} [\cos(\omega_C - \omega_M)t - \cos(\omega_C + \omega_M)t \\ + \cos(\omega_C - \omega_M)t + \cos(\omega_C + \omega_M)t] \quad \text{V} \tag{7-45}$$

The second and fourth terms of Eq. (7-45) represent signal components that are 180° out of phase with each other and of equal amplitude; thus the upper side frequency is eliminated from the output. The first and third terms are of equal amplitude and are in phase with each other, producing the following output expression:

$$v_o = kV_C V_M \cos(\omega_C - \omega_M)t \quad \text{V} \tag{7-46}$$

Based on the preceding analysis, the circuit represented by Fig. 7-38 will transmit the lower sideband, without need of filtering.

The balanced modulator sections of a SSB-SC modulator such as that of Fig. 7-38 must be very closely matched in terms of gain. Mismatching between the balanced modulators will result in less than maximum suppression of unwanted spectral components. Similarly, the phase-shifting sections of the system must provide 90° of phase shift to all applied signal frequency components, with no alteration of the relative amplitudes of these components. This operation is most difficult to perform on the modulating signal, which may be a complex signal with many frequency components.

EXAMPLE 7-6

The following input signals are present in Fig. 7-38:

$v_C = 5 \sin 10^6 t$ V $v_M = 2 \sin 10^3 t$ V

Determine the trigonometric expression for the output signal. Assume that both balanced modulators have $k = 2$.

Solution

The output is found through direct application of Eq. (7-46), yielding

$v_o = kV_C V_M \cos(\omega_C - \omega_M)t$ V

$= (2)(5)(2) \cos(10^6 - 10^3)t$

$= 20 \cos(9.99 \times 10^5)t$ V

7-6 ♦ THE 555 TIMER

Signal sources with highly predictable and controllable output characteristics are required in many electronic systems. There are a wide variety of linear ICs available that can generate various waveforms at frequencies ranging from less than 1 Hz to greater than 100 MHz. In this section, the operation and typical applications of the 555 timer are discussed. The 555 timer is a widely used general-purpose device that may be configured for use as a monostable multivibrator (one-shot) or an astable multivibrator.

555 Operation Principles

The 8-pin DIP pin designations for the 555 are shown in Fig. 7-39a. The components inside the dashed box in Fig. 7-39b represent the 555's equivalent internal circuitry. The external components shown are connected such that the 555 operates as an astable multivibrator.

The operation of the astable 555 can be determined by referring to the circuit of Fig. 7-39b and the waveforms shown in Fig. 7-40. Upon the initial application of power, the internal resistors divide the supply voltage such that $\frac{1}{3}V_{CC}$ is applied to the noninverting input of comparator 1 and $\frac{2}{3}V_{CC}$ is applied to the inverting input of comparator 2. Also at this time, capacitor C begins charging through external resistors R_1 and R_2. Under these conditions, the flip-flop will be reset, forcing the discharge transistor Q_1 into cutoff. While the capacitor is charging, the output of the 555 (pin 3) will be driven to approximately V_{CC}. Once v_C reaches $\frac{2}{3}V_{CC}$, the output of comparator 1 pulses high, resetting the flip-flop ($\overline{Q}$ goes high) and driving Q_1 into saturation. Now the capacitor begins discharging through R_2 and Q_1. Discharge continues until $v_C = \frac{1}{3}V_{CC}$, at which time the output of comparator 2 sets the flip-flop, cutting off Q_1, which in turn causes the capacitor to begin charging again. During the period of capacitor discharge, the output of the timer will be driven to ground potential. The load that is driven by the 555 may be referred to either ground or the power supply. The output of the 555 can typically sink or source 200 mA.

Because the capacitor charges through both R_1 and R_2 and discharges through R_2, the output of the 555 will be asymmetrical, with a duty cycle of some value greater than 50 percent. The output waveform parameters for the astable configuration of Fig. 7-39b are as follows:

$$T_H = 0.693C(R_1 + R_2) \tag{7-47}$$

$$T_L = 0.693CR_2 \tag{7-48}$$

$$T = T_H + T_L$$
$$= 0.693\,C(R_1 + 2R_2) \tag{7-49}$$

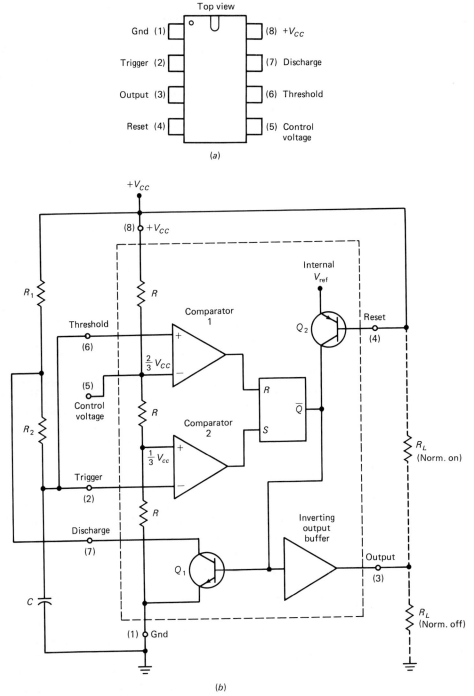

Fig. 7-39 The LM555 timer. (*a*) Pin designations; (*b*) equivalent internal circuitry (external components connected for astable operation). (*Courtesy of National Semiconductor Corp.*)

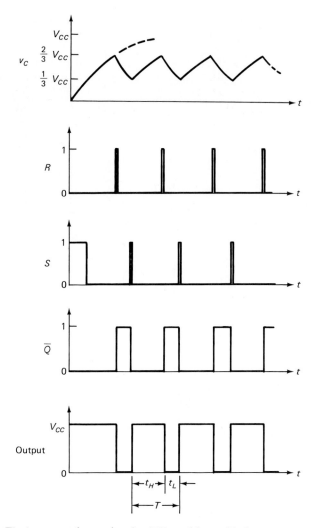

Fig. 7-40 Timing waveforms for the 555 astable multivibrator.

$$\text{PRR (or PRF)} = \frac{1}{T} = \frac{1.44}{C(R_1 + 2R_2)} \quad (7\text{-}50)$$

$$D = \frac{R_1 + R_2}{R_1 + 2R_2} \quad (7\text{-}51)$$

With rectangular waveforms or pulse trains like those produced at the output of the 555, generally either the term pulse repetition rate (PRR) or the term pulse repetition frequency (PRF) is preferred over frequency (*f*) when defining the reciprocal of the period. Like frequency, PRR and PRF have the units hertz. Notice that the value of the supply voltage has no effect on any of the output signal parameters. This is a very desirable characteristic. The 555 can be operated over a supply-voltage range of from 4.5 to 18 V.

EXAMPLE 7-7

The 555 timer in Fig. 7-41 is to produce an output waveform with a pulse repetition rate of 5 kHz. Determine the required value for R_1.

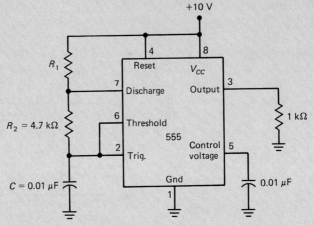

Fig. 7-41 Circuit for Example 7-7.

Solution

Equation (7-50) must be solved for R_1, producing

$$R_1 = \frac{1.44}{(PRR)(C)} - 2R_2$$

$$= 19.4 \text{ k}\Omega$$

In the circuit of Fig. 7-41, a 0.01-μF capacitor was connected from the control voltage input to ground. The capacitor may not be required in all circumstances; however, it makes the 555 more immune to noise.

The 555 has a maximum oscillating frequency of approximately 500 kHz. If a large electrolytic timing capacitor is used with the 555, periods of several hours are possible, with an upper limit determined by the leakage of the capacitor.

The 555 as a Voltage-Controlled Oscillator

In various astable configurations, the application of a control voltage to pin 5 of the 555 allows either the pulse width (duration) or the pulse position of the timer to be modulated. These types of pulse modulation are abbreviated PWM and PPM, respectively, and two 555 configurations that will produce such modulation are shown in Fig. 7-42a and b. In these circuits, an analog signal is applied to the control voltage input, causing a proportional variation in some particular pulse parameter. Representative input signals and output

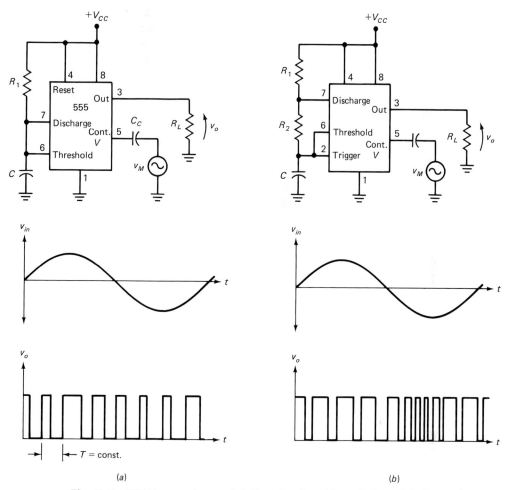

Fig. 7-42 555 timer pulse modulation circuits with typical modulating and output waveforms. (*a*) PWM modulator; (*b*) PPM modulator.

pulse trains for the PWM and PPM circuits are also shown in Fig. 7-42. Pulse modulation is widely used in the transmission of digital data over fiber optic cables. In such applications, the modulator drives a light source, such as an LED or a laser diode. A more sophisticated form of modulation called *pulse code modulation* (PCM) is also commonly used in such data communications applications. The principles of PCM are beyond the scope of this discussion.

7-7 ♦ PHASE-LOCKED LOOPS

Linear integrated circuit technology has been of fundamental importance in the widespread application of phase-locked loop (PLL) circuitry. PLLs are frequently used in digital and analog communication and control system applications. This section presents the basic principles of operation of PLLs and their typical applications.

Phase-Locked Loop Operation

A phase-locked loop operates in a manner such that the frequency and phase of the output of a voltage-controlled oscillator are synchronized with a second reference signal. The major subsections that comprise the typical PLL are shown in Fig. 7-43.

The operation of the PLL can be described in the following manner. Let us assume that initially, a reference signal is not present. Under these conditions, the VCO will operate at what is called its *free-running frequency* ω_o. Thus, the PLL is said to be operating in the free-running mode. The free-running frequency of the VCO is often adjustable, for reasons that are explained later. Now, let us assume that a reference signal is applied to the PLL.

If the applied reference signal is within a certain range of frequencies centered about ω_o, the PLL will begin to track the signal. This is called the phase-locked or tracking mode of operation. The range of frequencies through which the PLL will make the transition from the free-running mode to the tracking mode is called the capture range. Once in the capture mode, a difference in phase between the reference signal and the VCO signal will generate a voltage that is proportional to the phase difference (phase error) between the two signals. This error voltage is applied to the control input of the VCO, which in turn forces the frequency of the oscillator to match that of the reference signal. Assuming that no phase shift is introduced in the feedback path, the VCO-generated feedback signal will lead the reference signal by 90° once phase lock has occurred, for most PLL designs.

Once the PLL is locked onto the reference signal, it will track that signal until the reference signal exceeds the limits of what is called the PLL's lock range. The lock range of a PLL must be the same as or wider than the capture range for that particular PLL. Usually, the lock range will be wider. A graphical representation of the PLL's operation as described above is shown in Fig. 7-44, where the VCO control voltage versus the reference signal frequency is plotted. Typically, the lock range of a PLL may range from ±1 percent to

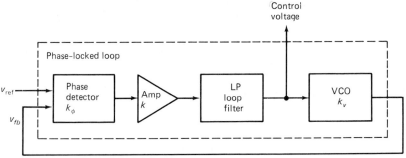

Fig. 7-43 Block diagram for the basic phase-locked loop.

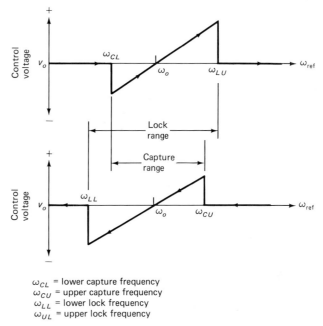

ω_{CL} = lower capture frequency
ω_{CU} = upper capture frequency
ω_{LL} = lower lock frequency
ω_{UL} = upper lock frequency

Fig. 7-44 Conceptual diagrams showing PLL capture and lock ranges.

±60 percent of the VCO free-running frequency, depending on the particular device being used. The various sections that comprise the PLL will now be described.

The Phase Comparator

As stated previously, the phase comparator, or phase detector, must produce an output voltage that is proportional to the difference in phase between two input signals. One of the simplest types of phase comparators used is the exclusive OR (XOR) gate. The logic symbol for a two-input XOR gate and its truth table are shown in Fig. 7-45. The important thing to notice here is that the output of the XOR gate is high only when the two inputs are logically opposite. This characteristic makes the XOR gate well suited for use as a phase detector.

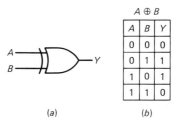

Fig. 7-45 The XOR gate. (*a*) Logic symbol; (*b*) truth table.

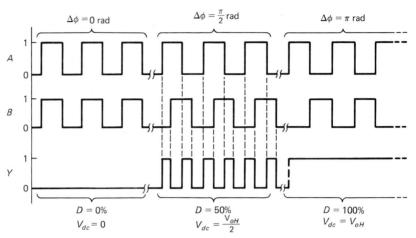

Fig. 7-46 XOR phase comparator input and output voltage waveforms.

The output of an XOR phase detector is illustrated for varying degrees of phase shift between input signals of the same frequency in the timing diagram of Fig. 7-46. When the input signals are in phase, the output is continuously low. This is equivalent to zero duty cycle. For a $\pi/2$ rad (90°) phase shift between the inputs, the output will be a 50 percent duty cycle pulse train of twice the frequency of the input signals. The average dc voltage produced under these conditions is 1/2 of the logic high output voltage. When the input signals differ in phase by π rad (180°), the output of the XOR gate is high continuously, which is equivalent to 100 percent duty cycle. It is clear that it is the XOR phase detector's output duty cycle, and hence its average output voltage, that is proportional to the phase difference.

A plot of the average output voltage versus phase difference for the XOR detector is shown in Fig. 7-47. This type of phase detector will operate correctly only if the phase difference is in the range from 0 to π rad. The slope of the transfer characteristic curve is used to define the conversion gain k_ϕ of the phase detector. Usually, the units volts per radian are used with this parameter.

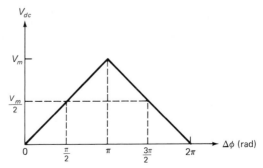

Fig. 7-47 Plot of XOR phase comparator conversion gain versus input phase difference.

There are several other types of phase detectors that are used. Some provide higher conversion gain, a wider phase response range, or both. Alternative phase comparator designs may utilize edge-triggered flip-flops or linear circuits, such as balanced modulators. Phase detectors that are designed around digital devices generally require rectangular pulse-type inputs, while linear phase detectors may operate with either digital or continuously varying analog signals. The use of a Schmitt trigger would allow an analog signal such as a sine wave to be squared up for application to a digital phase detector.

EXAMPLE 7-8

A 74LS86 quad XOR gate is to be used to build a phase detector. Given $V_{oL} = 0$ V and $V_{oH} = 4.5$ V, determine the conversion gain of the circuit in volts per radian and determine V_{dc} for an input phase differential of 2.36 rad.

Solution

The XOR phase detector produces zero output voltage when $\Delta\phi = 0$ rad and reaches maximum average output voltage ($V_{dc} = V_{oH}$) when $\phi = \pi$ rad. So, for the value of V_{oH} given,

$$k_\phi = \frac{\Delta V_{dc}}{\Delta\phi}$$

$$= \frac{4.5 \text{ V}}{\pi \text{ rad}}$$

$$= 1.43 \text{ V/rad}$$

An input phase difference of 2.36 rad produces

$$V_{dc} = k_\phi \Delta\phi$$

$$= 1.43 \text{ V/rad} \times 2.36 \text{ rad}$$

$$= 3.37 \text{ V}$$

It should be mentioned that in Example 7-8, the XOR gate will produce an average output voltage of 3.37 V regardless of whether the phase of the B input signal leads or lags that of the A input by 2.36 rad. In some instances, this characteristic limits the usefulness of this type of phase comparator.

It can be shown mathematically that the output of an XOR phase comparator is related to the product of the two input signals. In the most general sense, this is true of all phase comparator circuits. Because of the multiplicative nature of the phase comparator, when a PLL is in phase lock (the reference and feedback signals are of the same frequency), and the output of the phase comparator will consist of a dc component that is proportional to the cosine of the phase difference between the two input signals and an ac component that is twice the frequency of the input signals. The dc output component is used to control the VCO, and this is where the loop filter comes into play.

The Loop Filter

The function of the loop filter is to remove or greatly attenuate high-frequency components that are present in the output of the phase comparator, providing the VCO with a control voltage that is proportional to the instantaneous phase difference between the two signals present at the phase comparator's input.

Aside from removing high-frequency components from the output of the phase comparator, the characteristics of the loop filter play a major part in the determination of the dynamic behavior of the PLL. That is, the loop filter determines how quickly the PLL can track a changing reference frequency. In addition, when an XOR phase comparator is used, the loop filter also determines the width of the capture range.

One of the most commonly implemented types of loop filter is the first-order *RC* low-pass section, as shown in Fig. 7-48a. The low-pass network shown in Fig. 7-48b is referred to as a *lag-lead network*. The actual type of

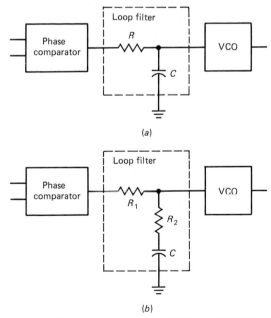

Fig. 7-48 Typical loop filter configurations. (a) Simple *RC* LP section; (b) lag-lead network.

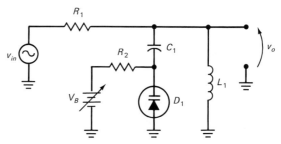

Fig. 7-49 Voltage-controlled parallel tuned circuit using a varactor diode.

loop filter and the component values that would be used with a given PLL may usually be determined based on information presented in the device data sheet. An example of a typical loop filter design is presented in the next section, where a specific PLL is discussed.

The VCO

There are many variations used in the design of the VCO. For example, a circuit similar to the 555 timer could be used as a voltage-controlled oscillator. For higher-frequency operation (into the hundreds of megahertz), varactor-diode-tuned oscillators are usually used. Varactor diodes are special PN junction diodes that are designed to exhibit relatively large, well-defined junction capacitances. The magnitude of the reverse bias applied to the varactor controls the depletion region width and hence the junction capacitance C_J. A voltage-controlled tuned circuit is shown in Fig. 7-49. In this circuit, C_1 is used to isolate the bias voltage source from the tank circuit. If $C_1 \gg C_{J(max)}$, then the resonant frequency of the tank circuit is

$$f_o = \frac{1}{2\pi\sqrt{LC_J}}$$

Resistor R_2 will usually be several megohms in value, in order to prevent loading of the tank and a reduction in Q. A tuned circuit such as this would be used to form an LC oscillator. In a PLL, the diode bias voltage would be produced by the phase comparator and the loop filter.

Although the VCO is usually an integral part of the PLL IC, there are commercially available monolithic VCO chips. An example of such a device is the Motorola MC1648.

PLL Applications

PLLs are used in many areas of electronics. Here, we shall briefly examine several popular PLL applications. Many other applications for PLLs also exist, however, and the reader is encouraged to explore the literature for further information.

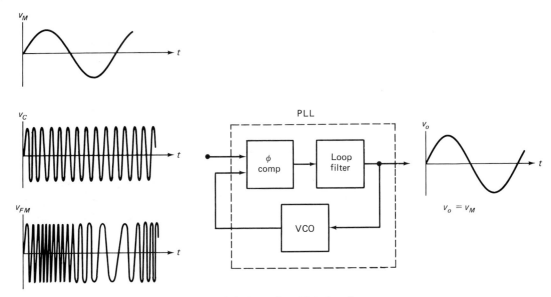

Fig. 7-50 PLL demodulation of an FM signal.

FM Demodulation

One of the most familiar PLL applications is the demodulation of frequency-modulated (FM) signals. In the generation of FM, a high-frequency carrier signal is shifted (modulated) in frequency in proportion to the instantaneous amplitude of a lower-frequency modulating signal.

The waveforms shown at the left side of Fig. 7-50 illustrate how an FM signal is derived. As usual, the inputs to the modulator are designated v_M and v_C, while the FM signal is v_{FM}. The v_{FM} signal is applied to one of the phase comparator inputs. In order for the PLL to lock onto the signal, the VCO free-running frequency must be adjusted such that the FM carrier frequency is within the capture range of the PLL. Once phase lock occurs, the frequency shift of the FM signal causes the VCO control voltage to vary in such a way that the VCO tracks the input. Because the FM signal varies in frequency in proportion to the modulating signal amplitude, the VCO control voltage must vary in a like manner; thus the loop filter output will be a replica of the original modulating signal.

Synchronous AM Demodulation

The PLL may also be used to demodulate amplitude-modulated signals. This application is shown in Fig. 7-51. In this application, the PLL is used to produce an unmodulated version of the incoming carrier frequency component. The phase-shift circuit (Hilbert transformer) is required because the PLL's VCO output will lead the input by 90°. Thus, the phase shifter returns the phase of the VCO to that of the incoming carrier signal. The AM signal

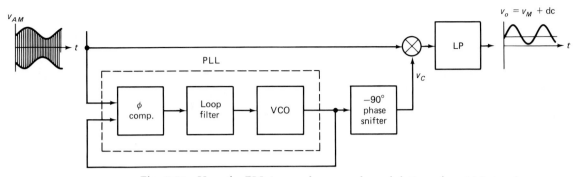

Fig. 7-51 Use of a PLL in synchronous demodulation of an AM signal.

and the VCO output are multiplied together (possibly by a balanced modulator), which produces an output that is a replica of the original modulating signal. The mathematical analysis of the operation of this circuit is left as an exercise for the reader.

Phase-Shifting Applications

Both the preceding application and the balanced modulator SSB-SC application required the use of a phase-shifting circuit. Sometimes simple *RC* sections can be used as phase shifters. However, such circuits produce a phase shift that varies with frequency. Also, because *RC* sections are actually filters, the attenuation of such a circuit will also vary with frequency. These characteristics make *RC* phase-shifting networks impractical for use over a wide range of frequencies. The PLL provides a solution to these problems.

The basic PLL circuit of Fig. 7-52 will produce an output signal that leads the input in phase by 90°. This phase shift will be constant as long as the PLL remains locked onto the input signal.

Frequency Synthesis

Many communication systems applications require that several different, very accurately controlled signal frequencies be available for transmission and reception on different channels. A familiar example is a 40-channel CB transceiver. Prior to the advent of PLLs, about the only way to meet such

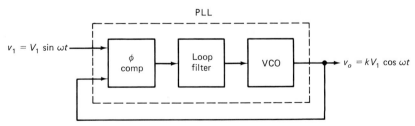

Fig. 7-52 PLL phase shifter.

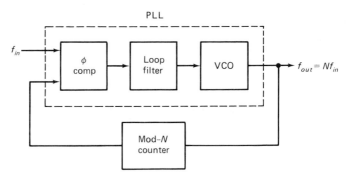

Fig. 7-53 Frequency synthesizer using a PLL and mod-N counter.

requirements would be to use (in the case of the 40-channel CB) 40 different crystal oscillators, each tuned to a specific frequency. Obviously, this would be a rather expensive proposition. Fortunately, the PLL provides a simple solution to this problem.

The block diagram in Fig. 7-53 illustrates the implementation of a PLL-based frequency synthesizer. The circuit works as follows. The reference input frequency is determined by a very stable signal source, such as a crystal oscillator. The output of the VCO is returned to the phase comparator via a mod-N digital counter. Now, in order for the PLL to lock onto the reference signal, the VCO must operate at N times the frequency of the reference. Thus, if f_{in} = 1 MHz and N = 10, the VCO must oscillate at 10 MHz in order for the PLL to remain in lock. If a programmable counter is used, the frequency of the VCO may be set to nearly any arbitrary multiple of the input signal frequency.

The advantages of the PLL frequency synthesizer are: (1) its output frequency will be as stable as that of the reference frequency source, and (2) it will generally be much less expensive than the multiple oscillator approach.

EXAMPLE 7-9

Refer to Fig. 7-53. Assume that the reference frequency is 250 kHz. It is desired that the VCO operate at a frequency of 6.25 MHz. Determine the required modulus of the counter.

Solution

The output frequency is given by $f_{out} = Nf_{in}$. Solving for N yields

$$N = \frac{f_{out}}{f_{in}}$$

$$= \text{mod } 25$$

Commercially Available PLLs

There are many different PLLs available in monolithic IC form. In this section, we shall briefly examine two such devices, the 565 PLL and the 567 tone decoder.

The 565 PLL

The 565 is a general-purpose PLL that is designed for operation with frequencies from 0.001 Hz up to 500 kHz. The internal block diagram for the 565 is shown in Fig. 7-54. The phase comparator used in this device is a balanced-modulator-type circuit. In most applications, pin 2 is used as the reference signal input, while pin 3 is grounded. These terminals provide access to the bases of two transistors similar to the signal inputs shown for the balanced modulator of Fig. 7-33b. An input signal applied to pin 2 should be between 10 and 1000 mV$_{rms}$ for best operation. The third phase comparator input (pin 5) terminates the VCO feedback loop. An internal amplifier provides two outputs, the demodulated output and the reference output. An external capacitor is connected from the demodulated output to the positive supply voltage. This capacitor and the internal 3.6-kΩ resistor form the loop

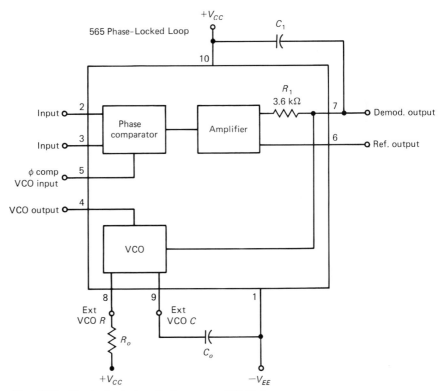

Fig. 7-54 Equivalent circuit for a type 565 PLL.

filter, which provides the control voltage for the VCO. The output of the VCO is normally connected to the third phase detector input (pin 5).

The main design equations for the 565 are as follows:

$$\text{VCO free-running frequency } f_o = \frac{1}{3.7\, R_o C_o} \qquad (7\text{-}52)$$

$$\text{Loop gain } A_{\text{loop}} = \frac{33.6 f_o}{V_{CC} + |V_{EE}|} \quad \text{Hz/V} \qquad (7\text{-}53)$$

$$\text{Tracking range} = \pm \frac{8 f_o}{V_{CC} + |V_{EE}|} \qquad (7\text{-}54)$$

The 567 Tone Decoder

The internal block diagram for the 567 tone decoder is shown in Fig. 7-55. The 567 is a PLL that is designed specifically to detect the presence of a frequency that is within its capture range. The main output of the 567 is the open collector of an internal transistor, available at pin 8. Normally, a pullup resistor or a load such as an LED would be connected to this pin. The PLL operates such that when an input signal is captured, the output transistor is driven into saturation. When the 567 is operated from a 5-V supply, and a collector pullup resistor is used (typically around 20 kΩ), the output is compatible with TTL logic devices.

Resistor R_1 and capacitor C_1 determine the free-running frequency of the current-controlled oscillator (CCO). R_1 should be in the range of 2 kΩ to 20 kΩ for best operation. Capacitor C_2 and the internal 3.9-kΩ resistor form the loop filter. The value of C_2 is chosen based on the expected input signal amplitude and the desired PLL percent bandwidth (%BW), using Eq. (7-56). %BW is the PLL's capture range, expressed as a percent of f_o. For example, if f_o = 100 kHz and BW = 10 percent, then the PLL will lock onto signals that fall within ± 10 percent of 100 kHz (frequencies from 90 kHz to 110 kHz will be captured by the PLL). Capacitor C_3 filters the output of the phase comparator that drives the output transistor. Typically, this capacitor is chosen to be about twice C_2 in value.

The major 567 design equations are as follows:

$$\text{Free-running frequency } f_o = \frac{1.1}{R_1 C_1} \qquad (7\text{-}55)$$

$$\%\text{BW} = 1070 \sqrt{\frac{V_{in(\text{rms})}}{f_o C_2}} \qquad (7\text{-}56)$$

where BW is expressed as a percentage of f_o for V_{in} < 200 mV$_{\text{rms}}$, and the value of C_2 is in microfarads.

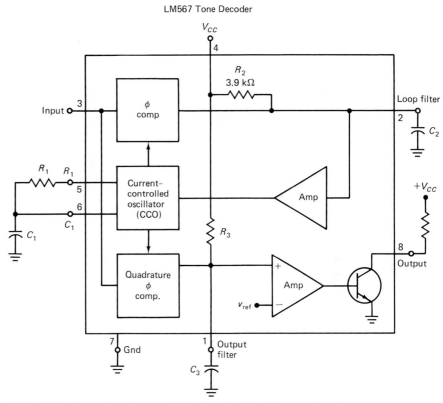

Fig. 7-55 Equivalent internal circuitry for the 567 tone decoder.

A typical application in which the 567 would be used is in the decoding of "touch-tone" encoded signals. Touch-tone encoding is a technique in which the pressing of a given button (closing of a switch or switches) on a telephone generates a signal that consists of a unique combination of two frequency components. Table 7-1 presents the frequencies that are used to represent the buttons present on the typical touch-tone telephone.

Table 7-1 TOUCH-TONE FREQUENCIES

Key	Frequencies (Hz)	Key	Frequencies (Hz)
1	697, 1209	7	852, 1209
2	697, 1336	8	852, 1336
3	697, 1477	9	852, 1477
4	770, 1209	0	941, 1336
5	770, 1336	*	941, 1209
6	770, 1477	#	941, 1477

EXAMPLE 7-10

Using 567 decoders and any required digital circuits, design a circuit that will produce a logic high output if the input to the circuit is the touch-tone signal for the digit 1. For a given 567 decoder, assume a required bandwidth of 5 percent. A 5-V power supply is to be used. The VCO timing capacitors are to be 0.1 µF in value, and $V_{in} = 100$ mV$_{rms}$.

Solution

The digit 1 is represented by simultaneous transmission of 697-Hz and 1209-Hz signals; thus, this application requires that two 567s be used, with free-running frequencies of 697 Hz and 1209 Hz. Pullup resistors must be used, and when the correct frequencies are present at the input, both outputs will be driven low. In order to generate a logic 1 output when the correct input is present, the outputs of the 567s must be NORed together. One gate of a 7402 quad NOR gate will be used here. The complete decoding circuit is shown in Fig. 7-56.

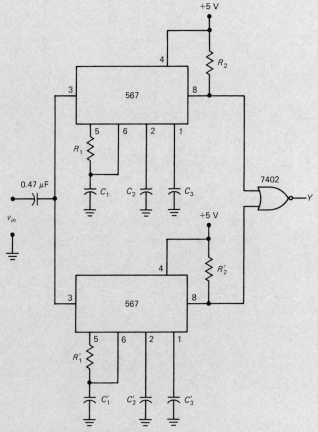

Fig. 7-56 Circuit for Example 7-10.

The required component values are determined by rearranging Eqs. (7-55) and (7-56), producing

$$R_1 = \frac{1.1}{f_o C_1} \quad \text{and} \quad C_2 = \frac{V_{in(rms)}\left(\frac{1070}{\%BW}\right)^2}{f_o}$$

Performing the necessary calculations yields

$R_1 = 15.8 \text{ k}\Omega \qquad R_1' = 8.3 \text{ k}\Omega$

$C_2 = 6.6 \text{ }\mu\text{F} \qquad C_2' = 3.8 \text{ }\mu\text{F}$

$C_3 = 15 \text{ }\mu\text{F} \qquad C_3' = 7.6 \text{ }\mu\text{F}$

where the unprimed values are for the 697-Hz section and the primed values are for the 1209-Hz section.

Chapter Review

Monolithic ICs are available that are designed to perform nearly any imaginable function in electronic circuits and systems. Many of these specialized ICs are very closely related to conventional op amps, while others are quite different.

A comparator is an amplifier that is used to signal the existence of a voltage that exceeds a certain threshold, called a setpoint or trip point. Positive feedback may be employed to the comparator, providing hysteresis. Hysteresis makes the circuit more immune to noise. Comparators may also be realized using standard op amps.

Analog switches are FET-based ICs that are designed to replace their mechanical counterparts in low-power switching applications. Analog switches are typically used to implement multiplexers, demultiplexers, and track-and-hold (T/H) circuits. Track-and-hold circuits are also available in monolithic form, and they are typically used in data acquisition applications.

Current difference amplifiers (CDAs), also called Norton and automotive op amps, are designed to be used in situations where only a unipolar supply voltage is available. CDAs require the designer to bias up the device for a given operating point. Capacitive coupling is required with the CDA, because the biasing results in a high dc potential at the amplifier's output.

Operational transconductance amplifiers (OTAs) form another class of monolithic devices. OTAs are basically voltage-controlled current sources (VCIS or VCCS). The voltage gain of the OTA is set by the amplifier bias current (I_{ABC}) and the load resistance. Because I_{ABC} is externally controlled, the OTA may be used as a voltage-controlled gain block or cell. OTAs may be used to realize linear amplification, as well as many other functions traditionally performed by op amps. OTAs are also very well suited for use in amplitude-modulation applications.

Monolithic balanced modulators are used to perform various types of amplitude-modulation functions. Most often, balanced modulators are used in radio communication applications, although they are also used in other areas of signal processing as well. The output of a balanced modulator is proportional to the product of two input signals.

The 555 timer is a widely used general-purpose monolithic multivibrator. When configured as an astable multivibrator, the 555 can produce pulse trains from less than 1 Hz to around 500 kHz. The period of oscillation of the 555 is relatively immune to supply voltage variations, and a load may be referenced either to ground or to the supply rail. Both the pulse repetition rate and the duty cycle of the 555's output are controlled by external resistors and capacitors.

Phase-locked loops (PLLs) make up another class of popular monolithic ICs. In PLL operation, a voltage-controlled oscillator (VCO) tracks the phase and frequency of an externally applied reference signal. Differences in phase between the VCO and reference signal result in an error voltage being produced, which forces the VCO to change frequency so that the phase error is eliminated. PLLs are frequently used in modulation, demodulation, frequency synthesis, signal encoding and decoding, and control system applications, as well as many others.

Questions

7-1. What type of comparator is used to indicate when a voltage is within a specific range?

7-2. Would the transfer function of a comparator generally be described as linear or nonlinear? What about the transfer function of an op amp with negative feedback?

7-3. Why are CDAs inappropriate for use in applications that require the processing of dc levels?

7-4. Why would the noninverting input terminal of a CDA be well suited for use as a summing junction?

7-5. When an FET is used as an analog switch, would it be operated in the ohmic region or the saturation region of its transfer characteristics?

7-6. Describe how an OTA could be used to implement an integrator. *Hint*: Consider that the output is effectively a current source.

7-7. What OTA parameter is affected by the modulation of the amplifier bias current?

7-8. Could a balanced modulator be used to perform the (mathematical) squaring function in an analog computer circuit?

7-9. What effect does the insertion of a mod-N counter in the feedback loop of a PLL have on the output of the VCO when the PLL is phase-locked?

7-10. What is the function of the PLL in a synchronous demodulator?

Problems

7-1. Refer to Fig. 7-3. Given $V_{ref} = 3.5$ V, $+V_{sat} = 4.75$ V, $-V_{sat} = 0$ V, $A_{OL} \to \infty$, and $v_{in} = 6 \sin 5000t$ V, determine the period and duty cycle of the comparator's output.

7-2. Refer to Fig. 7-9. Assume $+V_{sat} = 5.0$ V, $-V_{sat} = -5.0$ V, $R_1 = 15$ kΩ, and $R_2 = 10$ kΩ. Determine the upper and lower trip points and the size of the deadband in volts.

7-3. Given the circuit of Fig. 7-9 and the parameters stated in Prob. 7-2, determine the period and duty cycle of the output given $v_{in} = 5 \cos 377t$ V.

7-4. Refer to Fig. 7-9. Given $+V_{sat} = 10$ V, $-V_{sat} = -10$ V, and $R_2 = 1$ kΩ, determine the value of R_1 such that $V_{UT} = |V_{LT}| = 4$ V.

7-5. Given the values of R_1 and R_2 from Prob. 7-4, if $+V_{sat} = 10$ V and $-V_{sat} = 0$ V, determine V_{UT} and V_{LT}.

7-6. Refer to Fig. 7-57. **(a)** Determine V_{UT} and V_{LT}. **(b)** Given $v_{in} = 2t$ V, sketch the input and output voltage waveforms for the interval from $t = 0$ to 5 s. Label all pertinent times and voltages.

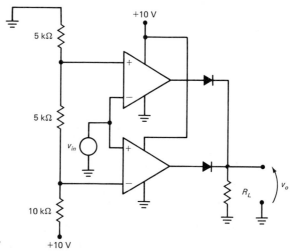

Fig. 7-57

7-7. Refer to Fig. 7-16. Assume that the analog switches are ideal ($R_{open} \to \infty$ and $R_{closed} \to 0$). Sketch the output voltage, given $V_1 = 10$ mV, $V_2 = 20$ mV, $\cdots$, $V_8 = 80$ mV, and a 3-bit binary up-counter driving the 74LS138 (starting at count = 000). $R = 1$ kΩ and $R_F = 10$ kΩ.

7-8. Refer to Fig. 7-17. The T/H circuit has the following parameters: $C_S = 50$ pF and for A_2, $R_{in} = 10$ GΩ. Assuming that at the instant the circuit enters the hold mode, $V_{CS} = 5.00$ V, determine V_o: **(a)** after 500 ms has passed, **(b)** after 1 s has passed.

7-9. Refer to Fig. 7-21. Given $V_{CC} = 12$ V, determine the required values for R_M, R_F, and R_1 such that $I_M = 50$ μA, $V_{oQ} = 6.0$ V, and $A_V = -5$.

7-10. Repeat Prob. 7-9 for the same design requirements except that $V_{oQ} = 3.0$ V.

7-11. Refer to Fig. 7-22. $V_{CC} = +18$ V, $R_1 = 2.2$ kΩ, $R_M = 150$ kΩ, and $R_F = 63.3$ kΩ. Determine I_M, A_v, and the quiescent voltage present at the amplifier's output terminal.

7-12. Based on the conditions presented in Prob. 7-11, what is the maximum unclipped peak-to-peak output voltage that could be produced by the circuit?

7-13. Refer to Fig. 7-24a. Given $V_{CC} = V_{EE} = \pm 12$ V, $R_1 = R_2 = 100$ Ω, $R_{ABC} = 50$ kΩ, and $R_L = 100$ kΩ, determine I_{ABC}, g_m, and A_v.

7-14. Given the conditions of Prob. 7-13, determine the value of R_{ABC} that will result in $A_v = -10$.

7-15. Refer to Fig. 7-24b. Given $V_{CC} = V_{EE} = \pm 10$ V and $R_1 = R_2 = 1$ kΩ, determine the required values for R_{ABC} and R_L such that $g_m = 2500$ μS and $A_v = 15$.

7-16. Refer to Fig. 7-58. Determine the trigonometric expressions for i_{abc} and v_o. Determine the modulation index of the circuit and sketch the frequency domain representation for v_o.

7-17. Determine the maximum rate of change of the output of Fig. 7-58, based on the input signals given.

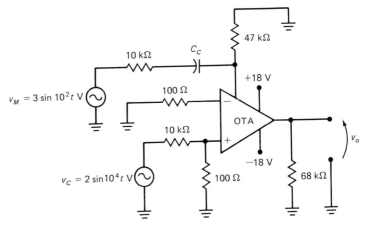

Fig. 7-58

7-18. Determine the trigonometric expressions for the output of a balanced modulator with $k = 1$ for each pair of input signals listed below.

(a) $v_1 = 1 \sin \omega t$ V, $v_2 = 1 \sin \omega t$ V

(b) $v_1 = 2 \sin \omega t$ V, $v_2 = 2 \sin (\omega t + \frac{\pi}{2} \text{ rad})$ V

(c) $v_1 = 3 \sin \omega t$ V, $v_2 = 3 \sin 2\omega t$ V

7-19. Refer to Fig. 7-37. The DSB-SC input signal is given by $v_1 = 1 \cos 9.9 \times 10^5 t - 1 \cos 1.01 \times 10^6 t$ V and the BFO output is given by $1 \sin 10^6 t$ V. Assuming that the LP filter has a brickwall response with $\omega_C = 20$ krad/s, determine the expression for v_o. (Assume that the filter contributes no phase shift.)

7-20. Refer to Fig. 7-38. Determine the trigonometric expression for v_o, assuming that the phase-shifting sections of the system each produce a phase shift for $-90°$. The input signals are $v_C = V_C \sin \omega_C t$ V and $v_M = V_M \sin \omega_M t$ V.

7-21. A 555 timer is configured for operation as an astable multivibrator with $R_1 = 2.2$ kΩ and $C = 0.022$ μF. Determine the required value of R_1 such that PRR = 5.6 kHz. Determine the duty cycle of the output.

7-22. An astable-configured 555 timer has $R_1 = R_2 = 10$ kΩ. Determine the required value for C such that PRR = 1 kHz.

7-23. A CMOS XOR phase comparator has $V_{oH} = 9$ V and $V_{oL} = 0$ V. Determine the average dc output voltage produced by the gate for the following input phase differentials.

(a) $\Delta\phi = 20°$

(b) $\Delta\phi = 30°$

(c) $\Delta\phi = \frac{\pi}{4}$ rad

(d) $\Delta\phi = \frac{3\pi}{2}$ rad

7-24. Refer to Fig. 7-49. The varactor diode junction capacitance is defined by $C_J = C_o/\sqrt{1 + 2V_R}$, where C_o is the junction capacitance when $V_R = 0$ V. Determine f_o, assuming that $C_1 \gg C_J$, $C_o = 20$ pF, $L_1 = 25$ μH, and $V_B = -1.0$ V.

7-25. Refer to Fig. 7-49. Given $L_1 = 68$ μH and $C_o = 300$ pF, determine the bias voltage required such that $f_o = 4.3$ MHz.

7-26. Refer to Fig. 7-51. Given $v_{AM} = V_C \sin \omega_C t + V_x \cos (\omega_C - \omega_M)t - V_x \cos (\omega_C + \omega_M)t$ V, determine the expression for v_o. Assume that $k = 1$ for the multiplier.

7-27. Refer to Fig. 7-59. Determine the component values required for $f_o = 200$ kHz. Determine A_{loop} and the lock range.

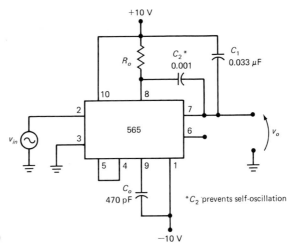

Fig. 7-59

7-28. Refer to Fig. 7-60. This circuit is an example of a servomechanism. The PLL, the external VCO, and the power amplifier provide automatic regulation of the speed of the motor, as set by the command (speed) potentiometer. The dc motor turns a chopper wheel that has 60 equally spaced holes around its circumference. As the motor turns the chopper, the interruption of the LED emission causes voltage pulses to be produced at the collector of a phototransistor. Thus, the feedback signal pulse repetition rate is directly proportional to the speed of the motor. Assuming that the external VCO is adjusted to oscillate at 500 Hz, determine the rotational speed of the motor in rpm.

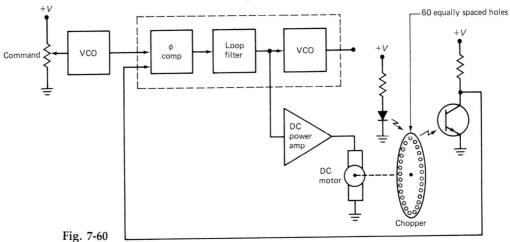

Fig. 7-60

7-29. Given a 567 tone decoder with $C_1 = 0.033$ μF, determine the required values of the remaining components such that $f_o = 7.35$ kHz and BW = 10 percent. Assume that $V_{in} = 100$ mV$_{\text{rms}}$.

CHAPTER 8
DATA CONVERSION DEVICES

Digital computers are finding use in more and more areas of electronic technology every day. The advent of powerful, yet low-cost microprocessors and related support devices has been at the forefront of this trend. Microcomputers are routinely being used to perform many signal processing and system control tasks that until recently had been solely the domain of linear circuits.

Digital devices and systems are designed to work with data in a discrete form, that is, in the form of voltage levels that represent binary numbers. However, the world outside the computer itself is primarily an analog one, as most natural phenomena are continuous in nature. Conversely, digital control of a continuous device, such as a motor, requires the conversion of discrete data into an analog form, such as a voltage or current. In this chapter, we shall discuss the operation, design, and limitations of digital-to-analog (D/A) and analog-to-digital (A/D) converters, which are used to provide an interface between the discrete world of digital systems and the continuous world within which they are used.

8-1 ♦ D/A CONVERSION FUNDAMENTALS

The function of a D/A converter, or DAC, is to accept a group of bits from a computer or some other digital device and convert that bit pattern into an equivalent analog voltage level. Generally, the bit pattern presented to the D/A converter is interpreted as being a binary number.

The output of the D/A converter should be able to assume a different level for each unique digital input that is applied. The output of a D/A converter may be either a voltage or a current, depending on the internal construction of the device. Actual D/A circuit operation is discussed later in this section, but before reaching this level, we must first lay a little groundwork.

Resolution and Full-Scale Output

The number of different output levels that can be produced by a given D/A converter is related to the number of input lines that the converter has by the following relation:

$$N = 2^n \tag{8-1}$$

where N is the number of different output levels that can be produced and n is the number of input bits that the converter has.

EXAMPLE 8-1

The D/A converter of Fig. 8-1 accepts 10-bit binary numbers as input. Determine the number of different output levels that it should be possible for this device to produce.

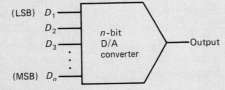

Fig. 8-1 Symbol for an n-bit D/A converter.

Solution

Direct application of Eq. (8-1) yields

$$N = 2^{10}$$
$$= 1024$$

The number of different levels that can be produced at the output of a D/A converter is used to define the resolution of that device. The more inputs a D/A converter has, the higher (or finer) its resolution will be. Resolution may be expressed as 1 part in N, where N is defined by Eq. (8-1). For example, the D/A converter of Example 8-1 could be said to have a resolution of 1 part in 1024. Resolution may also be expressed as a percentage. The following equation quantifies the relationship between resolution and the number of inputs the D/A converter has.

$$\text{Percent resolution} = 1/2^n \times 100\% \tag{8-2}$$

where, again, n is the number of bits in the input word.

Applying Eq. (8-2) to the converter of Example 8-1, it may be determined that the 10-bit D/A converter has a resolution of 0.098 percent. The practical

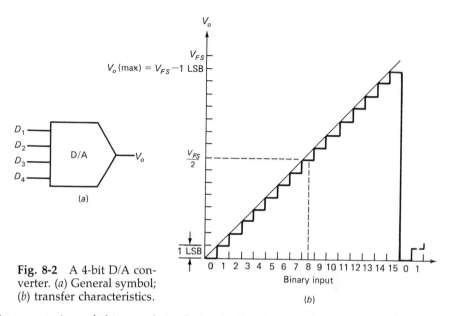

Fig. 8-2 A 4-bit D/A converter. (*a*) General symbol; (*b*) transfer characteristics.

interpretation of this result is that, ideally, the smallest possible change in the output of the converter that can be caused by a change of the input will be 0.098 percent of the full-scale output.

Full-scale output is the voltage (V_{FS}) or current (I_{FS}) level produced at the output of a hypothetical D/A converter that has infinite resolution (infinitely many inputs), with binary 1's applied to all inputs. Since a real D/A converter obviously cannot have infinitely many inputs, its output can never reach the ideal full-scale output level. For example, consider the 4-bit D/A converter of Fig. 8-2*a*. Starting at zero (all inputs low), if we apply successively higher 4-bit binary codes to the inputs, the stair-step transfer curve of Fig. 8-2*b* is produced. Notice that there are 16 distinct output voltage levels or steps possible (including 0 V), and there are 15 risers. In order to achieve full-scale output, 17 (which is not a power of 2) steps would be required. Thus we can see from this illustration that at maximum V_o, the D/A converter will fall short of V_{FS} by one increment, or step. The smallest change in the output that can be caused by the inputs is 1 LSB, or step. This change is termed 1 LSB because it occurs when the least significant bit of the input changes states.

Ideally, the output LSB size is constant. In Fig. 8-2*b*, this means that all risers are of equal height. Thus, the increase in the output for each of the steps is equal, and is determined by the number of steps (the resolution) and the full-scale voltage V_{FS} through the following relationship:

$$\text{Step size} = 1 \text{ LSB} = \frac{V_{FS}}{2^n} \tag{8-3}$$

where n is the number of binary inputs to the converter and V_{FS} is the full-scale voltage of an ideal equivalent D/A converter.

Using Eq. (8-3) it is possible to determine the D/A converter's output voltage (or possibly current) for a given binary input.

EXAMPLE 8-2

The 4-bit D/A converter of Fig. 8-2 has $V_{FS} = 10.00$ V. Determine V_o given the following binary inputs:

(a) 0001_2 (1_{10})

(b) 0100_2 (4_{10})

(c) 1000_2 (8_{10})

(d) 1111_2 (15_{10})

Solution

The LSB (step) size is calculated using Eq. (8-3), giving

$$1 \text{ LSB} = \frac{V_{FS}}{2^n}$$

$$= \frac{10 \text{ V}}{16}$$

$$= 0.625 \text{ V}$$

The output voltage for a given input word is found by multiplying the LSB voltage by the numerical value of that particular word. This procedure yields

(a) 1×0.625 V $= 0.625$ V

(b) 4×0.625 V $= 2.500$ V

(c) 8×0.625 V $= 5.000$ V

(d) 15×0.625 V $= 9.375$ V

Examination of the results of the preceding example shows that the output of the D/A converter is $V_{FS}/2$ when the MSB of binary input is high and the remainder of the input bits are low. Also notice that when the numerical value of the input is increased or decreased, the output voltage increases or decreases (ideally) in direct proportion. The resolution of a D/A converter may be used as a general indicator of its potential for accuracy. In a more conventional sense, resolution is analogous to precision.

Accuracy

Accuracy and resolution are not the same thing. For example, a 16-bit D/A converter would generally be considered to have high resolution (1 part in 65,536), but it is not necessarily true that the output is an accurate representation of a given input. Under normal conditions, the output of the D/A converter should be accurate to within $\pm\frac{1}{2}$ LSB (some converters are accurate to $\pm\frac{1}{4}$ LSB). However, there are many possible sources of error in the typical D/A converter circuit that can cause further deviation from the ideal. Rather than attempting to list these causes of errors, the effects of the various errors on the output of the converter will be summarized. Figure 8-3 illustrates the effects of the various errors on the transfer characteristics of a D/A converter with infinite resolution. Notice that the transfer function of a perfect D/A converter is exactly linear and is continuous (it has infinitely many infinitesimal steps). Of course, in practice such ideal characteristics cannot be realized.

The absolute error of the output of the D/A converter is given by the following equation:

$$e_{abs} = Y - X \tag{8-4}$$

where Y is the expected output and X is the actual output. The output error may also be expressed in relative terms, using

$$e_{rel} = \frac{e_{abs}}{Y} \tag{8-5}$$

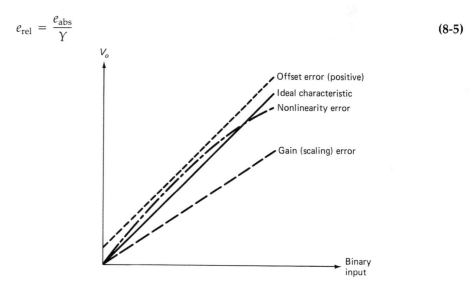

Fig. 8-3 The effects of various errors on the transfer characteristics of a D/A converter with infinite resolution.

The various errors (other than those inherent in finite resolution) that can affect the accuracy of a D/A converter will now be discussed.

Offset Error

An offset error causes the output of the converter not to be zero when the binary input is zero. This results in a constant shift of the output either above or below the expected value over the entire range of binary inputs. If offset is the only type of error present, the absolute output error will be constant regardless of the input. However, the relative error at the output decreases as the binary input is increased.

Gain Error

A gain error will produce output step sizes that are either larger or smaller than the expected LSB. Gain error causes the output to deviate further from the expected value as the binary input is increased. Thus the absolute error increases as the input increases. Gain error is also sometimes referred to as scaling error.

Linearity Error

This type of error is caused by nonlinearities inherent in the D/A converter circuitry. Nonlinearity results in the transfer characteristic of a perfect D/A converter not being a straight line; that is, the output is not a linear function of the binary input. Of course, real D/A converters are inherently nonlinear as a result of finite resolution, which causes the discrete stair-step output. Given this inherent nonlinearity, the effect of further nonlinearity on such a converter is a varying LSB step size. That is, the step size may increase or decrease as the binary input changes. Temperature fluctuations and gradients caused by internal and external influences are a major cause of nonlinearity. Figure 8-4 illustrates the effect of nonlinearity on the output of a 3-bit D/A converter. In this example, the gain of the circuit increases, for some unknown reason, as the input is increased.

Settling Time

The ideal D/A converter would respond instantaneously to an input. A practical D/A converter requires a finite amount of time to change output states and settle to within some well-defined limit of the final output state. This time interval is specified as the *settling time* of the converter. Typically, settling time is defined as the time required for the converter to come within $\pm \frac{1}{2}$ LSB (or some percentage, such as 99.5 percent of the final output level). Under worst-case conditions, maximum settling time is required when the converter output makes a transition from minimum to maximum or vice versa. Figure 8-5 illustrates the concept of settling time. Here, the final value

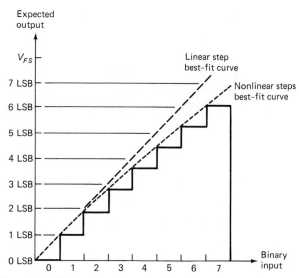

Fig. 8-4 Effect of nonlinearity (gain decreasing as the input code increases) in the transfer characteristics of a 4-bit D/A converter.

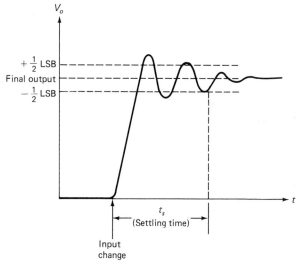

Fig. 8-5 Determination of D/A settling time.

is assumed to be $\pm \frac{1}{2}$ LSB. The presence of overshoot indicates that, overall, the system is underdamped.

Settling time limits the rate at which the D/A can make successive conversions. For example, consider a certain D/A converter that has a worst-case settling time of 1 ms. In order to obtain a meaningful output under all possible changes of the input word, the value of the binary input must not change states at a rate greater than once every millisecond. In practice, an even longer interval between successive inputs would probably be used.

EXAMPLE 8-3

A certain 5-bit D/A converter has $V_{FS} = 10.00$ V, with a guaranteed accuracy of $\pm\frac{1}{2}$ LSB. Determine the expected (ideal) output, the upper and lower limits of the possible output, and the percent error of the output at each extreme for the following binary inputs: **(a)** 0001, **(b)** 0010, **(c)** 1000.

Solution

The LSB size is first calculated, yielding

1 LSB = 312.50 mV

A given output cannot differ from the expected output by more than $\pm\frac{1}{2}$ LSB, where $\frac{1}{2}$ LSB = 156.25 mV. Using this information and Eq. (8-5), the following results are obtained.

(a) Expected output = 312.50 mV
   Upper output limit = 468.75 mV
   Lower output limit = 156.25 mV
   Maximum output error = ± 50 percent

(b) Expected output = 625.00 mV
   Upper output limit = 781.25 mV
   Lower output limit = 468.75 mV
   Maximum output error = ± 25 percent

(c) Expected output = 5.000 V
   Upper output limit = 5.156 V
   Lower output limit = 4.844 V
   Maximum output error = ± 3.12 percent

The preceding example shows that the relative error at the output of a D/A converter decreases as the input increases. It can be shown that for a 5-bit converter with the error specifications presented in the example, with maximum input (11111), the output error will be ± 1.61 percent. Of course, this assumes that the maximum absolute error is within $\pm\frac{1}{2}$ LSB.

In relative (percent error) terms, the effect of an offset on the output of a converter will be greatest (actually infinite) when all inputs are zero. Output offset error is constant in absolute terms.

A gain error (with no additional types of error present) will result in a constant relative or percent error over the entire range of the output. However, the absolute output error increases from zero with zero input, in proportion to the input. Both gain and offset errors can usually be minimized rather easily.

Nonlinearity errors generally have an unpredictable effect on the output of the converter, and are most difficult to eliminate. Generally, all types of errors will exist to some degree in an actual circuit.

8-2 ♦ D/A CONVERSION CIRCUITS

Digital-to-analog converters are available in monolithic IC form, and they may also be designed using operational amplifier circuits. This section examines the performance, complexity, and cost tradeoffs of the various implementations of D/A converters.

Weighted Resistor Summing D/A

One of the simplest types of D/A converters to analyze and design is the binary-weighted resistor summing-amp D/A converter. This is quite a name for such a simple (relatively speaking) circuit. As the name implies, a summing amplifier forms the heart of this type of converter. A 4-bit summing D/A converter is shown in Fig. 8-6. The input resistors are chosen such that, proceeding from the LSB (S_1), each successive input is amplified two times as much as its predecessor. Thus, the gains relative to the inputs are related by powers of 2. Application of the usual op amp analysis methods to the summing D/A converter yields the following generalized output voltage expression:

$$V_o = -V_{ref}\left(\frac{D_1 R_F}{R_1} + \frac{D_2 R_F}{R_2} + \cdots + \frac{D_n R_F}{R_n}\right) \qquad (8\text{-}6)$$

where $D_1, D_2, \cdots, D_n$ represent the states of the switches (0 = open and 1 = closed).

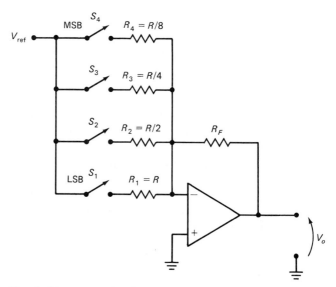

Fig. 8-6 Simple binary-weighted resistor summing-amp D/A converter.

EXAMPLE 8-4

The circuit of Fig. 8-6 has the following component values: $R = 80\ \text{k}\Omega$ and $R_F = 8\ \text{k}\Omega$. Assuming that $V_{\text{ref}} = -5.00\ \text{V}$, determine the LSB size, $V_{o(\text{max})}$, and V_{FS}.

Solution

The op amp will produce $V_o = 1\ \text{LSB}$ when S_1 is closed and the remaining switches are open. Applying Eq. (8-6) produces

$$V_o = 1\ \text{LSB} = -V_{\text{ref}}\frac{R_F}{R_1}$$

$$= (5\ \text{V})(0.1)$$

$$= 0.5\ \text{V}$$

The maximum output (produced when all switches are closed) may be determined in several different ways. Using Eq. (8-6) again, we obtain

$$V_{o(\text{max})} = -V_{\text{ref}}\left(\frac{8\ \text{k}\Omega}{80\ \text{k}\Omega} + \frac{8\ \text{k}\Omega}{40\ \text{k}\Omega} + \frac{8\ \text{k}\Omega}{20\ \text{k}\Omega} + \frac{8\ \text{k}\Omega}{10\ \text{k}\Omega}\right)$$

$$= 5\ \text{V}(0.8 + 0.4 + 0.2 + 0.1)$$

$$= (5)(1.5)$$

$$= 7.5\ \text{V}$$

Alternatively, we know that if the input is 1000, the output will be $V_{FS}/2$ V, and the maximum output is 1 LSB less than the full-scale output. Using this knowledge, we can determine both V_{FS} and $V_{o(\text{max})}$ as follows:

$$\frac{V_{FS}}{2} = -V_{\text{ref}}\frac{8\ \text{k}\Omega}{10\ \text{k}\Omega}$$

$$= (5)(0.8)$$

$$= 4.0\ \text{V}$$

Therefore

$$V_{FS} = (2)(4.0\ \text{V})$$

$$= 8.0\ \text{V}$$

and

$$V_{o(\text{max})} = V_{FS} - 1\ \text{LSB}$$

$$= 8.0\ \text{V} - 0.5\ \text{V}$$

$$= 7.5\ \text{V}$$

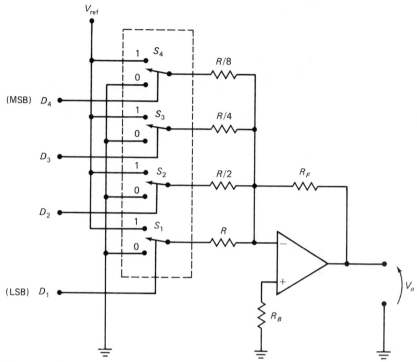

Fig. 8-7 Summing amp D/A converter modified to reduce offset errors.

There are several problems associated with the summing D/A converter of Fig. 8-6. First of all, because the resistance seen by the inverting input terminal changes as the input switches are set for various input words, a fixed bias current compensation resistor on the noninverting input will not completely cancel the resulting offset component. This error becomes more critical as the resolution of the converter is increased by summing more inputs. Second, open inputs are more susceptible to noise pickup.

Both of the previously mentioned problems can be solved by using the circuit of Fig. 8-7. In this circuit, the SPST switches have been replaced with SPDT units. Now, when a given bit is low, the corresponding input resistor is connected to ground. This provides the op amp with a constant effective inverting input resistance, allowing accurate selection of R_B, and it also helps reduce noise pickup problems. Equation (8-6) applies to this circuit as well as to that of Fig. 8-6. The switches used to select a given input word may be mechanical—BCD thumbwheel switches, for example—but if the converter is to be controlled directly by a digital circuit, analog switches would be used. Analog switches offer the advantage of high switching speeds and direct digital logic compatibility.

The circuits of both Figs. 8-6 and 8-7 are impractical for high-resolution applications. The main reason for this statement is the fact that a wide range of input resistor values must be used. For example, if we were to expand Fig. 8-7 to form an 8-bit converter, the input resistors would span a 128 to 1 ratio. It is extremely difficult to produce resistors with such radically differing values that will exhibit similar enough temperature tracking characteristics. In addition, obtaining or fabricating this wide assortment of precision resistors is also somewhat expensive and difficult, especially using IC design techniques.

The R-2R Ladder D/A Converter

One of the most popular D/A converter designs is based on the use of what is called an R-2R ladder. A 4-bit R-2R ladder D/A is shown in Fig. 8-8, from which it is obvious why the term *ladder* is used in the name. The advantage of this circuit over its summing amplifier counterpart is that only two different precision resistor values are required for any number of binary inputs. Also, because the resistors differ by only a factor of 2, they can be made to exhibit similar temperature tracking characteristics.

A straightforward, but somewhat tedious analysis of the ladder yields the following expression for the output of the 4-bit R-2R ladder converter:

$$V_o = (V_{\text{ref}}) \left(1 + \frac{R_F}{R_1}\right)\left(\frac{D_4}{2} + \frac{D_3}{4} + \frac{D_2}{8} + \frac{D_1}{16}\right) \tag{8-7}$$

We now have what appears to be a converter that can be expanded to nearly any arbitrary resolution. In terms of the input resistor network this

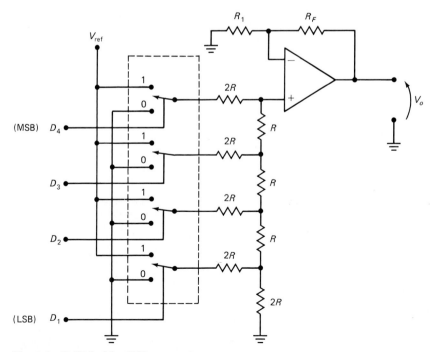

Fig. 8-8 R-2R ladder D/A converter.

may be true; however, increasing resolution requires greater accuracy from the op amp. Consider a 4-bit D/A converter with V_{FS} = 10.00 V. In order to maintain $\pm\frac{1}{2}$-LSB accuracy, the output must be within 312.5 mV of the ideal expected value. Such accuracy is well within the capabilities of the average op amp. Now, consider a 16-bit D/A with the same rated accuracy and full-scale output voltage. The output of such a circuit would have to be within 76.29 μV of the ideal expected output. Such accuracy requires the use of an extremely high-performance (expensive) op amp, and careful compensation for environmental influences. These are important considerations in any D/A converter circuit.

EXAMPLE 8-5

Refer to Fig. 8-8. The R-2R ladder consists of 10-kΩ and 20-kΩ resistors, V_{ref} = 2.00 V, and R_1 = 10 kΩ. Determine the value required for R_F such that V_{FS} = 10.00 V.

Solution

Possibly the most straightforward way to solve this problem is to set the MSB high (which produces $V_o = V_{FS}/2$ = 5.00 V) and solve Eq. (8-7) for R_F. Keep in mind that all terms of the input factor except the first will be zero, which simplifies the expression substantially. We may write

$$V_o = \frac{V_{FS}}{2} = (V_{ref})\left(1 + \frac{R_F}{R_1}\right)\left(\frac{1}{2}\right)$$

Multiplying by 2,

$$V_{FS} = (V_{ref})\left(1 + \frac{R_F}{R_1}\right)$$

$$\frac{V_{FS}}{V_{ref}} = 1 + \frac{R_F}{R_1}$$

$$R_F = \frac{R_1 V_{FS}}{V_{ref}} - R_1$$

Now, substituting the known values produces

$$R_F = \frac{10 \text{ k}\Omega \times 10 \text{ V}}{2 \text{ V}} - 10 \text{ k}\Omega$$

$$= 40 \text{ k}\Omega$$

Monolithic D/A Converters

Typical D/A converter applications require at least 8-bit resolution. While the op amps used in the previous designs are relatively inexpensive and small, the use of discrete resistors in the construction of D/A circuits tends to use up precious circuit board real estate. Also, significant temperature gradients tend to exist between such discrete resistors, which can result in offset and drift problems. For these reasons, monolithic and hybrid IC D/A converters are most frequently used in practice.

Hybrid converters are generally quite expensive, and are used in the most exacting applications. Monolithic D/A converters are relatively inexpensive, offer very good performance, and are available in a wide variety of resolutions. Typically, monolithic D/A converters are available with 6-, 8-, 10-, and 12-bit resolution. As we shall see, accuracy is another matter.

Introduction to the DAC0808 Family

The National Semiconductor DAC0808, DAC0807, and DAC0806 are members of a very popular family of 8-bit monolithic D/A converters. All three of these devices share the same operating principles, packaging, and pin designations. The pin diagram for the DAC0808/7/6 is shown in Fig. 8-9.

The major parameter that differentiates the three versions of this device is accuracy. The DAC0808 has full 8-bit accuracy, $\pm\tfrac{1}{2}$ LSB, while the DAC0807 effectively has 7-bit accuracy. Thus, in terms of absolute error, the output of the DAC0807 may vary $\pm\tfrac{3}{4}$ LSB from the expected output level. Finally, the DAC0806 has 6-bit accuracy; therefore, its output may vary ± 1 LSB from the expected output value. This should help drive home the earlier discussion relating resolution and accuracy, for here we see an 8-bit D/A converter that in one variation (the DAC0806) is no more accurate than a 6-bit D/A with $\pm\tfrac{1}{2}$ LSB accuracy. For convenience, from this point on we shall refer to the

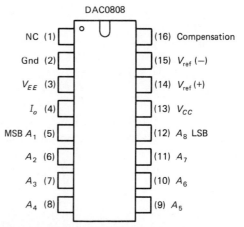

Fig. 8-9 Pin designations for the DAC0808.

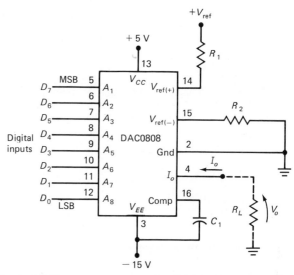

Fig. 8-10 Typical D/A converter using the DAC0808.

DAC0808/7/6 as a DAC0808. Typically, the DAC0808 has a settling time of 150 ns. The inputs are TTL and CMOS compatible, and it can operate with power-supply voltages that range from ± 4.5 V to ± 18 V.

DAC0808 Operation and Applications

A typical D/A circuit using the DAC0808 is shown in Fig. 8-10. The output current for this configuration is given by

$$I_o = -\frac{V_{\text{ref}}}{R_1}\left(\frac{A_1}{2} + \frac{A_2}{4} + \frac{A_3}{8} + \cdots + \frac{A_8}{256}\right) \tag{8-8}$$

where the A inputs are either 1 or 0, as determined by the data present on the data (D) lines.

Resistor R_1 is chosen based on the available reference voltage and the desired full-scale output current. The output of the DAC0808 is guaranteed to sink -1.9 mA; thus, typically, $I_{FS} = -1.000$ mA. Although the $+5$-V logic supply may be used as a reference, this practice is not recommended in situations in which full accuracy is desired. In these cases, a very stable external voltage reference is recommended. Resistor R_2 is chosen to equal the external Thevenin resistance present at the $V_{\text{ref}(+)}$ input. This provides offset compensation for the converter. For the circuit used in Fig. 8-10, $R_2 = R_1$. Compensation capacitor C_1 is used to prevent ringing and overshoot of the output. Typically, $C_1 = 0.001$ µF.

Since the output of the converter represents a controlled current sink, an external load must be provided. In most applications, it is preferable that the output of a D/A converter be a voltage. Simply connecting a resistor between the output of the DAC0808 and ground will cause an output voltage to be produced. The output voltage is given by

$$V_o = I_o R_L \qquad (8\text{-}9)$$

Substituting Eq. (8-9) into (8-8) yields

$$V_o = -\frac{R_L V_{\text{ref}}}{R_1}\left(\frac{A_1}{2} + \frac{A_2}{4} + \frac{A_3}{8} + \cdots + \frac{A_8}{256}\right) \qquad (8\text{-}10)$$

Notice that the output voltage increases negatively with the binary input. This occurs because the output sinks current from the ground-referenced load.

EXAMPLE 8-6

The circuit in Fig. 8-10 has $V_{\text{ref}} = 5.000$ V. Determine the required values for R_1, R_2, and R_L such that $I_{FS} = -500.00$ μA and $V_{FS} = 10.000$ V.

Solution

We begin by solving Eq. (8-8) for R_1, with a binary input of 10000000_2, which of course will produce $I_o = I_{FS}/2 = -250.00$ μA. This yields

$$R_1 = 10 \text{ k}\Omega$$

Now, since $I_{FS} = -500.00$ μA and $V_{FS} = 10.000$ V, we apply Ohm's law to determine R_L [use Eq. (8-9)]. This produces

$$R_L = 20 \text{ k}\Omega$$

In order to maintain high accuracy, the output of a circuit such as that of Fig. 8-10 must drive a constant resistance. This can be a problem if we need to connect other devices or instruments to the output terminal. One solution to this loading problem would be to buffer the output voltage with a non-inverting op amp voltage follower, which will have inherently high input impedance. Another potential problem is that the output of Fig. 8-10 is a negative voltage. Most likely, we would need to have a positive-going output; thus, an inverting amplifier might also be required. Fortunately, there is a simple and effective solution to these problems.

The connection of an op amp I/V converter (transresistance amplifier) to the output of the DAC0808 provides both the current-to-voltage conversion and the phase inversion required. Such a circuit would be configured as

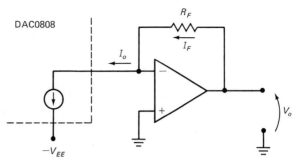

Fig. 8-11 Transresistance amplifier used to produce a voltage output.

shown in Fig. 8-11. The output voltage produced in this case is given by

$$V_o = \frac{R_F V_{ref}}{R_1}\left(\frac{A_1}{2} + \frac{A_2}{4} + \frac{A_3}{8} + \cdots + \frac{A_8}{256}\right) \qquad (8\text{-}11)$$

Using this approach, the desired output polarity is obtained, and output loading effects are essentially eliminated.

There are literally hundreds of applications in which D/A converters can be used. One possibility is in the design of a positional servo, as shown in Fig. 8-12. In this system, the output of a D/A converter is compared with a voltage produced by a potentiometer whose wiper position is controlled by a motor. The difference between these voltages (the error voltage V_e) is amplified and applied to the motor, causing it to rotate such that the feedback voltage approaches the reference voltage supplied by the D/A converter. Each input code corresponds to a unique wiper position.

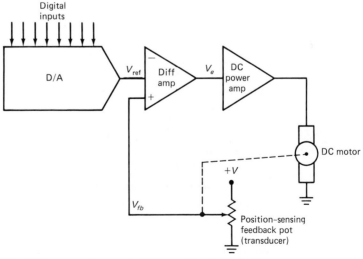

Fig. 8-12 D/A converter-controlled positional control servo.

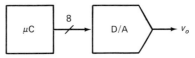

Fig. 8-13 Microcomputer control of a D/A converter.

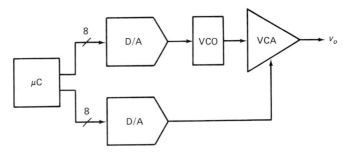

Fig. 8-14 A microprocessor-controlled signal generator.

Note the similarity between the general topology of this system and that of Fig. 7-60. Applications for these types of systems include automatic pilot and inertial guidance systems, robotics, and automatic gun director systems. The position of the motor in Fig. 8-12 could be adjusted by a computer, which would output positional information to the D/A converter.

If a microcomputer is used to drive the inputs of a D/A, as in Fig. 8-13, the converter can be used to synthesize various different waveforms. For example, by continuously incrementing and outputting the contents of an 8-bit register, a sawtooth wave could be generated. Conversely, decrementing the same register would result in a sawtooth with symmetry reversal. An approximation of a sinusoidal waveform could be produced by programming the computer to output values from a lookup table in memory to the converter.

Another approach to waveform generation that uses a microcomputer and D/A converters is shown in Fig. 8-14. In this example, one converter provides the control voltage to a VCO, while a second converter controls the gain of a voltage-controlled amplifier (VCA). This circuit would provide 256 different frequencies at 256 different amplitudes. Additional D/A converters could be added for even more versatility.

8-3 ♦ A/D CONVERSION

Computers and other digital systems are used extensively for the analysis and processing of signals that are actually continuously variable. A digital voltmeter is an example of a device that processes an analog voltage and produces a numerical representation of that voltage on its display. Sophisticated image processing systems convert the continuous shading and colors of images into a digital form that is then processed by a computer, in order

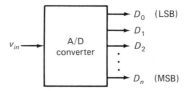

Fig. 8-15 Symbol for an *n*-bit A/D converter.

to enhance the image. For the computerized flight control systems of a jet fighter, the continuously variable outputs produced by transducers that monitor altitude, speed, course, and other flight parameters must be converted into digital form for processing. These are just a few applications which require the use of A/D converter circuits. In this section we shall discuss the basics of A/D conversion circuits and their operation.

A/D Conversion Fundamentals

The basic block diagram symbol for an A/D converter is shown in Fig. 8-15. The function of an A/D converter is to sample some analog level (usually voltage) and produce a quantized digital (usually binary) representation of that level at its output. As with the D/A converter, the number of bits in the output of the A/D defines the resolution and the potential for accuracy of the A/D converter.

Full-Scale Range

The transfer characteristics shown in Fig. 8-16 are those that would be produced by a perfect 4-bit A/D converter. A 4-bit A/D converter can resolve an analog input voltage into 1 of 16 unique output code words.

The input of a given A/D converter can accommodate only a limited range of values, for example, 0 to 10 V. The span of input values that can be resolved by the converter (0 to 10 V in this case) is called the *full-scale range (FSR)*. The converter's input range is divided into 2^n divisions, where n is the number of bits in the output code word. The input range division of an A/D converter is analogous to the resolution of a D/A converter, as expressed by Eq. (8-3); thus we have

$$\text{Input range division} = 1\ \text{LSB} = \frac{\text{FSR}}{2^n} \qquad (8\text{-}12)$$

The output of an A/D converter will usually be in standard 8-4-2-1 binary form, although other codes, such as binary-coded decimal (BCD), can also be produced. The horizontal axis of the A/D transfer function plot is scaled in fractions of the FSR. Somewhat similar to the case of the D/A converter, for which the full-scale output voltage cannot be realized, the FSR input

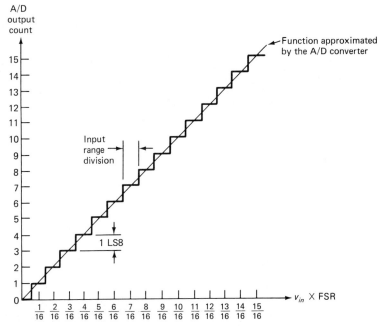

Fig. 8-16 Transfer characteristics for an ideal 4-bit A/D converter.

voltage of an A/D converter cannot be resolved. That is, a full-scale input applied to an A/D converter will normally result in an overrange condition.

A/D Error Specifications

Even within a finite range, an analog quantity can take on any of an infinite number of different values. The A/D converter samples an analog input and produces a quantized output code word that represents that input. This action in itself introduces the possibility of output error. This error, plus other forms of error, will now be discussed.

QUANTIZATION ERROR Because of the finite number of bits at the output of the converter, there will always be some uncertainty as to the actual value of the input. The most certain that we can be is that the input is within $\pm\frac{1}{2}$ LSB. This uncertainty associated with a given conversion is called *quantization error* or *quantization noise*. At the minimum, *any* A/D converter will exhibit at least $\pm\frac{1}{2}$-LSB quantization error. Figure 8-17a shows the plot of quantization error versus input voltage for an A/D converter with $\pm\frac{1}{2}$-LSB accuracy.

It can be quite difficult to produce high-resolution A/D converters that exhibit only the inherent $\pm\frac{1}{2}$-LSB quantization error. Figure 8-17b illustrates the error plot for a converter that has a quantization error of $\pm\frac{3}{4}$ LSB. In general, the locations of the error peaks for such converters will be random, and so we must assume that for any given input, the corresponding output code may be in error by as much as $\pm\frac{3}{4}$ LSB. Some A/D converters exhibit as much as ± 1 LSB of quantization error.

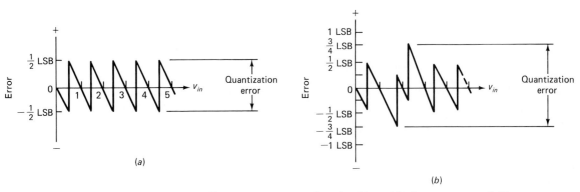

Fig. 8-17 A/D converter error plots for (a) an ideal converter and (b) a converter with $\pm\frac{3}{4}$-LSB accuracy.

EXAMPLE 8-7

A certain 6-bit A/D converter is specified as having $\pm\frac{1}{2}$-LSB accuracy. If the FSR is 0 to 10 V and the output is in normal binary format, determine: (a) input range division, (b) The possible range of input voltages that would cause the output code word to be 011000.

Solution

(a) Input range division $= \dfrac{\text{FSR (V)}}{2^n \text{ (div)}}$

$= \dfrac{10}{2^6}$

$= 156.25 \text{ mV/div}$

(b) Ideally, the input voltage represented by the given output code is found as follows:

$011000_2 = 24_{10}$

$V_{in} = 24 \times \text{input range division}$

$= 24 \times 1 \text{ LSB}$

$= 24 \times 156.25 \text{ mV}$

$= 3.75 \text{ V}$

However, because the accuracy is $\pm\frac{1}{2}$ LSB, the actual input voltage may be any value in the range of 3.75 V $\pm\frac{1}{2}$ LSB; therefore

$V_{in} = 3.75 \text{ V} \pm 78.125 \text{ mV}$

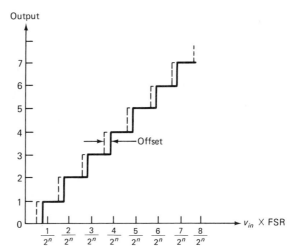

Fig. 8-18 Effect of offset error (positive) on the transfer characteristics of an n-bit A/D converter.

In signal processing applications, it is sometimes desirable to express the quantization noise in decibel form, called the *signal-to-quantization-noise ratio (SQNR)*. This is defined as

$$\text{SQNR} = 20 \log \frac{\text{FSR (V)}}{1 \text{ LSB (V)}} \quad \text{dB} \tag{8-13}$$

For the A/D converter of Example 8-7, SQNR = 36 dB. For an A/D converter with $\pm \frac{1}{2}$-LSB accuracy, Eq. (8-13) may be simplified to the following form:

$$\text{SQNR} = 6_n \quad \text{dB}$$

where n is the number of bits in the output.

OFFSET ERROR Because of the nature of the devices used to construct A/D converters, other types of error, aside from quantization error, may exist. Offset is one such error. Basically, an offset error will result in a shifting of the input range divisions by some fixed amount. Figure 8-18 illustrates the effect of a positive offset of approximately $\frac{1}{4}$ LSB on the transfer characteristics of an A/D converter. The dashed portion of the plot represents the transfer characteristics of a converter with no offset.

GAIN ERROR A gain error will result in a binary output that is not related to the input as a fraction of the FSR of the converter. This is shown in Fig. 8-19, where the ideal transfer characteristics are again shown as dashed lines. The presence of a gain error effectively results in a FSR that is smaller than expected if the gain is too high, or a FSR that is larger than expected if the gain is too low.

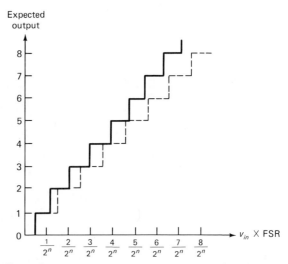

Fig. 8-19 Effect of gain error (positive) on the transfer characteristics of an n-bit A/D converter.

NONLINEARITY ERROR Nonlinearity results in the best-fit curve for the transfer characteristic plot deviating from a straight line. This is shown in Fig. 8-20. Thus, we see that the $\pm\frac{1}{2}$-LSB nonlinearity that is inherent in any A/D converter is not considered in this plot. Some causes for nonlinearity are variations in offset or gain as the input changes.

DIFFERENTIAL NONLINEARITY Differential nonlinearity is the difference between the theoretical and actual size of a given input range division. An A/D converter with accuracy of $\pm\frac{1}{4}$ LSB has an input range division of 1 LSB, and the differential nonlinearity is less than 1 LSB. A converter with accuracy of less than this, say $\pm\frac{3}{4}$ LSB, has a differential nonlinearity of greater than 1 LSB. In terms of the transfer characteristics, this means that it is possible that two input range divisions that should be separated by a

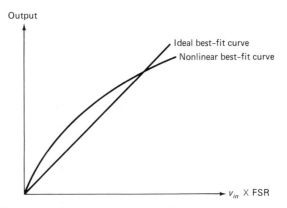

Fig. 8-20 Effect of nonlinearity on the transfer characteristics of an A/D converter.

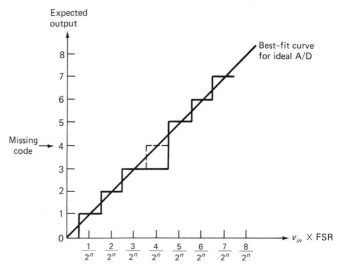

Fig. 8-21 Example of a missing code caused by differential nonlinearity.

central input range division could overlap. This results in a skipped code at the output, as shown in Fig. 8-21. If the specifications for a given A/D converter state that there are no missing codes, the converter has less than 1 LSB differential nonlinearity.

Conversion Time

A practical A/D converter requires some finite nonzero time to complete a conversion. This time interval, called the *conversion time* T_C, is illustrated in Fig. 8-22. As shall be shown in the next section, not all A/D converters have

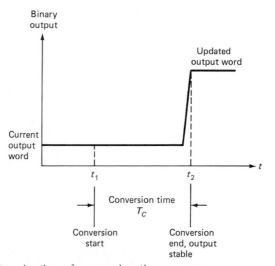

Fig. 8-22 Determination of conversion time.

a constant conversion time. Most often, T_C is specified under worst-case conditions. Conversion time specs for a specific A/D converter do not account for delays that may be caused by other support devices. The reciprocal of conversion time is termed *conversion rate*.

8-4 ♦ A/D CONVERSION CIRCUITS

There are many different ways to implement A/D conversion circuits. Many A/D converters are designed using D/A converters. Ramp converters and successive-approximation converters fall into this category. Other types of A/D converters include integrating, parallel, and tracking converters. Most of these types of converters and the main sections that comprise them are readily available in IC form.

Ramp Converters

Perhaps the A/D converter whose operation is easiest to understand is the ramp-type converter. A block diagram for a 4-bit ramp converter is shown in Fig. 8-23.

The ramp converter derives its name from the way in which it performs a conversion. The following steps outline this operation. For the purposes of this discussion, let us assume that the D/A converter section of the circuit has $V_{FS} = 10.00$ V, and that $V_{in} = 6.870$ V. Refer to the timing diagram of Fig. 8-24a at each step of the sequence. The output of the internal D/A converter for each corresponding clock period is shown in Fig. 8-24b.

1. The $\overline{\text{START}}$ input is driven low. This clears the counter, causing the output of the D/A converter $V_{fb} = 0$ V.

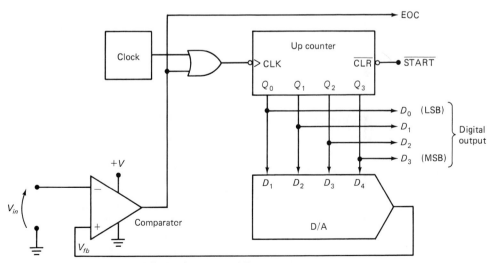

Fig. 8-23 Simplified 4-bit ramp A/D converter.

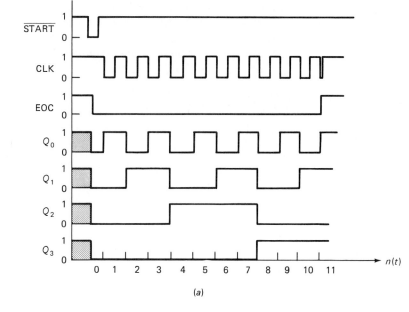

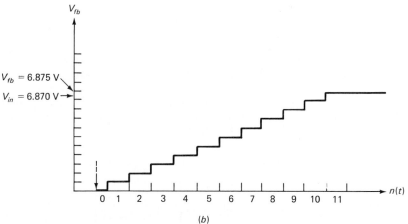

Fig. 8-24 Timing diagram for the 4-bit ramp converter.

2. Because $V_{in} > V_{fb}$, the output of the comparator is driven low, enabling the OR gate to pass clock pulses to the up counter. The output of the comparator is also used to indicate that a conversion is now in progress (EOC = 0).

3. The counter increments once for each falling edge of the clock signal, forcing the output of the D/A converter to increase by 1 LSB on each falling clock edge. This action continues until the output of the D/A converter is greater than the input voltage being sampled. In this example, the count will increase to 1011_2, or 11_{10}, when V_{fb} just exceeds V_{in}.

4. Once $V_{fb} > V_{in}$, the output of the comparator is driven high, which disables the clock input of the counter. A high logic state on the EOC output indicates that the conversion is complete and the output count is proportional to the analog input voltage. Another conversion may be initiated by pulsing the $\overline{START}$ input low.

There are several very important points relating to the operation of the ramp converter that should be discussed. First, notice that the decimal equivalent of the binary output is not a direct indication of the input voltage. Rather, the decimal value of the output corresponds to the fraction of V_{FS} of the internal D/A converter that is present at the input. For this type of circuit, the final output count will always be a slight overestimation of the input voltage, as the D/A output must exceed V_{in} in order to stop the count. A more detailed analysis of this circuit reveals that at best, the output uncertainty is -1 LSB (assuming an ideal comparator). A second important point to note is that conversion time will vary for different input voltages. For example, if $V_{in} = 1.00$ V, then the conversion would be complete after only 2 clock cycles. For ramp converters with higher resolution, the conversion time differences for different inputs can vary over a very wide range. There is also a conversion time uncertainty of ± 1 clock cycle in the conversion. This occurs because the $\overline{START}$ pulse is not synchronized with the clock. Thus, if $\overline{START}$ goes high just after a falling clock edge, we must wait nearly 1 cycle before the counter is incremented. Conversely, if $\overline{START}$ goes high just before a falling clock edge, the counter will increment almost immediately.

EXAMPLE 8-8

Refer to Fig. 8-23. Assume that the D/A converter has $V_{FS} = 10.00$ V, the clock PRF = 10 kHz, and the low-to-high transition of the $\overline{START}$ input is synchronized with the rising edge of the clock. Determine: **(a)** the maximum conversion time, **(b)** the output count and conversion time for $V_{in} = 2.85$ V.

Solution

(a) The maximum conversion time is determined by multiplying the clock period (1/10 kHz = 0.1 ms) by 2^n, where n is the number of bits in the output code word. This produces

$$T_{C(max)} = 0.1 \text{ ms} \times 2^4$$
$$= 1.6 \text{ ms}$$

(b) We must determine the first output count that produces $V_{fb} > 2.85$ V. To do this, we divide V_{in} by the value of 1 LSB and round up to the next highest integer. This produces the decimal equivalent of the binary output. Here, we obtain

$$\frac{2.85 \text{ V}}{0.625 \text{ V}} = 4.56$$

Output count = 5_{10} = 0101_2

Successive-Approximation A/D Converters

The most widely used general-purpose A/D converters are designed around a digital circuit called a *successive-approximation register (SAR)*. Like the ramp converter, the SAR A/D converter contains a D/A converter. As will be shown, however, the SAR converter offers several advantages over the ramp converter.

SAR Converter Operation

The block diagram for an 8-bit SAR A/D converter is shown in Fig. 8-25. The SAR is used to produce estimations of the input voltage, beginning with FSR/2 and working down to the LSB, one bit position at a time. The output of the internal D/A converter is compared with the input voltage after each internal state change. If the setting of a particular bit results in $V_{fb} > V_{in}$,

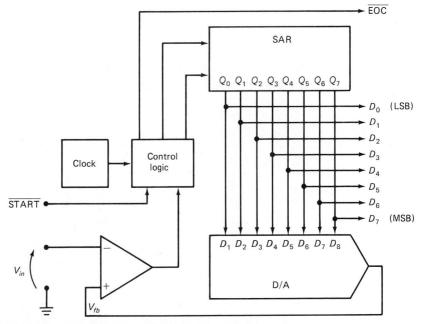

Fig. 8-25 Simplified 8-bit SAR A/D converter.

then that bit is cleared (discarded). If the setting of a particular bit results in $V_{fb} < V_{in}$, that bit is latched high.

Figure 8-26 shows a plot of the output of the D/A converter (V_{fb}) for each step of the conversion. The following steps explain the conversion process. Let us assume that the D/A converter has V_{FS} = 10.00 V, which produces FSR = 10.00 V, and V_{in} = 5.40 V.

0. The $\overline{\text{START}}$ input is pulsed low. This clears the SAR, producing V_{fb} = 0.00 V. The converter is now initialized.
1. The MSB of the SAR (Q_7) is set, producing V_{fb} = 5.00 V. Since $V_{fb} < V_{in}$, the output of the comparator is driven low. The control logic senses this and latches Q_7 high.
2. The next most significant bit of the SAR (Q_6) is set. This produces V_{fb} = 7.50 V. Since $V_{fb} > V_{in}$, the output of the comparator goes high, causing the control logic to clear Q_6 of the SAR.
3. Now, Q_5 of the SAR is set, producing V_{fb} = 6.25 V. Again, $V_{fb} > V_{in}$ and the comparator output goes high, causing the clearing of Q_5.

This keep/discard testing process is repeated for each successive bit until, after the eighth test, the conversion is complete. The technique used is similar to that which a person might use to efficiently guess a number between 0 and 255 that someone else is thinking of. Using this approach, it would take no more than eight guesses to hit the correct number.

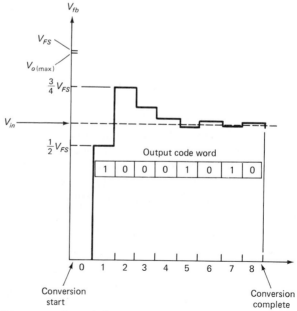

Fig. 8-26 Representation of the successive approximation process for the 8-bit converter.

Additional SAR Converter Considerations

One of the advantages of the SAR A/D converter is that the conversion time is constant, regardless of the input voltage. For the circuit of Fig. 8-25, it takes eight keep/discard operations to quantize the input. [Of course, the time required to initialize the converter (step 0 in Fig. 8-26) must also be included in the conversion time determination.] Compare this with the ramp converter, for which conversion time varies widely for input voltages at either extreme of the FSR, especially when high accuracy (many output bits) is required.

Another advantage of the SAR converter over the ramp converter is that for the SAR converter, the conversion time increases in direct proportion to the number of output bits, whereas resolution increases exponentially. For the ramp converter, both resolution and worst-case conversion time increase exponentially with the number of output bits. Thus, on the average, the SAR converter will greatly outperform a ramp converter of the same resolution in terms of conversion time.

The accuracy of the SAR converter (and the ramp converter) depends on the accuracy of the internal D/A converter and the comparator. For the best possible overall accuracy of the A/D converter ($\pm \frac{1}{2}$ LSB), the total inaccuracies of the internal D/A converter and comparator must be less than $\pm \frac{1}{2}$ LSB. This means that the D/A section should have $\pm \frac{1}{4}$-LSB accuracy, and the comparator must resolve input differentials ($V_{fb} - V_{in}$) to less than $\frac{1}{4}$ LSB as well. For the 8-bit A/D, with FSR = 10.00 V, the D/A must be accurate to within ± 9.76 mV, and the comparator must reliably respond to input differentials of less than ± 9.76 mV. Because of these strict requirements, the internal amplifiers must have very good offset and drift specifications. For converters with more than about 8-bit outputs, such accuracy can be difficult to attain, with the result that these devices are rather expensive. The accuracy/cost tradeoffs must be weighed carefully in the design process.

Up to this point, we have assumed that the input voltage has remained constant while a conversion is being performed. Because of the nature of the SAR conversion process, this is a necessity. Consequently, track-and-hold (T/H) amplifiers are usually placed before the input of the converter, as shown in Fig. 8-27. The use of T/H circuitry is a must when time-varying input voltages are to be sampled and quantized. The use of a T/H amplifier allows the input signal to be effectively frozen in time at a particular sample point. Any additional time delay introduced by the T/H amplifier must be considered when determining overall conversion time. More will be said about sampling later in this chapter.

Dual-Slope A/D Converters

The dual-slope converter is an example of an intergrating A/D converter. A simplified block diagram for an 8-bit dual-slope A/D converter is shown in Fig. 8-28. This circuit is designed to work with input voltages that are opposite in polarity to the reference voltage. The heart of the circuit is an

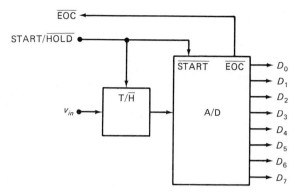

Fig. 8-27 Track-and-hold amplifier added to the A/D converter allows time-varying inputs to be quantized.

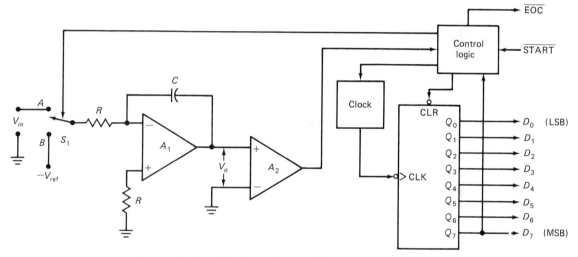

Fig. 8-28 Simplified dual-slope A/D converter.

integrator, formed by amplifier A_1. Amplifier A_2 is a comparator that uses ground as its reference. The output of the circuit is taken from an 8-bit up counter. Also, for this circuit, FSR = $|V_{ref}|$. The operation of the circuit is summarized in the following two steps. Refer to Fig. 8-29 during this discussion.

1. The $\overline{\text{START}}$ input is pulsed low. This clears the counter, sets S_1 to position A, and gates on the clock. This marks the beginning of the sample interval t_1. This interval is always of constant duration. The output of the integrator ramps negatively at a constant rate, given by

$$\Delta V_{o1} = \frac{-V_{in} t_1}{RC} \qquad (8\text{-}14)$$

During this time, the output of the comparator is low, and $\overline{\text{EOC}} = 1$.

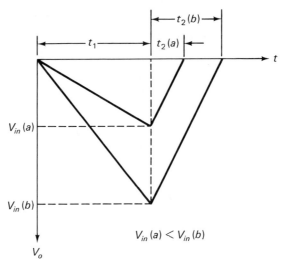

Fig. 8-29 Dual-slope integrator response to two different input voltages.

2. When the counter rolls over to zero, t_1 has expired, and S_1 is set to position B. This starts time interval t_2. During this time, the reference voltage forces the output of the integrator to ramp back toward 0 V. The output of the integrator is now given by

$$\Delta V_{o2} = \frac{-V_{ref}t_2}{RC} \qquad (8\text{-}15)$$

When the output of the integrator just exceeds 0 V (by less than $\frac{1}{2}$ LSB, for $\pm\frac{1}{2}$-LSB accuracy), the output of the comparator goes high, inhibiting the clock and causing $\overline{EOC}$ to go low; this signals the end of the conversion.

From the preceding description, we see that the input voltage sample period t_1 is constant, and dependent on the clock frequency. The reference period t_2 is variable, depending on both the reference voltage and the voltage to which the integrator capacitor charged during the sample period. The value of the unknown input voltage is determined in the following way. First, we may equate Eqs. (8-14) and (8-15):

$$\frac{V_{in}t_1}{RC} = \frac{V_{ref}t_2}{RC}$$

Solving for V_{in}, we obtain

$$V_{in} = -V_{ref}\frac{t_2}{t_1} \qquad (8\text{-}16)$$

Now, since t_1 and t_2 are determined by the counts accumulated during the conversion, for this circuit we may write

$$V_{in} = -V_{ref}\frac{N}{2^n} \qquad (8\text{-}17)$$

where n is the number of output bits and N is the count accumulated during the period t_2.

A digital circuit or microprocessor would process the t_2 count as required by the application. An example of an application that would require such processing is the direct display of the input voltage in decimal form using seven-segment displays.

In general, dual-slope converters can be made to have extremely high accuracy. Also, because of the integration operation, input noise tends to be suppressed. This occurs because, on the average, a random noise voltage will tend to have equal area about 0 V, and such signals integrate to zero. The accuracy of the integrator is critical to the overall accuracy of the converter. Once again, very low offset amplifier circuitry is required, and the feedback capacitor must have low leakage and a low temperature coefficient. Integrating A/D converters are sensitive to significant short-term variations in clock frequency. In this context, *short-term* means frequency variations that occur over the course of a conversion. This should be rather obvious, as the output of the integrator is strongly time-dependent. Long-term frequency drift has little effect on the accuracy of the integrating converter.

Integrating A/D converters are relatively slow compared with SAR converters; therefore, they are usually used in devices such as digital multimeters, where conversion speed is not a major concern.

EXAMPLE 8-9

Refer to Fig. 8-28. Given $R = 10$ kΩ, $C = 0.1$ μF, PRF$_{clk}$ = 100 kHz, V_{ref} = -5.00 V, and V_{in} = 3.00 V, determine the following: **(a)** The output count after the conversion is completed. **(b)** The indicated input voltage, based on the output count.

Solution

(a) First, we solve Eq. (8-17) for N, producing

$$N = 256 \times \frac{3.00 \text{ V}}{5.00 \text{ V}}$$

$$= 153.6$$

Fractional output bits are not possible; therefore, we must conclude that the count is halted before it reaches 154. Thus, we truncate the above result, producing

$$N = 153 = 1001\ 1001_2$$

(b) The indicated output voltage is found by direct application of Eq. (8-17), producing

$$V_{in} = 5\text{ V}\frac{153}{256}$$
$$= 2.99\text{ V}$$

Parallel A/D Converters

For applications which require extremely fast conversion, parallel, or flash, converters are used. The block diagram for a 3-bit parallel A/D converter is shown in Fig. 8-30. The main section of this circuit is the chain of comparators, which in this case have their noninverting inputs connected in parallel to the input terminal; hence the name *parallel converter*. The reference voltage is applied to a resistive voltage divider that provides the inverting input of each comparator with a power of 2 fraction of V_{ref}. In this case, the lower comparator (A) effectively has $V'_{ref} = 0.125V_{ref}$, comparator B has $V'_{ref} = 0.250V_{ref}$, and so on until we reach comparator G, where $V'_{ref} = 0.875V_{ref} = \frac{7}{8}V_{ref}$. Thus, FSR = V_{ref}.

To get a good understanding of how the parallel converter works, let us assume that initially, $V_{in} = 0.00$ V. This results in the outputs of all the comparators being driven low. Now, if the input voltage begins increasing positively, comparator A will change states (ideally) when $V_{in} = \frac{1}{8}V_{ref}$. Further increases in V_{in} will result in consecutively more comparators changing output states. The chart in Fig. 8-31 illustrates this sequence of events. Notice that there are eight different comparator output states possible. Decoding logic is required to generate an equivalent binary output.

Because the input to the parallel converter is sampled simultaneously by all the comparators, the only delay in producing a binary output is the switching time of a single comparator and the propagation delay of the decoding logic. High-speed comparators and fast logic, such as ECL or gallium arsenide devices, allow extremely fast conversion. Parallel converters have been produced that have conversion times that are measured in hundreds of picoseconds.

The main disadvantage associated with parallel converters is the large number of comparators required for moderately high resolution. The number of comparators N required to provide n output bits is given by

$$N = 2^n - 1 \tag{8-18}$$

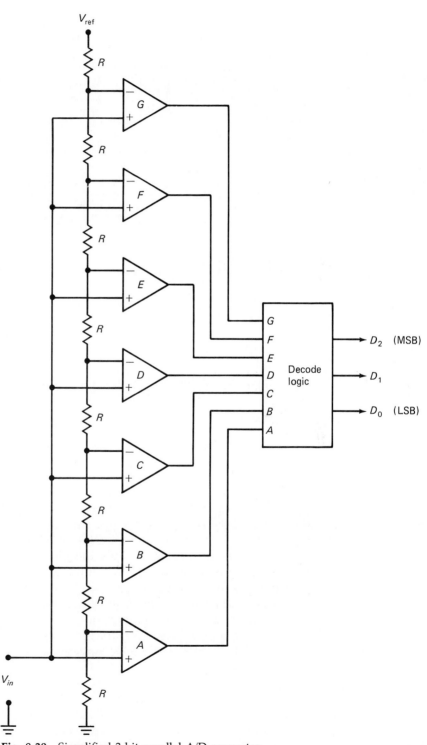

Fig. 8-30 Simplified 3-bit parallel A/D converter.

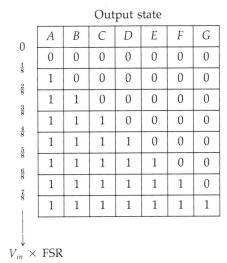

Fig. 8-31 3-bit parallel converter comparator output states versus input voltage.

A parallel A/D converter with 10-bit resolution would require 1023 comparators, and 1024 resistors in the voltage divider. Clearly, this would be a difficult circuit to produce. For these reasons, parallel converters are used only in applications in which conversion speed is of paramount importance. One application is the conversion of digital color and luminosity codes into analog voltages for the display of computer-generated graphics.

Tracking A/D Converters

A variation of the ramp converter is the tracking A/D converter. The basic structure of one variety of tracking converter is shown in Fig. 8-32a. Once a conversion is started, the up/down counter begins counting up. When the output of the D/A just exceeds the input voltage, the output of the comparator goes high, causing the counter to begin counting down. If V_{in} is constant, the output of the counter will toggle such that V_{fb} oscillates about V_{in}. If the input is time-varying, the output of the counter will track the input voltage as it increases and decreases; thus the term *tracking converter*. Conversion may be halted by driving the $\overline{\text{STOP}}$ input low.

Tracking converters are sometimes called *continuous converters* because they do not stop after a given conversion is complete. Such converters have much faster conversion times than ramp converters, making them better suited for higher-speed applications. Also, tracking converters do not require T/H amplifiers at their inputs, as do SAR converters.

Monolithic A/D Converters

There is a wide range of monolithic A/D converters available, ranging from 6 bits to 16 bits in accuracy, designed using all the techniques that were just discussed. Converters are available with both binary and BCD outputs. An-

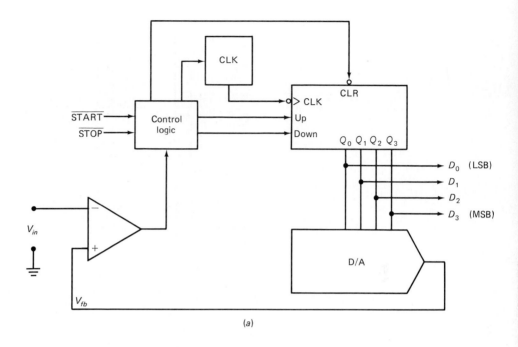

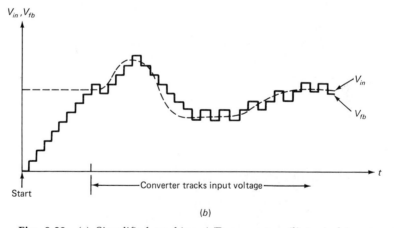

Fig. 8-32 (a) Simplified tracking A/D converter; (b) typical input and feedback signals.

alog multiplexers, similar to that in Fig. 7-16, are sometimes integrated into A/D converters to allow multiple input sources to be digitized as required.

The ADC0801 A/D Converter

The National Semiconductor ADC0801 is a popular 8-bit SAR converter that is available in several degrees of accuracy. Additional members of this family include the ADC0802, ADC0803, ADC0804, and ADC0805. The pin designations for the ADC0801 are shown in Fig. 8-33.

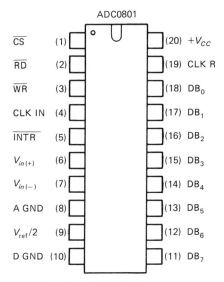

Fig. 8-33 Pin designations for the ADC0801.

The ADC0801 is designed primarily to be controlled by a microprocessor or microcomputer; hence the chip select ($\overline{CS}$), read ($\overline{RD}$), and write ($\overline{WR}$) inputs and the interrupt request ($\overline{INTR}$) output. The digital inputs and outputs are compatible with TTL and MOS logic devices.

Either the ADC0801 may operate using its own internal clock, or an external clock signal may be applied. The frequency of the internal clock is set by an external resistor and capacitor. The internal clock frequency is approximately

$$f_{clk} = \frac{1.1}{RC} \tag{8-19}$$

where $100 \text{ kHz} \leq f_{clk} \leq 1.46 \text{ MHz}$

Typically, $R_{clk} = 10 \text{ k}\Omega$. If an external TTL clock is used, pin 19 is left open, and the clock signal is applied to pin 4.

Separate analog and digital grounds (A GND and D GND) are provided for high noise immunity. The +5-V V_{CC} supply line is used to generate the reference voltage, resulting in FSR = 5.00 V. This is suitable for many applications; however, for highest accuracy, an external +2.50-V reference should be applied to the $V_{ref}/2$ input (pin 9). If this option is not used, pin 9 should be left open.

The actions of the ADC0801 and its required control signals are described in the following steps.

1. $\overline{CS}$ and $\overline{WR}$ are driven low. At this time, the output lines are in a high-impedance state, and $\overline{INTR}$ is high.

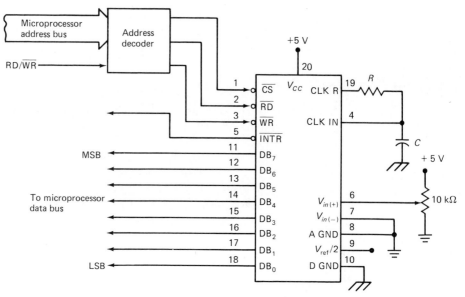

Fig. 8-34 Interfacing the ADC0801 to an 8-bit microprocessor.

2. $\overline{CS}$ and $\overline{WR}$ are returned to their inactive states (high). Conversion begins when $\overline{WR}$ returns high.

3. When the conversion is complete, the $\overline{INTR}$ output is driven low. This signal may cause the controlling CPU to suspend operation and read the converter, or the microcomputer may periodically test the status of this line (polling) to determine the status of the converter.

4. $\overline{CS}$ and $\overline{RD}$ are activated (driven low). The binary representation of V_{in} is driven onto output lines DB_7 (MSB) through DB_0 (LSB), and $\overline{INTR}$ is driven high.

5. When $\overline{RD}$ is released, the output data lines return to the high-impedance state. The converter is now ready to begin another conversion.

Figure 8-34 shows a typical application in which the ADC0801 is used to digitize the voltage drop produced by a potentiometer.

8-5 ♦ DATA ACQUISITION AND DIGITAL SIGNAL PROCESSING

Digital signal processing (DSP) techniques have been gaining use and sophistication nearly as rapidly as microprocessor technology. There are many reasons for the increase in DSP usage. However, the main reason is that digital processing can simulate any operation that can be performed by a linear system, and some operations that cannot be realized using linear techniques as well. DSP has largely been made possible by the availability of relatively low-cost, yet powerful microprocessors, high-density memory devices, and specialized DSP devices. This section presents a brief introduction to some of the fundamentals of DSP, and the applications of D/A and A/D converters in the data acquisition process.

Effects of Sampling

Recall that, ideally, the time during which a signal is sampled for a conversion is infinitesimally short. The use of a T/H circuit allows this situation to be closely approximated. The T/H circuit has two inputs, one for the analog signal and another for the control of the T/H mode. Normally, a signal will be sampled at a fixed rate, called the *sample frequency* f_{sa}. Under these conditions, the output of the T/H will contain the original signal frequency (or frequencies) plus the sum and difference of the signal frequency and the sample frequency, located at integer multiples of the sample frequency. This is shown in Fig. 8-35.

During the processing of the digitized signal information, the higher-frequency components that were created by the sampling process may be compensated for. Notice that in Fig. 8-35, the sample frequency is more than twice as high as the signal frequency. This situation will always result in the creation of frequency components that are higher than the frequency of the signal being sampled. As will be shown, this is a very important fact.

The Nyquist Sampling Theorem

The Nyquist sampling theorem states that in order to produce an accurate representation of a periodic signal, the frequency at which the signal is sampled must be at least twice the frequency of that signal. Mathematically, this is written

$$f_{sa} \geq 2f_{in} \qquad (8\text{-}20a)$$

or, alternatively, we may write

$$f_{in} \leq \frac{f_{sa}}{2} \qquad (8\text{-}20b)$$

where f_{sa} is the sample frequency and f_{in} is the highest signal frequency to be sampled. For convenience, let us call $f_{sa}/2$ the Nyquist frequency.

Equation (8-20) says that for complex signals that consist of several frequency components, the sample frequency must be at least twice as high as the highest-frequency component to be quantized. In general, it is desirable to sample at a rate that is much higher than the highest frequency to be processed. This is called *oversampling*. As will be seen, oversampling results in a more accurate digital representation of the original signal.

Aliasing

If the conditions of the Nyquist sampling theorem are not met, the signal is said to be *undersampled*, and a phenomenon known as *aliasing* will occur.

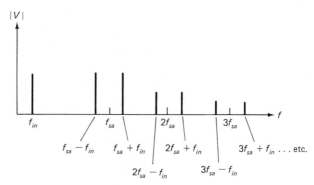

Fig. 8-35 T/H output spectrum for $f_s \ll f_{sa}$.

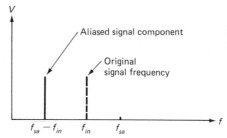

Fig. 8-36 T/H output spectrum produced by sampling at $f_{sa} < 2f_s$ results in aliasing error.

Aliasing is the translation of signal frequency components that lie above the Nyquist frequency down below their original frequencies. This occurs because the difference between f_{sa} and f_{in} is lower than f_{in}. This phenomenon is illustrated in Fig. 8-36, where $f_{sa} < 2f_{in}$. Normally, this is an undesirable situation, as it appears that the original signal was actually composed of two sinusoidal components, which was not the case.

A more intuitively obvious illustration of aliasing error is shown in Fig. 8-37. This is an extreme case, with the signal frequency just slightly less than the sample frequency. The resultant aliasing in this case is obvious. It should be mentioned that undersampling is sometimes used in digital sampling oscilloscopes to translate signals that would normally be too high in frequency to process down to frequencies that are within the capability of the scope to display.

In order to reduce the effects of aliasing on the sample data, a low-pass filter is used to band-limit the incoming signal. Such a filter is called an *anti-aliasing filter*. Figure 8-38a shows the connection of an anti-aliasing filter to an A/D converter.

Figure 8-38b shows the frequency response of the ideal anti-aliasing filter, where we have a brick-wall response and a cutoff frequency of $f_{sa}/2$. A practical low-pass filter will pass frequency components beyond f_c, and therefore the corner frequency of such a filter will be set below $f_{sa}/2$. Figure 8-39 shows the possible response curves for first- and second-order anti-aliasing

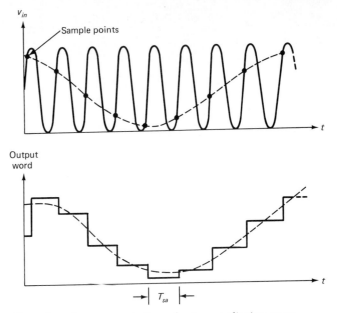

Fig. 8-37 Time domain representation of extreme aliasing error.

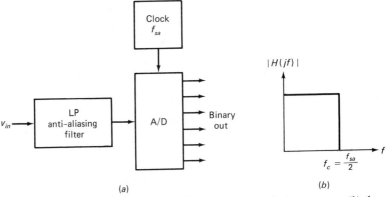

Fig. 8-38 The anti-aliasing filter. (*a*) Connection to the converter; (*b*) frequency response of the ideal filter.

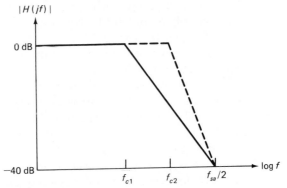

Fig. 8-39 Idealized response of practical anti-aliasing filters of first and second order.

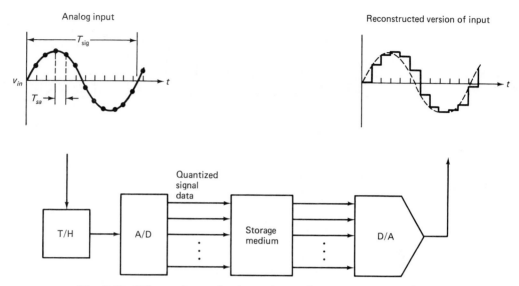

Fig. 8-40 Effects of quantization noise on the reconstruction of a digitized signal.

filters. As a general rule of thumb, the response of the filter should be down by at least 40 dB at $f_{sa}/2$. Here, we see that the higher the order of the anti-aliasing filter, the wider the range of input signal frequencies that can be accommodated. Passive *RC* filters or active filters may be used in the design of the anti-aliasing filter.

Quantization Noise and Reconstruction

Quantized (or digitized) signals may be processed and analyzed immediately, or the data may be stored for later use. Often, the digital data that represents the quantized signal must be converted back into its original analog form, as in the case of music that is stored on an audio compact disk (CD). Figure 8-40 illustrates this process. The upper right reconstructed version of the original signal shows the effects of the digitization process in the form of quantization noise.

Increasing the resolution of the A/D converter (and the D/A converter) will reduce the quantization noise and increase the signal-to-quantization-noise ratio (SQNR) of the output. As was shown in Fig. 8-35, quantization causes harmonics of f_{in} to appear in the sampled signal. In order to reconstruct the input signal more accurately, the output of the D/A converter should be processed through a low-pass filter called a *reconstruction filter*, as shown in Fig. 8-41. The reconstruction filter smooths out the discontinuities in the output of the D/A converter.

Once again, because of the finite rolloff rate of a practical filter, total elimination of the quantization noise is not possible. As a rule of thumb, the response of the reconstruction filter should be down by 40 dB at $(1/T_o) - f_{in(max)}$, where $1/T_o$ is the rate at which data are supplied to the D/A converter.

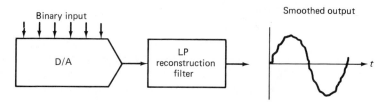

Fig. 8-41 Frequency response of a practical reconstruction filter.

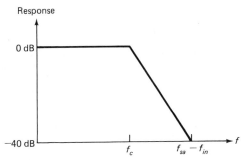

Fig. 8-42 Time domain representation of input (top trace) and output (bottom trace) of an 8-bit A/D–D/A conversion system, with $f_{in} = 1$ kHz and $f_{sa} = 1/T_o = 12.5$ kHz.

The corner frequency of the filter should be the same as, or preferably higher than, the highest signal frequency that is to be reproduced. The response curve for such a reconstruction filter is shown in Fig. 8-42. Normally, the sample readout frequency $1/T_o$ is the same as the original A/D sample frequency f_{sa}. This means that the frequency response is determined at the outset by the sampling and quantizing circuitry.

Oversampling the input signal allows lower-order anti-aliasing and reconstruction filters to be used, while still providing good rejection of the higher-frequency harmonics in the signal.

EXAMPLE 8-10

Refer to the system shown in Fig. 8-40. The A/D and D/A converters both have 10-bit resolution, $\pm\frac{1}{2}$-LSB accuracy, and $f_{sa} = 1/T_o = 50$ kHz. A second-order anti-aliasing filter with $f_c = 2$ kHz, unity gain, and Butterworth response is used at the input. Determine the following.

(a) The attenuation of the anti-aliasing filter at $f_{sa}/2$.

(b) The highest signal frequency that may be processed without producing aliasing.

(c) The SQNR of the A/D and D/A converters.

(d) The specifications required for a reconstruction filter.

> **Solution**
>
> (a) To determine the attenuation of the anti-aliasing filter at $f_{sa}/2$ (25 kHz), we refer to Eq. (6-16b). Setting $A = 1$, we obtain
>
> $|H(jf)| = -43.9$ dB
>
> (b) The converter is actually capable of quantizing signal frequencies up to 25 kHz, but the anti-aliasing filter limits the input bandwidth to 2 kHz.
>
> (c) Since the A/D and D/A converters have the same resolution and accuracy, they both have the same SQNR.
>
> SQNR = 60 dB
>
> (d) Because the A/D converter input signal frequency is limited to 2 kHz, the corner frequency of the reconstruction filter should also be 2 kHz. Clearly, a second-order Butterworth filter will suffice in this case, as the gain of the filter at $f_{sa} - f_{in(max)}$ (48 kHz) is -55 dB.

A more systematic method of determining the order of a Butterworth filter that is required to provide a given attenuation at a given frequency can be devised. Recall that the normalized amplitude response of an nth-order Butterworth LP filter is given by

$$|H(jf)| = \frac{1}{\sqrt{1 + (f_{in}/f_c)^{2n}}} \qquad (8\text{-}21)$$

where n is the order of the filter, f_c is the corner frequency of the filter, and f_{in} is the frequency of interest. Equation (8-21) may be solved for n, which leaves us an expression that can be used to determine the required filter order, given a specific corner frequency and a desired attenuation (gain). The reader is encouraged to solve Eq. (8-21) on his or her own.

The upper trace of the oscilloscope display in Fig. 8-43 shows a sinusoidal signal with $f_{in} = 1$ kHz that was applied to an actual system (with 8-bit resolution) like that of Fig. 8-40. The unfiltered, reconstructed output waveform is shown in the lower trace. In this case, the sampling rate and the readout rate are both 12.5 kHz.

The effects of quantization noise on the output are readily apparent in the time domain display of Fig. 8-43. The effects of quantization noise on the frequency domain representation of the signal are shown in Fig. 8-44, where 8-44a is the actual spectrum for the input signal and 8-44b is the spectrum of the reconstructed output.

340 CHAPTER 8 DATA CONVERSION DEVICES

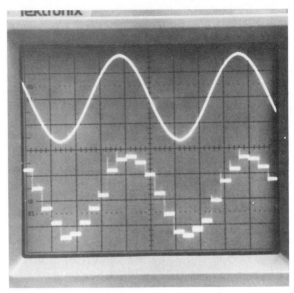

Fig. 8-43 Time domain representations for (*a*) the sinusoidal input signal and (*b*) the reconstructed output signal, without filtering.

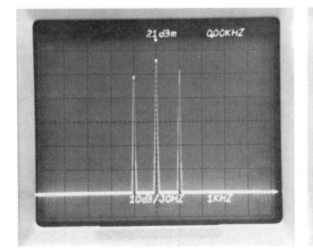
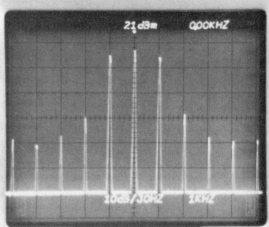

Fig. 8-44 Input signal spectrum (*a*) and spectrum of the reconstructed output signal (*b*).

The spectrum analyzer display is interpreted as shown in Fig. 8-45, where 0 frequency (dc) is located at the center of the screen. Each major division, moving out from the center, spans a distance of 1 kHz. The negative frequency components are produced as a result of the time-to-frequency-domain transformation (Fourier transform). In Fig. 8-44*b*, the important thing to notice is that quantization has produced output frequency components that were not in the original input. The undesired components lie at integer multiples of the input signal frequency.

8-5 DATA ACQUISITION AND DIGITAL SIGNAL PROCESSING

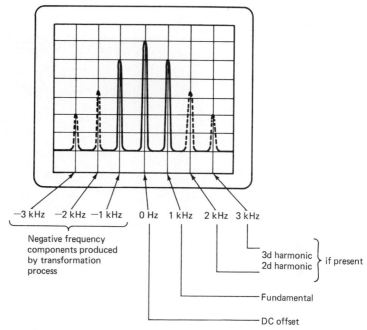

Fig. 8-45 Interpretation of the spectrum analyzer display.

Quantization noise was reduced by placing a second-order LP reconstruction filter with $f_c = 1.2$ kHz at the output of the D/A converter. This resulted in the time domain display of Fig. 8-46, where the smoothing effect of the reconstruction filter can easily be seen in the lower trace.

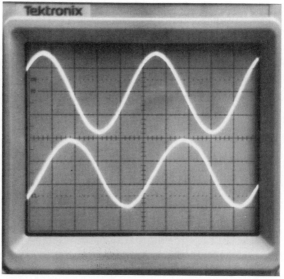

Fig. 8-46 System I/O relationships after the addition of a reconstruction filter.

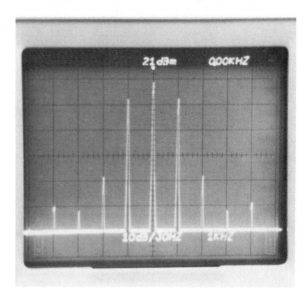

Fig. 8-47 Frequency domain representation of system output after the addition of a reconstruction filter.

The frequency domain representation of the filtered output is shown in Fig. 8-47. Notice that the undesired frequency components have been attenuated. Because the vertical (amplitude) axis of the display is in decibels, not linear units, the attenuation of the noise does not appear to be significant. However, this is not the case, as was apparent in Fig. 8-46.

Chapter Review

The output of a D/A converter is a voltage or current level that is proportional to the value of its binary input. The resolution of a D/A converter is determined by the number of input bits that the device has. The accuracy of a D/A converter may be specified as a fraction of 1 LSB, with ± 1 LSB to $\pm \frac{1}{4}$ LSB being common. Lower accuracy is usually specified by equivalence to a lower-resolution converter. The maximum rate at which a D/A converter can operate is determined by its settling time. D/A converters are usually designed using op amps and binary weighted resistors or R-2R ladder networks. All D/A converters require a stable and accurate reference voltage.

The output of an A/D converter is a binary code word whose numerical value is proportional to the value of an input voltage. The FSR of an A/D converter is the range of input voltages that can be quantized. At least $\pm \frac{1}{2}$ LSB of quantization uncertainty will always exist for any A/D converter. Conversion time determines the rate at which conversions can be performed by a given A/D. The various types of A/D converters available include ramp, SAR, integrating, parallel, and tracking converters. Track-and-hold amplifiers are often used in conjunction with A/D converters.

According to the Nyquist sampling theorem, the sample rate at which a given A/D converter is operated must be at least twice as high as the signal being quantized; otherwise aliasing error will exist. Anti-aliasing filters are used to band-limit signals applied to A/D converters to prevent aliasing. Reconstruction filters are used to remove high-frequency quantization noise from digitally reproduced signals.

Questions

8-1. What is the major source of inaccuracy associated with the D/A converter of Fig. 8-6?

8-2. What D/A converter parameter limits the rate at which the input code may be updated?

8-3. In terms of LSB size and V_{FS}, define the maximum output voltage that should be produced by a given D/A converter.

8-4. Theoretically, what is the minimum quantization uncertainty that can be obtained with any A/D converter?

8-5. What is the term applied to the act of assigning a finite value to a continuous variable?

8-6. Which type of D/A converter is most sensitive to variations of the input voltage during a conversion?

8-7. What type of error will exist if an analog signal is undersampled?

8-8. In general, what filter type would be used in the design of an anti-aliasing or reconstruction filter?

8-9. Which type of A/D converter provides the lowest conversion time?

8-10. In general, what effect does increasing the resolution of an A/D converter have on SQNR?

Problems

8-1. Determine the percent resolution and the LSB size for a 12-bit D/A converter with $V_{FS} = 5.00$ V.

8-2. Determine the minimum number of input bits required for a D/A converter with $V_{FS} = 10.00$ V, such that 1 LSB ≤ 80 mV.

8-3. A certain 6-bit D/A converter has $V_{fs} = 10.00$ V. Determine the ideal values of V_o for the following inputs:

(a) 010000

(b) 100000

(c) 100001

(d) 011111

8-4. Refer to Fig. 8-6. Given $R = 100$ kΩ and $V_{ref} = -10.00$ V, determine the required value of R_F such that $V_{FS} = 10.00$ V.

8-5. Refer to Fig. 8-7. Given $R_F = 50$ kΩ and $R = 200$ kΩ, determine V_{ref} such that $V_{FS} = 10.00$ V, and determine the necessary value for R_B.

8-6. Refer to Fig. 8-8. Given $R_F = 50$ kΩ, $R = 12.5$ kΩ, and $V_{ref} = 5.00$ V, determine the required value for R_1 such that $V_{FS} = 10.00$ V.

8-7. Given the conditions of Prob. 8-6, determine the effective resistances present at the input terminals of the op amp.

8-8. A DAC0808 has $V_{ref} = 5.00$ V. Determine the required values for R_1 and R_2 such that $I_{FS} = -1.50$ mA.

8-9. Refer to Fig. 8-10. Given $V_{ref} = 2.50$ V and $R_1 = R_2 = 2.50$ kΩ, determine I_{FS} and the value of R_L that will result in $V_{FS} = 10.00$ V.

8-10. An op amp I/V converter is to be placed at the output of a DAC0808 with $V_{ref} = 10.00$ V. Determine the necessary values for R_1, R_2, and R_F such that $I_{FS} = 2.00$ mA and $V_{FS} = 10.00$ V.

8-11. Refer to Fig. 8-12. Assume that an 8-bit D/A converter with accuracy of $\pm \frac{1}{4}$ LSB and $V_{FS} = 10.00$ V is used. Determine the binary input required to produce $5.30 \leq V_{fb} \leq 5.34$ V.

8-12. A certain application requires an A/D converter to resolve input voltages to within ± 100 mV. Assuming $\pm \frac{1}{2}$-LSB accuracy and FSR = 10.00 V, determine the required resolution of the A/D converter in bits.

8-13. A certain 8-bit A/D converter has an input range division of 31.250 mV. Determine the FSR of this converter.

8-14. Determine the minimum number of output bits required of an ideal A/D converter such that SQNR ≥ 70 dB.

8-15. Refer to Fig. 8-23. Assuming that the start of conversion is synchronized with the rising edge of the clock, PRF$_{clk}$ = 1 kHz, and all propagation delays are negligible, determine the minimum and maximum possible conversion times.

8-16. Refer to Fig. 8-25. The converter requires 3 clock cycles for each keep/discard test and for the initialization period. Given PRF$_{clk}$ = 500 kHz, and assuming that the converter is read and reset 6 μs after a given conversion is completed, determine the maximum sample frequency (f_{sa}) that could be obtained.

8-17. Refer to Fig. 8-25. Determine the value of V_{in} (ideally) that will produce an output of 10110000. (FSR = 10.00 V)

8-18. Refer to Fig. 8-23. Devise a digital circuit that could be added to this circuit that will output logic 1 if the input voltage results in an overrange condition (i.e., $V_{in} \geq$ FSR).

8-19. Refer to Fig. 8-28. Given $V_{ref} = -10.00$ V, $R = 100$ kΩ, $C = 0.1$ μF, $PRF_{clk} = 25.6$ kHz, and $V_{in} = 8.00$ V, determine the binary output that would be produced and the time required for the conversion. Assume that propagation delays are negligible, and $t_1 = 2^n \times T_{clk}$.

8-20. Repeat Prob. 8-19 for $V_{in} = 2.00$ V.

8-21. A certain A/D converter has $f_{sa} = 20$ kHz. Determine the highest input signal frequency that may be sampled without the introduction of aliasing error. Determine the aliased frequency component that would be produced if the input voltage had $f_{in} = 15$ kHz.

8-22. An A/D converter is required to quantize audio signals that range from 20 Hz to 20 kHz, with an oversampling factor of 10. Determine the necessary sample frequency. Determine the order required for a Butterworth anti-aliasing filter with $f_c = 20$ kHz such that response is down by at least 40 dB at $f_{sa}/2$.

8-23. An audio CD player uses a 16-bit D/A converter. Data are transferred to the converter, 16 bits at a time, at a 500-kHz rate. If a second-order Butterworth filter with $f_c = 20$ kHz is used as a reconstruction filter, determine the attenuation of the lowest-frequency quantization noise component.

8-24. A certain D/A converter has FSR = 5.00 V, with $0.00 \leq V_{in} \leq 5.00$ V. Design an op amp circuit that will allow signals in the range -2.50 V $\leq V_{in} \leq +2.50$ V to be digitized by the converter. Assume that a -5.00-V reference source is available.

CHAPTER 9
VOLTAGE REGULATORS

Up to this point, little has been said about the power sources that are required to operate the circuits that have been presented. Nearly all electronic circuits (digital and analog) require a stable dc power source for correct operation. Discrete semiconductors, ICs, and combinations of both types of devices are used to regulate power-supply outputs. This chapter provides an introduction to the basic principles of operation of voltage regulator circuits and devices.

9-1 ♦ UNREGULATED SUPPLIES

Direct-current power supplies may be classified as either single-polarity or bipolar. A typical unregulated single-polarity supply is shown in Fig. 9-1a, while an unregulated bipolar supply is shown in 9-1b. Both of these circuits employ full-wave bridge rectifiers and capacitive filtering.

Under no-load conditions, the output voltages of the circuits in Fig. 9-1 will be approximately equal to the peak amplitude of the rectified voltage. This output is called the *no-load voltage* V_{NL}. Application of a load will cause the output voltage to drop. Increasing the load (reducing the load resistance) causes the output voltage to drop even further. At some point, the supply will be fully loaded. That is, the load is drawing the maximum current that the supply is designed to handle. The output voltage under these conditions is called the full-load voltage, V_{FL}. The stability of the output voltage over this range of load conditions is expressed in terms of percent voltage regulation by the following equation.

$$\% \text{ VR}_{\text{load}} = \frac{V_{NL} - V_{FL}}{V_{FL}} \times 100 \qquad (9\text{-}1)$$

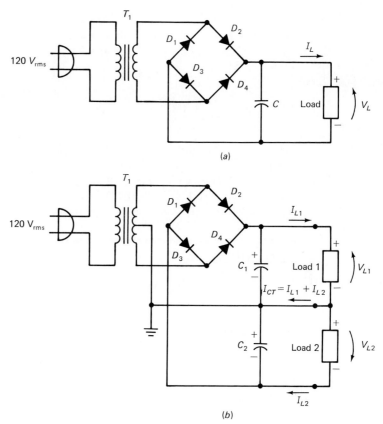

Fig. 9-1 Typical unregulated power supplies: (*a*) single-polarity; (*b*) bipolar.

A low value of VR_{load} indicates that the output voltage will remain relatively constant, regardless of the load. This is generally a very desirable characteristic. A typical plot of V_L versus I_L is shown in Fig. 9-2.

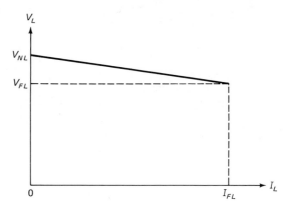

Fig. 9-2 Typical plot of load voltage versus load current.

The decrease in output (load) voltage depends on the effective output resistance of the power-supply circuitry and the resistance of the load being driven. Voltage regulation can also be defined in terms of these parameters as follows:

$$\% \text{ VR}_{\text{load}} = \frac{R_o}{R_L} \times 100 \tag{9-2}$$

where R_o is the effective internal resistance of the power supply. The internal resistance of the supply can be reduced by using a transformer with a low-resistance (heavy gage) secondary winding, and by using diodes in the bridge that exhibit very low bulk resistance. The use of a large filter capacitor also helps keep output variations to a minimum.

EXAMPLE 9-1

A certain power supply has $V_{NL} = 15.0$ V and $V_{FL} = 13.0$ V. Determine the percent regulation. If the output voltage drops to 13.0 V with a 100-Ω load, determine the internal resistance of the supply and the load current.

Solution

The percent regulation is found by direct application of Eq. (9-1):

$$\% \text{VR} = \frac{15 \text{ V} - 13 \text{ V}}{13 \text{ V}} \times 100$$

$$= 15.4\%$$

The internal resistance of the supply is found by rearranging Eq. (9-2), which produces

$$R_{\text{int}} = R_L \frac{\% \text{VR}}{100}$$

$$= 100 \text{ } \Omega \times 0.154$$

$$= 15.4 \text{ } \Omega$$

The full-load current is found by application of Ohm's law:

$$I_{FL} = \frac{V_{FL}}{R_{FL}}$$

$$= \frac{13.0 \text{ V}}{100 \text{ } \Omega}$$

$$= 130 \text{ mA}$$

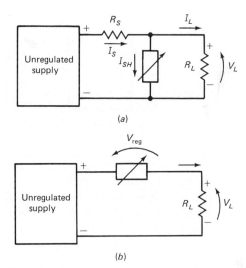

Fig. 9-3 Types of linear voltage regulators: (a) shunt regulator; (b) series regulator.

9-2 ♦ LINEAR VOLTAGE REGULATORS

In the preceding discussion it was stated that the regulation of a power supply can be improved by several means of lowering the effective output resistance. Probably the most effective method would be through the use of a larger transformer. Unfortunately, there are several major drawbacks to this approach, the most important of which is increased expense. The use of a voltage regulator provides a solution to this problem.

Linear voltage regulators fall into two categories, the shunt type and the series type. Simplified representations for both shunt and series regulators are shown in Fig. 9-3a and b, respectively.

Shunt Regulators

In the shunt regulator, the regulating element is in parallel with the load. Under no-load conditions, the current drawn from the unregulated supply, I_S, flows through a series resistor R_S and the regulating element; thus, $I_{SH} = I_S$. When a load is applied, I_S divides between the load and the regulator. Thus, the regulator effectively gives up current in order to maintain a constant voltage drop. As long as sufficient current continues to flow through the regulator, the output voltage will remain essentially constant. Zener diodes are probably the most commonly encountered shunt-type voltage regulators. A typical zener-regulated supply is shown in Fig. 9-4.

The effective output resistance of the zener regulator is approximately the same as the dynamic resistance r_z of the zener diode used in the circuit. Typically, zener dynamic resistances range from about 0.5 Ω to around 25 Ω.

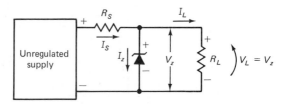

Fig. 9-4 Zener-diode-based shunt regulator.

An important characteristic of any shunt regulator is that under normal operating conditions, the current drawn from the unregulated supply is constant. Thus, under no load, full load, or something in between, the components making up the unregulated section of the supply are under constant load. Also, the shunt regulating element will dissipate maximum power under no-load conditions. The reader is encouraged to verify this independently.

Series Regulators

The series regulator of Fig. 9-3b operates in such a way that when the load current changes (due to a changing load resistance), the voltage drop across the regulating element varies, causing the load voltage to remain constant. Typically, a power transistor is used as a series regulating element; in such applications it is called a series pass transistor. A simple series regulator is shown in Fig. 9-5. In this circuit, R_1 provides base current to Q_1, the series pass transistor. The output voltage is sampled by the voltage divider formed by resistors R_3 and R_4. When the output voltage increases such that $V_{R4} > V_z + V_{BE}$, transistor Q_2 begins to conduct. This robs Q_1 of base current, effectively increasing its C-E resistance, and causes a decrease in V_L. At some point, equilibrium is achieved, and a constant load voltage results. If the load current should increase (possibly because of a decrease in load resistance), V_{CE1} increases, causing a decrease in V_L. This causes a decrease in V_{R4}, which

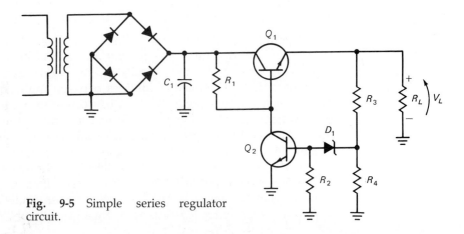

Fig. 9-5 Simple series regulator circuit.

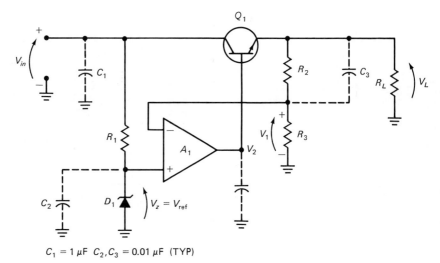

Fig. 9-6 Improved series regulator circuit.

in turn decreases the base drive on Q_2, causing more base current to be supplied to Q_1. The net effect of this action is a constant load voltage.

Ideally, the load voltage produced by Fig. 9-5 is given by

$$V_L = \frac{(V_z + V_{BE})(R_4 + R_3)}{R_4} \qquad (9\text{-}3)$$

Based on Eq. (9-3), we see that the output voltage can be varied by adjusting either R_3 or R_4.

The performance of the series regulator can be improved dramatically by using a high-gain amplifier to sense changes in the load voltage. This modification is shown in Fig. 9-6. In this circuit, a zener diode is used to develop a stable reference voltage, to which the sampled output voltage V_1 is compared. Let us assume that the output is unloaded and in equilibrium. Now, when a load is applied, V_L begins to decrease. This causes a decrease in V_1. Because $V_1 < V_{ref}$, the output of A_1 goes positive, turning Q_1 on harder. This decreases V_{CE} and increases V_L to its original level. That is, the load voltage is held such that $V_1 = V_{ref}$. The regulation of this circuit is primarily dependent on the stability of the internal reference voltage, and to a lesser extent on the gain of A_1. In general, higher gain means better regulation. The output voltage for this regulator is given by

$$V_o = V_{ref}\left(1 + \frac{R_2}{R_3}\right) \qquad (9\text{-}4)$$

Because of the high gain of the amplifier, oscillation may occur; thus, the regulator may require that one or more compensation capacitors be used. Typical placement of these capacitors is shown as the dashed portions of the circuit.

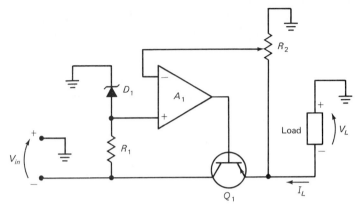

Fig. 9-7 Negative regulator.

A negative series regulator is shown in Fig. 9-7. The basic circuit operation is the same as that of Fig. 9-6, with the main difference being the use of a PNP series pass transistor. Notice also that a potentiometer is used to provide output voltage adjustment. This arrangement could also have been used in the other series regulators.

The main source of power dissipation within a series regulator is the series pass transistor. This power dissipation is given by

$$P_C = V_{CE}I_C \tag{9-5}$$

where V_{CE} is the difference between the output load voltage and the unregulated input voltage, and $I_C = I_L$. The lower the power dissipation of the regulator, the higher will be the efficiency of the power supply. Regulator efficiency is defined as

$$\eta = \frac{P_L}{P_L + P_{reg}} \tag{9-6a}$$

where P_{reg} is the power dissipated by the regulator and P_L is the power dissipated by the load. Because in most cases, most of the regulator power dissipation occurs in the pass transistor, we may also write

$$\eta = \frac{P_L}{P_L + P_C} \tag{9-6b}$$

EXAMPLE 9-2

The circuit of Fig. 9-6 has $V_{in} = 25$ V. Determine the efficiency of the regulator under the following conditions:

(a) $V_L = 15$ V, $I_L = 2$ A

(b) $V_L = 5$ V, $I_L = 2$ A

Solution

(a) $V_{CE1} = V_{in} - V_L$
$= 25$ V $- 15$ V
$= 10$ V

$P_C = I_C V_{CE}$
$= 2$ A $\times 10$ V
$= 20$ W

$P_L = I_L V_L$
$= 2$ A $\times 15$ V
$= 30$ W

$\eta = \dfrac{P_L}{P_L + P_C}$

$= \dfrac{30\text{ W}}{50\text{ W}}$

$= 60\%$

(b) $V_{CE1} = V_{in} - V_L$
$= 25$ V $- 5$ V
$= 20$ V

$P_C = I_C V_{CE}$
$= 2$ A $\times 20$ V
$= 40$ W

$P_L = I_L V_L$
$= 2$ A $\times 5$ V
$= 10$ W

$\eta = \dfrac{P_L}{P_L + P_C}$

$= \dfrac{10\text{ W}}{50\text{ W}}$

$= 20\%$

The preceding example indicates that for a given current, series regulators are most efficient when the load voltage is near maximum. A high-current, low-voltage load will cause the series pass transistor to dissipate substantial power. For this reason, the pass transistor will normally be mounted on a heat sink, often with forced-air cooling. In some cases, two or more pass transistors may be connected in parallel, as shown in Fig. 9-8. This reduces the power dissipation and collector currents for the individual transistors, resulting in lower junction temperatures. The emitter swamping resistors are used to provide negative feedback, causing the transistors to share the load current equally. This is necessary, because mismatch between devices could cause one transistor to carry more current than the other, possibly causing it to overheat and self-destruct. Typically, the values of such emitter resistors range from 0.1 Ω to 10 Ω.

Overcurrent Protection

An overload condition, such as a short circuit at the output of a power supply, can mean the destruction of the circuit. Usually, it will be the series pass transistor that will fail, as it is the main power-dissipating element in

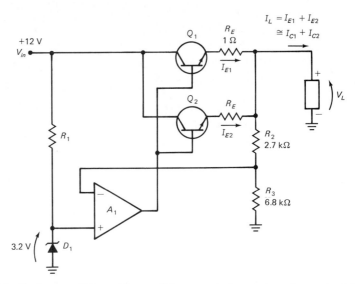

Fig. 9-8 Series regulator with parallel-connected pass transistors for higher-current operation.

the series regulator circuit. Occasionally, other supply components such as the rectifier or even the power transformer may be destroyed when overload conditions exist for long periods of time. In any case, current limiting is employed in many regulator designs, to help prevent these occurrences. A positive series regulator with current limiting is shown in Fig. 9-9. In this circuit, current sensing resistor R_{SC} and a second transistor Q_2 are added to implement overcurrent protection. The operation of the circuit under normal load conditions is the same as that of the previously discussed circuits, with R_{SC} and Q_2 having no effect on the output.

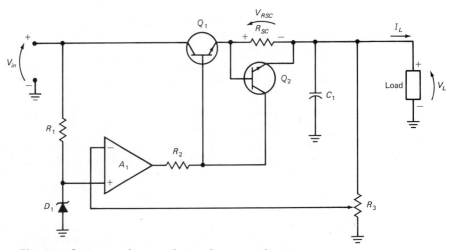

Fig. 9-9 Series regulator with simple current limiting.

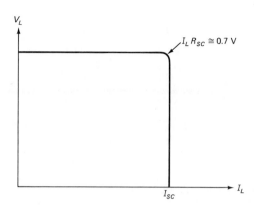

Fig. 9-10 Plot of V_L versus I_L for a regulator with simple current limiting.

Let us assume that a short circuit is placed across the output of the circuit in Fig. 9-9. This causes increases in I_{C1} and I_L, which in turn increases the voltage drop across R_{SC}. When I_L reaches a level where V_{RSC} is about 0.7 V, the base-emitter junction of Q_2 becomes forward-biased. Now Q_2 goes into conduction, diverting base current from Q_1. This tends to drive Q_1 toward cutoff, and results in a constant short-circuit current I_{SC} being drawn from the supply. The short-circuit current is approximated by

$$I_{SC} = \frac{0.7 \text{ V}}{R_{SC}} \tag{9-7}$$

A plot of V_L versus I_L for the current-limiting regulator is shown in Fig. 9-10.

EXAMPLE 9-3

A certain series-regulated supply has $V_{in} = 25$ V and $V_o = 15$ V, and is current-limited at 4 A. Determine the power dissipation of the regulator with a load that draws 2 A, and with a short circuit at the output terminals.

Solution

Under both normal load and short-circuited conditions, the following equation applies:

$$P_{reg} = (V_{in} - V_o) I_L$$

For the 2-A load, $V_{in} = 25$ V and $V_o = 15$ V; therefore

$$P_{reg} = 20 \text{ W}$$

Under short-circuited conditions, $V_{in} = 25$ V, $V_o = 0$ V, and $L_L = 4$ A; therefore

$P_{reg} = 100$ W

Even with current limiting, the series pass transistor may heat up excessively under long-term short-circuit conditions. For these reasons, other overload protection methods are also used.

Foldback Current Limiting

The idea behind foldback limiting is to reduce the load current and the load voltage once overload conditions occur. Figure 9-11 illustrates the V-I characteristics of a typical foldback-limited power supply. Here we see that once a maximum load current I_{knee} is reached, the output voltage drops and the load current is reduced. This reduces the power dissipation of the regulator and the remaining power-supply components, and it also affords protection to the device being driven by the power supply.

A simple foldback-limited regulator is shown in Fig. 9-12. Under normal load conditions, R_2, R_3, and Q_2 have no effect on circuit operation. Under overload conditions, the voltage drop across R_4 will exceed V_{BE2} and the drop across R_2. This drives Q_2 into conduction, robbing Q_1 of base current and causing a decrease in output voltage. The decrease in V_o further reduces the drop across R_2, driving Q_2 further toward saturation and reducing load current and voltage even further. Thus, the characteristics of Fig. 9-11 are produced. The knee and short-circuit currents are given by

$$I_{knee} = \frac{V_o R_2}{R_4 R_3} + \frac{0.7 \text{ V } (R_2 + R_3)}{R_4 R_3} \qquad (9\text{-}8)$$

$$I_{SC} = \frac{0.7 \text{ V}}{R_4} \left(\frac{R_2 + R_3}{R_3} \right) \qquad (9\text{-}9)$$

Foldback-limited supplies are usually used in high-current applications.

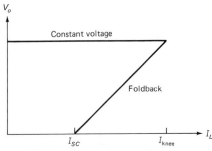

Fig. 9-11 Typical plot of V_L versus I_L for a regulator with foldback current limiting.

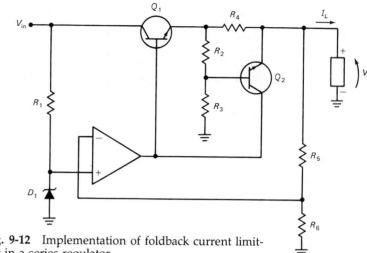

Fig. 9-12 Implementation of foldback current limiting in a series regulator.

9-3 ♦ MONOLITHIC LINEAR VOLTAGE REGULATORS

The majority of voltage regulators in use are monolithic IC types. Low-power devices are typically housed in plastic or ceramic DIPs or TO-5 cans, while higher-power devices usually come in TO-220 or TO-3 packages. In some cases, provision is made for the connection of external pass transistors and current-sensing resistors. Some units offer the option of foldback limiting, while others have thermal shutdown capability. In this section, we shall examine several popular monolithic voltage regulators.

The 723

The 723 is a versatile series regulator that is available in a 14-pin DIP (plastic or ceramic) or in a TO-5 metal can package. The pin designations for the 723 DIP and its equivalent internal circuitry are shown in Fig. 9-13*a* and *b*. A few of the major specifications of the 723 are listed below.

V_{CC} to V_{EE} differential: 40 V continuous maximum
          9 V minimum

P_D plastic DIP: 1.25 W
 ceramic DIP: 1.25 W
 TO-5 can: 1 W

I_{SC}: 65 mA (typ.)
Line regulation: 0.01%/V
Load regulation: 0.03%

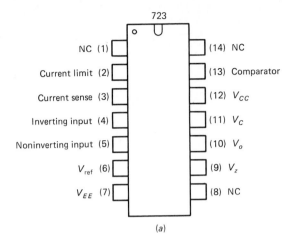

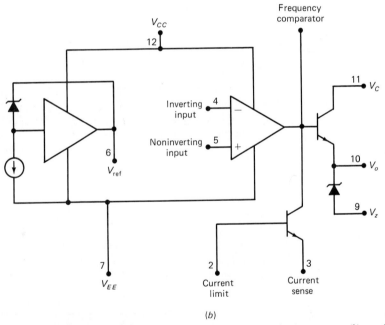

Fig. 9-13 The 723 monolithic voltage regulator: (*a*) pin designations; (*b*) equivalent internal circuitry.

The 723's specifications indicate that it is basically a low-power regulator. However, an external pass transistor can be used to increase current-handling capability. A new parameter, line regulation, has been introduced here. Line regulation is a measure of the regulator's ability to maintain a constant output voltage for variations of the unregulated input voltage. The 723's line regulation specification indicates that a 1-V change in V_{in} will produce a 0.01

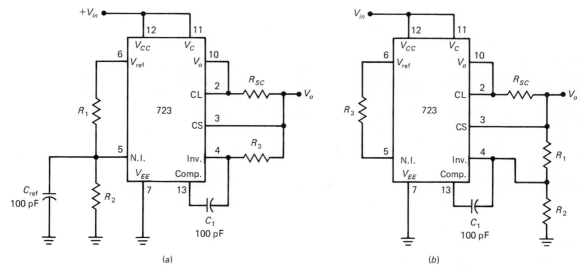

Fig. 9-14 Typical 723 configurations. (a) 723 connected to supply output voltages from 2 to 7 V; (b) 723 connected to supply output voltages from 7 to 37 V.

percent change in V_o. Most liner IC regulators have similar line regulation specifications.

Two typical regulator circuits using the 723 are shown in Fig. 9-14. For output voltages ranging from 2 to 7 V, the circuit of Fig. 9-14a is used. The output voltage is determined by the following equation:

$$V_o = V_{ref} \frac{R_2}{R_1 + R_2} \tag{9-10}$$

where $V_{ref} = 7.15$ V. For minimum output error, R_3 should be equal to the parallel combination of R_1 and R_2. Resistor R_2 should be limited to the following range of values:

$$10 \text{ k}\Omega < R_2 < 100 \text{ k}\Omega \tag{9-11}$$

The circuit of Fig. 9-14b is used to produce regulated voltages from 7 to 37 V. The output voltage is given by

$$V_o = V_{ref} \frac{R_1 + R_2}{R_2} \tag{9-12}$$

Resistor R_3 should be approximately equal to the parallel combination of resistors R_1 and R_2 for minimum output error, and R_2 is constrained by Eq. (9-11). For both circuits in Fig. 9-14, the short-circuit current is given by

$$I_{SC} = \frac{0.7 \text{ V}}{R_{SC}} \tag{9-13}$$

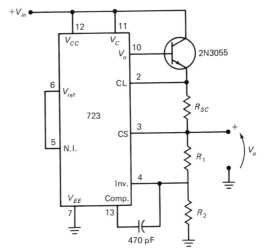

Fig. 9-15 Connection of an external pass transistor increases the current drive of the 723.

A series pass transistor may be connected to the 723 as shown in Fig. 9-15. The output voltage for this configuration is given by Eq. (9-12), and the short-circuit current is given by Eq. (9-13).

EXAMPLE 9-4

Refer to Fig. 9-14a. Given $R_2 = 27$ kΩ and $V_{in} = 18$ V, determine the required values for the remaining components such that $V_o = 5$ V and $I_{SC} = 50$ mA.

Solution

Solving Eq. (9-10) for R_1 yields

$R_1 = 11.6$ kΩ

$R_3 = R_1 \| R_2$
$ = 8.1$ kΩ

By application of Eq. (9-13), we obtain

$R_{SC} = 14$ Ω

The 117

The 117 is an adjustable three-terminal linear series voltage regulator that is available in TO-39, TO-92, TO-202, TO-220, and TO-3 packages. The 117 is adjustable over a range of about 2 to 37 V. Output current and power

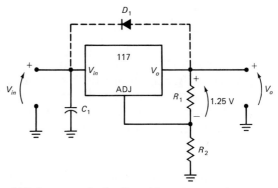

Fig. 9-16 The 117 three-terminal adjustable series regulator.

dissipation ratings depend on the case style used. At the high end, the TO-3 package can dissipate 20 W and has a load current rating of 1.5 A. At the opposite extreme, the TO-92 packaged device can dissipate 0.6 W and deliver a load current of 0.1 A. The 117 is designed to shut down automatically when the device temperature exceeds the design limit.

Figure 9-16 shows the 117 in a typical configuration. The output voltage is given by the following equation:

$$V_o = 1.25 \left(1 + \frac{R_2}{R_1}\right) \qquad (9\text{-}14)$$

Capacitor C_1 is recommended to help prevent oscillation, and is required when the regulator is located more than about 6 in. from the actual power supply. The output terminal of the 117 should always be at a lower potential than the input terminal. If a capacitive load is driven by the regulator, and the power supply is turned off, the load may hold the output terminal at a higher voltage than the input, possibly damaging the regulator. Diode D_1 is used to provide a discharge path, preventing such occurrences. The 117 develops a constant 1.25-V drop between the V_o and ADJ terminals. The ADJ terminal current, I_{ADJ}, is less than 100 µA, and causes negligible output voltage error.

Dual-Tracking Regulators

Bipolar power supplies can be regulated through the use of separate positive and negative regulator chips. However, linear IC voltage regulators are available that feature positive and negative regulators housed in a single case. Such devices are usually dual-tracking regulators. In a dual-tracking regulator, both outputs may be adjusted simultaneously by varying one resistor. Dual-tracking regulators maintain both supply rails at the same potential (with opposite polarity) relative to ground. Thus, load variations affecting one supply voltage will cause a similar change in the other output. In some cases, the output voltages may also be independently adjustable. An example of a dual-tracking regulator is the Motorola MC1568.

9-4 ◆ SWITCHING REGULATORS

It is apparent that in many applications, linear voltage regulators are quite inefficient. In the case of a series regulator, this inefficiency stems from the fact that the pass transistor is carrying the load current, which may be quite high, and is dropping the excess voltage not required by the load. Thus the series pass transistor is in the active region, and transistors dissipate the most power when they are in the active region.

In a switching regulator, the regulating element (transistor) is constantly switched between cutoff and saturation. In these two regions of operation, the transistor dissipates very little power (typically < 1 mW in cutoff and < 1 W in saturation). Because of this operation, switching regulators are quite efficient (often greater than 90 percent efficient), especially when high $V_{in} - V_o$ differentials are involved.

Switching Regulator Operation

There are three basic switching regulator configurations: step-down, step-up, and inverting. The basic circuit configuration for each type is shown in Fig. 9-17a, b, and c, respectively. Notice that each type requires the use of an inductor to filter the output voltage. The heart of the switching regulator is an oscillator with voltage-controlled pulse width. This is a PWM circuit like that discussed in Chap. 7. The output of the PWM switches a transistor between cutoff and saturation, with a duty cycle that is controlled by the difference between the feedback voltage V_{fb} and V_{ref}. Basically, if $V_{fb} > V_{ref}$, the duty cycle of the PWM is decreased until the reference and feedback voltages are equal. Conversely, if $V_{fb} < V_{ref}$, the duty cycle of the PWM is increased. Thus, the output voltage is proportional to the duty cycle of the PWM.

Typical PWM frequencies range from 1 kHz to 200 kHz. Because of the high ripple frequencies that are produced, the filter components may be relatively small and yet still achieve excellent ripple reduction. Most modern switching regulator designs employ power VMOS FETs as the switching elements. The major reason for this is that FETs can switch at higher rates than BJTs.

The Step-Down Configuration

Figure 9-17a illustrates the basic step-down switching regulator configuration. Transistor Q_1 supplies a pulse train to the filter composed of L_1 and C_1. Because of the inductor, the capacitor charges to the average of the input voltage waveform. The average output voltage is proportional to the duty cycle of the PWM output. Because of the relatively complex relationship among the frequency and the duty cycle of the PWM, the LC filter parameters, the output ripple voltage, and the load current, an expression for the output voltage will not be derived here. Diode D_1 provides a path for current when the magnetic field around L_1 collapses. The required values for the

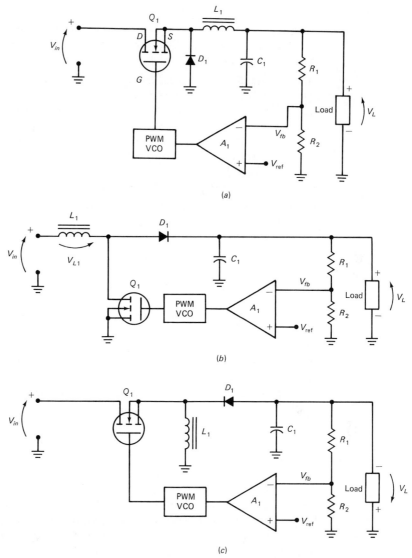

Fig. 9-17 The three basic switching regulator configurations: (a) step-down; (b) step-up; (c) inverting.

inductor and the capacitor are rather critical and are determined based on the switching frequency, allowed ripple, and allowed load current.

The Step-Up Configuration

The step-up configuration is shown in Fig. 9-17b. In this circuit, when Q_1 is turned on, a magnetic field builds up around the inductor. When the transistor is driven into cutoff, the inductor field begins to collapse, producing a voltage that is series-aiding with the input voltage. Capacitor C_1 charges to the sum of V_{L1} and V_{in}, and thus the load voltage is higher than the input voltage. The amplitude of the inductor voltage is proportional to the duty cycle of the switching waveform, which is in turn determined by the comparison of V_{fb} and V_{ref}.

The Inverting Configuration

Figure 9-17c shows an inverting switching regulator. Again, when the transistor is turned on, a magnetic field is built up around the inductor. Now, when the transistor is turned off, the inductor field begins to collapse. The inductor forces current to remain flowing in the direction which originally built up the magnetic field. Since the transistor is off, the inductor causes current to flow from ground, up through the load and D_1. The filter capacitor smoothes out the fluctuations in load voltage caused by the switching of Q_1. This circuit is useful where bipolar dc voltages are required and there is only a positive source available.

Additional Switching Regulator Considerations

The transistor in a switching regulator may be switching between essentially zero current and several amperes at a relatively high frequency. This switching action may produce higher harmonics (into the megahertz) of substantial power. For this reason, switching supplies are normally shielded in a metal or wire screen enclosure. Another technique used to prevent high-frequency noise from being radiated is the placement of ferrite beads around the output leads. Ferrite is a magnetic material that effectively absorbs energy at high frequencies. Also, to prevent noise from being impressed on the ac line, a filter may be placed at the input of the switching supply. These modifications are shown in Fig. 9-18.

There are integrated circuits that are specifically designed for use in switching regulator circuits. The application data for these devices contains information on selection of inductors and capacitors, based on design requirements. Often, inductors must be custom-wound or specially ordered for use in switching supplies. Magnetic core manufacturers can provide data relating to inductor design and construction.

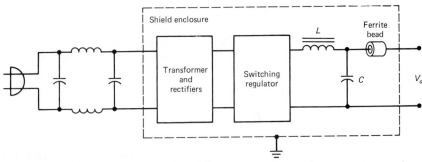

Fig. 9-18 Noise suppression techniques used with switching power supplies.

Chapter Review

Voltage regulators are used to provide a constant voltage to a load regardless of the current drawn by the load or variations of input and line voltage. Linear regulators are relatively simple and effective, but they tend to be inefficient. The majority of IC regulators are of the series type. Most voltage regulators employ some form of current limiting. Foldback limiting reduces the load current to a level that is safe for the power supply and the possibly defective load device.

Switching regulators are more complex than linear regulators, but they are usually much more efficient, weigh less, and take up less space than their linear equivalents. Switching regulators can provide an output voltage that is lower or greater than the input voltage, and they can produce an output with the opposite polarity from the input. Precautions must be taken to ensure that excessive noise is not radiated from a switching power supply.

Questions

9-1. In general, when will a linear series regulator operate most efficiently?

9-2. Which component of a series regulator will dissipate the most power?

9-3. What factor(s) determine the overall stability of a series regulator?

9-4. In general, under what conditions will a shunt-type regulator operate most efficiently?

9-5. State one advantage of the shunt regulator over an equivalent series regulator.

9-6. State two disadvantages of switching regulators, as compared with linear regulators.

9-7. List the three basic switching regulator configurations.

9-8. What pulse parameter controls the output voltage produced by a switching regulator?

9-9. What type of current limiting reduces short-circuit current to a safe level?

Problems

9-1. Refer to Fig. 9-1. Given load 1 = 500 Ω, load 2 = 250 Ω, V_{L1} = 12 V, and V_{L2} = 12 V, determine I_{L1}, I_{L2}, and I_{CT}.

9-2. A certain power supply has the following performance specifications: V_{NL} = 15 V, load regulation = 0.5 percent. Determine the full-load output voltage.

9-3. Refer to Fig. 9-3b. Given $V_{in} = 12$ V, $V_L = 5$ V, and $I_L = 2.5$ A, determine the power dissipated by the regulator and the efficiency of the regulator.

9-4. Refer to Fig. 9-6. $V_{in} = 18$ V, $V_z = 4.2$ V, $I_{zT} = 5.0$ mA, and $R_2 = 2.2$ kΩ. Determine the required values for R_1 and R_3 such that $V_L = 10$ V.

9-5. Given the conditions stated in Prob. 9-4, determine the power dissipation of Q_1 when $R_L = 50$ Ω.

9-6. Refer to Fig. 9-8. Given $V_{in} = 12$ V, $V_z = 3.2$ V, $R_2 = 2.7$ kΩ, $R_3 = 6.8$ kΩ, $R_E = 1$ Ω, and $R_L = 20$ Ω, determine V_o, I_L, I_{C1}, I_{C2}, P_{C1}, P_{C2}, P_L, the power dissipation of each emitter resistor, the total power dissipation of the regulator, and the efficiency of the regulator. Assume that op amp, zener, and output sampling losses are negligible.

9-7. Refer to Fig. 9-9. Determine the required value of R_{SC} such that $I_{SC} = 5$ A. Determine the power dissipation of R_{SC} under short-circuit conditions.

9-8. Refer to Fig. 9-12. Given $R_2 = 4.7$ kΩ, $R_3 = 1.2$ kΩ, $R_4 = 10$ Ω, and $V_o = 10$ V, determine I_{knee} and I_{SC}.

9-9. Repeat Prob. 9-8 using $R_2 = 6.2$ kΩ, $R_3 = 1.5$ kΩ, $R_4 = 5$ Ω, and $V_o = 5$ V.

9-10. Refer to Fig. 9-14b. Given $R_1 = 10$ kΩ, determine the required values for R_2, R_3, and R_{SC} such that $V_o = 12$ V and $I_{SC} = 25$ mA. Determine the minimum value of V_{in} that is necessary for correct regulator operation.

9-11. Refer to Fig. 9-15. The external pass transistor is required to deliver a current of 1 A to a load. Assuming that the 723 can source 65 mA, determine the minimum β required of the pass transistor.

9-12. A certain voltage regulator has a line regulation of 1 percent per volt. Given $V_o = 10.0$ V when $V_{in} = 20.0$ V, determine V_o for $V_{in} = 25$ V.

9-13. Refer to Fig. 9-16. Given $R_2 = 470$ Ω and $V_{in} = 25$ V, determine the required value for R_1 such that $V_o = 12$ V.

9-14. Refer to Fig. 9-19. The 117 is used as a constant current source, with $V_{in} = 20$ V and $R_1 = 3$ Ω. Determine the load current and the power dissipation of R_1. Assuming that the load is a 25-Ω resistor, determine the power dissipation of the load and the 117 regulator.

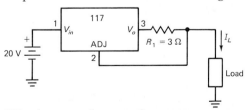

Fig. 9-19 A 117 voltage regulator used as a constant current source.

APPENDIX A
DATA SHEETS

The following data sheets are reprinted courtesy of National Semiconductor Corporation, 2900 Semiconductor Drive, Santa Clara, California 95051.

LM3045, LM3046, LM3086 Transistor Arrays

LM108A/LM208A/LM308A, LM308A-1, LM308A-2 Operational Amplifiers

LM741/LM741A/LM741C/LM741E Operational Amplifier

LH0036/LH0036C Instrumentation Amplifier

LM3080/LM3080A Operational Transconductance Amplifier

LM1596/LM1496 Balanced Modulator-Demodulator

LM555/LM555C Timer

LM567/LM567C Tone Decoder

Transistor/Diode Arrays

LM3045, LM3046, LM3086 Transistor Arrays

General Description

The LM3045, LM3046, and LM3086 each consist of five general purpose silicon NPN transistors on a common monolithic substrate. Two of the transistors are internally connected to form a differentially-connected pair. The transistors are well suited to a wide variety of applications in low power system in the DC through VHF range. They may be used as discrete transistors in conventional circuits however, in addition, they provide the very significant inherent integrated circuit advantages of close electrical and thermal matching. The LM3045 is supplied in a 14-lead cavity dual-in-line package rated for operation over the full military temperature range. The LM3046 and LM3086 are electrically identical to the LM3045 but are supplied in a 14-lead molded dual-in-line package for applications requiring only a limited temperature range.

Features

- Two matched pairs of transistors
  V_{BE} matched ±5 mV
  Input offset current 2μA max at I_C = 1 mA
- Five general purpose monolithic transistors
- Operation from DC to 120 MHz
- Wide operating current range
- Low noise figure   3.2 dB typ at 1 kHz
- Full military
  temperature range (LM3045) $-55°C$ to $+125°C$

Applications

- General use in all types of signal processing systems operating anywhere in the frequency range from DC to VHF
- Custom designed differential amplifiers
- Temperature compensated amplifiers

Schematic and Connection Diagram

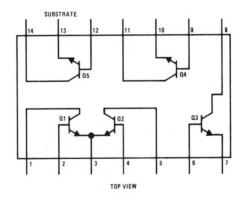

Dual-In-Line Package

TOP VIEW

Order Number LM3045J
See NS Package J14A
Order Number LM3046N
or LM3086N
See NS Package N14A

Absolute Maximum Ratings ($T_A = 25°C$)

	LM3045		LM3046/LM3086		
	Each Transistor	Total Package	Each Transistor	Total Package	Units
Power Dissipation:					
$T_A = 25°C$	300	750	300	750	mW
$T_A = 25°C$ to $55°C$			300	750	mW
$T_A > 55°C$			Derate at 6.67		mW/°C
$T_A = 25°C$ to $75°C$	300	750			mW
$T_A > 75°C$	Derate at 8				mW/°C
Collector to Emitter Voltage, V_{CEO}	15		15		V
Collector to Base Voltage, V_{CBO}	20		20		V
Collector to Substrate Voltage, V_{CIO} (Note 1)	20		20		V
Emitter to Base Voltage, V_{EBO}	5		5		V
Collector Current, I_C	50		50		mA
Operating Temperature Range	$-55°C$ to $+125°C$		$-40°C$ to $+85°C$		
Storage Temperature Range	$-65°C$ to $+150°C$		$-65°C$ to $+85°C$		
Lead Temperature (Soldering, 10 sec)	300		300		°C

Electrical Characteristics ($T_A = 25°C$ unless otherwise specified)

PARAMETER	CONDITIONS	LIMITS LM3045, LM3046			LIMITS LM3086			UNITS
		MIN	TYP	MAX	MIN	TYP	MAX	
Collector to Base Breakdown Voltage ($V_{(BR)CBO}$)	$I_C = 10\mu A$, $I_E = 0$	20	60		20	60		V
Collector to Emitter Breakdown Voltage ($V_{(BR)CEO}$)	$I_C = 1$ mA, $I_B = 0$	15	24		15	24		V
Collector to Substrate Breakdown Voltage ($V_{(BR)CIO}$)	$I_C = 10\mu A$, $I_{CI} = 0$	20	60		20	60		V
Emitter to Base Breakdown Voltage ($V_{(BR)EBO}$)	$I_E = 10\mu A$, $I_C = 0$	5	7		5	7		V
Collector Cutoff Current (I_{CBO})	$V_{CB} = 10V$, $I_E = 0$		.002	40		.002	100	nA
Collector Cutoff Current (I_{CEO})	$V_{CE} = 10V$, $I_B = 0$			.5			5	μA
Static Forward Current Transfer Ratio (Static Beta) (h_{FE})	$V_{CE} = 3V$ $\begin{cases} I_C = 10\text{ mA} \\ I_C = 1\text{ mA} \\ I_C = 10\mu A \end{cases}$	40	100 100 54		40	100 100 54		
Input Offset Current for Matched Pair Q_1 and Q_2 $|I_{IO1} - I_{IO2}|$	$V_{CE} = 3V$, $I_C = 1$ mA		.3	2				μA
Base to Emitter Voltage (V_{BE})	$V_{CE} = 3V$ $\begin{cases} I_E = 1\text{ mA} \\ I_E = 10\text{ mA} \end{cases}$		.715 .800			.715 .800		V
Magnitude of Input Offset Voltage for Differential Pair $|V_{BE1} - V_{BE2}|$	$V_{CE} = 3V$, $I_C = 1$ mA		.45	5				mV
Magnitude of Input Offset Voltage for Isolated Transistors $|V_{BE3} - V_{BE4}|$, $|V_{BE4} - V_{BE5}|$, $|V_{BE5} - V_{BE3}|$	$V_{CE} = 3V$, $I_C = 1$ mA		.45	5				mV
Temperature Coefficient of Base to Emitter Voltage $\left(\frac{\Delta V_{BE}}{\Delta T}\right)$	$V_{CE} = 3V$, $I_C = 1$ mA		-1.9			-1.9		mV/°C
Collector to Emitter Saturation Voltage ($V_{CE(SAT)}$)	$I_B = 1$ mA, $I_C = 10$ mA		.23			.23		V
Temperature Coefficient of Input Offset Voltage $\left(\frac{\Delta V_{IO}}{\Delta T}\right)$	$V_{CE} = 3V$, $I_C = 1$ mA		1.1					$\mu V/°C$

Note 1: The collector of each transistor of the LM3045, LM3046, and LM3086 is isolated from the substrate by an integral diode. The substrate (terminal 13) must be connected to the most negative point in the external circuit to maintain isolation between transistors and to provide for normal transistor action.

Electrical Characteristics (Continued)

PARAMETER	CONDITIONS	MIN	TYP	MAX	UNITS
Low Frequency Noise Figure (NF)	$f = 1$ kHz, $V_{CE} = 3$V, $I_C = 100\mu$A, $R_S = 1$ kΩ		3.25		dB
Low Frequency, Small Signal Equivalent Circuit Characteristics:					
Forward Current Transfer Ratio (h_{fe})			110 (LM3045, LM3046) (LM3086)		
Short Circuit Input Impedance (h_{ie})	$f = 1$ kHz, $V_{CE} = 3$V, $I_C = 1$ mA		3.5		kΩ
Open Circuit Output Impedance (h_{oe})			15.6		μmho
Open Circuit Reverse Voltage Transfer Ratio (h_{re})			1.8×10^{-4}		
Admittance Characteristics:					
Forward Transfer Admittance (Y_{fe})			$31 - j\,1.5$		
Input Admittance (Y_{ie})	$f = 1$ MHz, $V_{CE} = 3$V, $I_C = 1$ mA		$0.3 + j\,0.04$		
Output Admittance (Y_{oe})			$0.001 + j\,0.03$		
Reverse Transfer Admittance (Y_{re})			See curve		
Gain Bandwidth Product (f_T)	$V_{CE} = 3$V, $I_C = 3$ mA	300	550		
Emitter to Base Capacitance (C_{EB})	$V_{EB} = 3$V, $I_E = 0$		.6		pF
Collector to Base Capacitance (C_{CB})	$V_{CB} = 3$V, $I_C = 0$		.58		pF
Collector to Substrate Capacitance (C_{CI})	$V_{CS} = 3$V, $I_C = 0$		2.8		pF

Typical Performance Characteristics

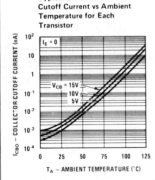

Typical Collector To Base Cutoff Current vs Ambient Temperature for Each Transistor

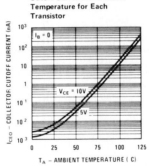

Typical Collector To Emitter Cutoff Current vs Ambient Temperature for Each Transistor

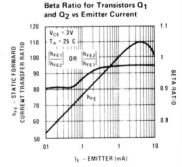

Typical Static Forward Current-Transfer Ratio and Beta Ratio for Transistors Q_1 and Q_2 vs Emitter Current

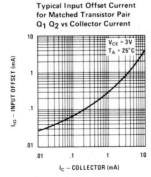

Typical Input Offset Current for Matched Transistor Pair $Q_1\,Q_2$ vs Collector Current

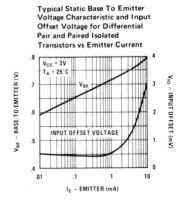

Typical Static Base To Emitter Voltage Characteristic and Input Offset Voltage for Differential Pair and Paired Isolated Transistors vs Emitter Current

Typical Performance Characteristics (Continued)

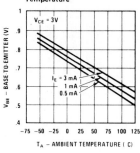

Typical Base To Emitter Voltage Characteristic for Each Transistor vs Ambient Temperature

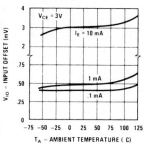

Typical Input Offset Voltage Characteristics for Differential Pair and Paired Isolated Transistors vs Ambient Temperature

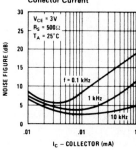

Typical Noise Figure vs Collector Current

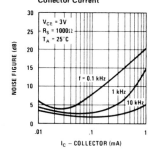

Typical Noise Figure vs Collector Current

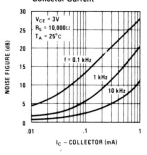

Typical Noise Figure vs Collector Current

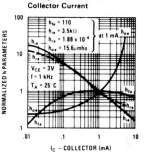

Typical Normalized Forward Current Transfer Ratio, Short Circuit Input Impedance, Open Circuit Output Impedance, and Open Circuit Reverse Voltage Transfer Ratio vs Collector Current

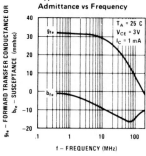

Typical Forward Transfer Admittance vs Frequency

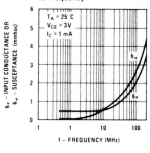

Typical Input Admittance vs Frequency

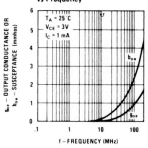

Typical Output Admittance vs Frequency

Typical Performance Characteristics (Continued)

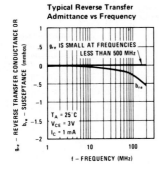

Typical Reverse Transfer Admittance vs Frequency

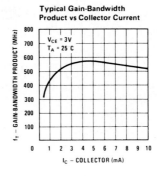

Typical Gain-Bandwidth Product vs Collector Current

National Semiconductor

Operational Amplifiers/Buffers

LM108A/LM208A/LM308A, LM308A-1, LM308A-2 Operational Amplifiers

General Description

The LM108/LM108A series are precision operational amplifiers having specifications about a factor of ten better than FET amplifiers over their operating temperature range. In addition to low input currents, these devices have extremely low offset voltage, making it possible to eliminate offset adjustments, in most cases, and obtain performance approaching chopper stabilized amplifiers.

The devices operate with supply voltages from ±2V to ±18V and have sufficient supply rejection to use unregulated supplies. Although the circuit is interchangeable with and uses the same compensation as the LM101A, an alternate compensation scheme can be used to make it particularly insensitive to power supply noise and to make supply bypass capacitors unnecessary. Outstanding characteristics include:

- Offset voltage guaranteed less than 0.5 mV
- Maximum input bias current of 3.0 nA over temperature
- Offset current less than 400 pA over temperature
- Supply current of only 300 µA, even in saturation
- Guaranteed 5 µV/°C drift.
- Guaranteed 1 µV/°C for LM308A-1

The low current error of the LM108A series makes possible many designs that are not practical with conventional amplifiers. In fact, it operates from 10 MΩ source resistances, introducing less error than devices like the 709 with 10 kΩ sources. Integrators with drifts less than 500 µV/sec and analog time delays in excess of one hour can be made using capacitors no larger than 1 µF.

The LM208A is identical to the LM108A, except that the LM208A has its performance guaranteed over a −25°C to 85°C temperature range, instead of −55°C to 125°C. The LM308A devices have slightly-relaxed specifications and performance guaranteed over a 0°C to 70°C temperature range.

Compensation Circuits

Standard Compensation Circuit

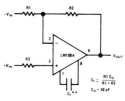

Alternate* Frequency Compensation

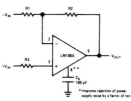

Feedforward Compensation

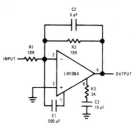

Typical Applications

Sample and Hold

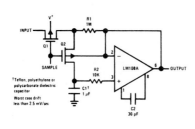

High Speed Amplifier with Low Drift and Low Input Current

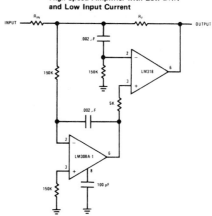

LM108A/LM208A
Absolute Maximum Ratings

Supply Voltage	±20V
Power Dissipation (Note 1)	500 mW
Differential Input Current (Note 2)	±10 mA
Input Voltage (Note 3)	±15V
Output Short-Circuit Duration	Indefinite
Operating Temperature Range LM108A	−55°C to 125°C
LM208A	−25°C to 85°C
Storage Temperature Range	−65°C to 150°C
Lead Temperature (Soldering, 10 sec)	300°C

Electrical Characteristics (Note 4)

PARAMETER	CONDITIONS	MIN	TYP	MAX	UNITS
Input Offset Voltage	$T_A = 25°C$		0.3	0.5	mV
Input Offset Current	$T_A = 25°C$		0.05	0.2	nA
Input Bias Current	$T_A = 25°C$		0.8	2.0	nA
Input Resistance	$T_A = 25°C$	30	70		MΩ
Supply Current	$T_A = 25°C$		0.3	0.6	mA
Large Signal Voltage Gain	$T_A = 25°C$, $V_S = ±15V$ $V_{OUT} = ±10V$, $R_L \geq 10 \, k\Omega$	80	300		V/mV
Input Offset Voltage				1.0	mV
Average Temperature Coefficient of Input Offset Voltage			1.0	5.0	µV/°C
Input Offset Current				0.4	nA
Average Temperature Coefficient of Input Offset Current			0.5	2.5	pA/°C
Input Bias Current				3.0	nA
Supply Current	$T_A = +125°C$		0.15	0.4	mA
Large Signal Voltage Gain	$V_S = ±15V$, $V_{OUT} = ±10V$ $R_L \geq 10 \, k\Omega$	40			V/mV
Output Voltage Swing	$V_S = ±15V$, $R_L = 10 \, k\Omega$	±13	±14		V
Input Voltage Range	$V_S = ±15V$	±13.5			V
Common Mode Rejection Ratio		96	110		dB
Supply Voltage Rejection Ratio		96	110		dB

Note 1: The maximum junction temperature of the LM108A is 150°C, while that of the LM208A is 100°C. For operating at elevated temperatures, devices in the TO-5 package must be derated based on a thermal resistance of 150°C/W, junction to ambient, or 45°C/W, junction to case. The thermal resistance of the dual-in-line package is 100°C/W, junction to ambient.

Note 2: The inputs are shunted with back-to-back diodes for overvoltage protection. Therefore, excessive current will flow if a differential input voltage in excess of 1V is applied between the inputs unless some limiting resistance is used.

Note 3: For supply voltages less than ±15V, the absolute maximum input voltage is equal to the supply voltage.

Note 4: These specifications apply for ±5V $\leq V_S \leq$ ±20V and −55°C $\leq T_A \leq$ 125°C, unless otherwise specified. With the LM208A, however, all temperature specifications are limited to −25°C $\leq T_A \leq$ 85°C.

LM308A, LM308A-1, LM308A-2

Absolute Maximum Ratings

Supply Voltage	±18V
Power Dissipation (Note 1)	500 mW
Differential Input Current (Note 2)	±10 mA
Input Voltage (Note 3)	±15V
Output Short-Circuit Duration	Indefinite
Operating Temperature Range	0°C to 70°C
Storage Temperature Range	−65°C to 150°C
Lead Temperature (Soldering, 10 sec)	300°C

Electrical Characteristics (Note 4)

PARAMETER	CONDITIONS	MIN	TYP	MAX	UNITS
Input Offset Voltage	$T_A = 25°C$		0.3	0.5	mV
Input Offset Current	$T_A = 25°C$		0.2	1	nA
Input Bias Current	$T_A = 25°C$		1.5	7	nA
Input Resistance	$T_A = 25°C$	10	40		MΩ
Supply Current	$T_A = 25°C$, $V_S = ±15V$		0.3	0.8	mA
Large Signal Voltage Gain	$T_A = 25°C$, $V_S = ±15V$, $V_{OUT} = ±10V$, $R_L \geq 10\ k\Omega$	80	300		V/mV
Input Offset Voltage	$V_S = ±15V$, $R_S = 100\Omega$				
LM308A				0.73	mV
LM308A-1				0.54	mV
LM308A-2				0.59	mV
Average Temperature Coefficient of Input Offset Voltage	$V_S = ±15V$, $R_S = 100\Omega$				
LM308A			2.0	5.0	μV/°C
LM308A-1			0.6	1.0	μV/°C
LM308A-2			1.3	2.0	μV/°C
Input Offset Current				1.5	nA
Average Temperature Coefficient of Input Offset Current			2.0	10	pA/°C
Input Bias Current				10	nA
Large Signal Voltage Gain	$V_S = ±15V$, $V_{OUT} = ±10V$, $R_L \geq 10\ k\Omega$	60			V/mV
Output Voltage Swing	$V_S = ±15V$, $R_L = 10\ k\Omega$	±13	±14		V
Input Voltage Range	$V_S = ±15V$	±14			V
Common-Mode Rejection Ratio		96	110		dB
Supply Voltage Rejection Ratio		96	110		dB

Note 1: The maximum junction temperature of the LM308A, LM308-1 and LM308-2 is 85°C. For operating at elevated temperatures, devices in the TO-5 package must be derated based on a thermal resistance of 150°C/W, junction to ambient, or 45°C/W, junction to case. The thermal resistance of the dual-in-line package is 100°C/W junction to ambient.

Note 2: The inputs are shunted with back-to-back diodes for overvoltage protection. Therefore, excessive current will flow if a differential input voltage in excess of 1V is applied between the inputs unless some limiting resistance is used.

Note 3: For supply voltages less than ±15V, the absolute maximum input voltage is equal to the supply voltage.

Note 4: These specifications apply for $±5V \leq V_S \leq ±15V$ and $0°C \leq T_A \leq 70°C$, unless otherwise specified.

Application Hints

A very low drift amplifier poses some uncommon application and testing problems. Many sources of error can cause the apparent circuit drift to be much higher than would be predicted.

Thermocouple effects caused by temperature gradient across dissimilar metals are perhaps the worst offenders. Only a few degrees gradient can cause hundreds of microvolts of error. The two places this shows up, generally, are the package-to printed circuit board interface and temperature gradients across resistors. Keeping package leads short and the two input leads close together help greatly.

Resistor choice as well as physical placement is important for minimizing thermocouple effects. Carbon, oxide film and some metal film resistors can cause large thermocouple errors. Wirewound resistors of evanohm or manganin are best since they only generate about 2 $\mu V/°C$ referenced to copper. Of course, keeping the resistor ends at the same temperature is important. Generally, shielding a low drift stage electrically and thermally will yield good results.

Resistors can cause other errors besides gradient generated voltages. If the gain setting resistors do not track with temperature a gain error will result. For example a gain of 1000 amplifier with a constant 10 mV input will have a 10V output. If the resistors mistrack by 0.5% over the operating temperature range, the error at the output is 50 mV. Referred to input, this is a 50 μV error. All of the gain fixing resistor should be the same material.

Offset balancing the LM308A-1 can be a problem since there is no easy offset adjustment incorporated into the circuit. These devices are selected for low drift with no offset adjustment to the internal circuitry, so any change of the internal currents will change the drift — probably for the worse. Offset adjustment must be done at the input. The three most commonly needed circuits are shown here.

Testing low drift amplifiers is also difficult. Standard drift testing technique such as heating the device in an oven and having the leads available through a connector, thermoprobe, or the soldering iron method — do not work. Thermal gradients cause much greater errors than the amplifier drift. Coupling microvolt signal through connectors is especially bad since the temperature difference across the connector can be 50°C or more. The device under test along with the gain setting resistor should be isothermal. The following circuit will yield good results if well constructed.

Offset Adjustment for Inverting Amplifiers

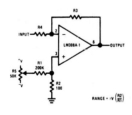

Offset Adjustment for Non-Inverting Amplifiers

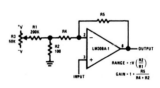

Offset Adjustment for Differential Amplifiers

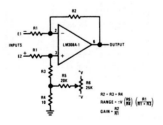

Drift Measurement Circuit

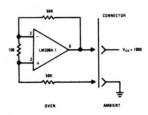

Schematic Diagram*

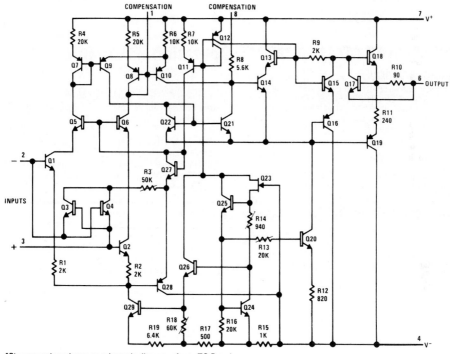

*Pin connections shown on schematic diagram refer to TO-5 package.

Connection Diagrams

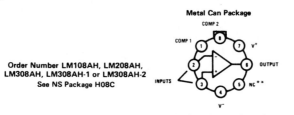

Metal Can Package

Order Number LM108AH, LM208AH,
LM308AH, LM308AH-1 or LM308AH-2
See NS Package H08C

**Unused pin (no internal connection) to allow for input anti-leakage guard ring on printed circuit board layout.

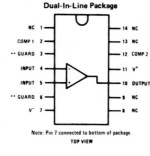

Dual-In-Line Package

Order Number LM108AJ, LM208AJ,
or LM308AJ
See NS Package J14A

Note: Pin 7 connected to bottom of package.
TOP VIEW

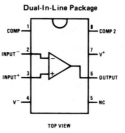

Dual-In-Line Package

Order Number LM108AJ-8,
LM208AJ-8 or LM308AJ-8
See NS Package J08A
Order Number LM208AN
or LM308AN
See NS Package N08B

TOP VIEW

377

Operational Amplifiers/Buffers

LM741/LM741A/LM741C/LM741E Operational Amplifier

General Description

The LM741 series are general purpose operational amplifiers which feature improved performance over industry standards like the LM709. They are direct, plug-in replacements for the 709C, LM201, MC1439 and 748 in most applications.

The amplifiers offer many features which make their application nearly foolproof: overload protection on the input and output, no latch-up when the common mode range is exceeded, as well as freedom from oscillations.

The LM741C/LM741E are identical to the LM741/LM741A except that the LM741C/LM741E have their performance guaranteed over a 0°C to +70°C temperature range, instead of −55°C to +125°C.

Schematic and Connection Diagrams (Top Views)

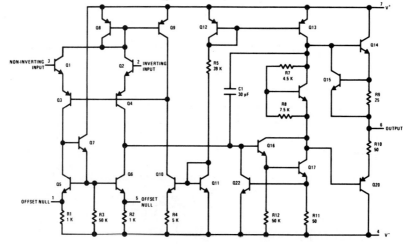

Metal Can Package

Order Number LM741H, LM741AH,
LM741CH or LM741EH
See NS Package H08C

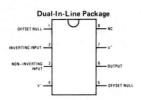

Dual-In-Line Package

Order Number LM741CN or LM741EN
See NS Package N08B
Order Number LM741CJ
See NS Package J08A

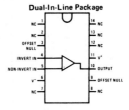

Dual-In-Line Package

Order Number LM741CN-14
See NS Package N14A
Order Number LM741J-14, LM741AJ-14
or LM741CJ-14
See NS Package J14A

Absolute Maximum Ratings

	LM741A	LM741E	LM741	LM741C
Supply Voltage	±22V	±22V	±22V	±18V
Power Dissipation (Note 1)	500 mW	500 mW	500 mW	500 mW
Differential Input Voltage	±30V	±30V	±30V	±30V
Input Voltage (Note 2)	±15V	±15V	±15V	±15V
Output Short Circuit Duration	Indefinite	Indefinite	Indefinite	Indefinite
Operating Temperature Range	−55°C to +125°C	0°C to +70°C	−55°C to +125°C	0°C to +70°C
Storage Temperature Range	−65°C to +150°C	−65°C to +150°C	−65°C to +150°C	−65°C to +150°C
Lead Temperature (Soldering, 10 seconds)	300°C	300°C	300°C	300°C

Electrical Characteristics (Note 3)

PARAMETER	CONDITIONS	LM741A/LM741E MIN	TYP	MAX	LM741 MIN	TYP	MAX	LM741C MIN	TYP	MAX	UNITS
Input Offset Voltage	$T_A = 25°C$										
	$R_S \leq 10 \, k\Omega$					1.0	5.0		2.0	6.0	mV
	$R_S \leq 50 \, \Omega$		0.8	3.0							mV
	$T_{AMIN} \leq T_A \leq T_{AMAX}$										
	$R_S \leq 50 \, \Omega$			4.0							mV
	$R_S \leq 10 \, k\Omega$						6.0			7.5	mV
Average Input Offset Voltage Drift				15							µV/°C
Input Offset Voltage Adjustment Range	$T_A = 25°C, V_S = \pm 20V$	±10				±15			±15		mV
Input Offset Current	$T_A = 25°C$		3.0	30		20	200		20	200	nA
	$T_{AMIN} \leq T_A \leq T_{AMAX}$			70		85	500			300	nA
Average Input Offset Current Drift				0.5							nA/°C
Input Bias Current	$T_A = 25°C$		30	80		80	500		80	500	nA
	$T_{AMIN} \leq T_A \leq T_{AMAX}$			0.210			1.5			0.8	µA
Input Resistance	$T_A = 25°C, V_S = \pm 20V$	1.0	6.0		0.3	2.0		0.3	2.0		MΩ
	$T_{AMIN} \leq T_A \leq T_{AMAX}$, $V_S = \pm 20V$	0.5									MΩ
Input Voltage Range	$T_A = 25°C$							±12	±13		V
	$T_{AMIN} \leq T_A \leq T_{AMAX}$				±12	±13					V
Large Signal Voltage Gain	$T_A = 25°C, R_L \geq 2 \, k\Omega$										
	$V_S = \pm 20V, V_O = \pm 15V$	50									V/mV
	$V_S = \pm 15V, V_O = \pm 10V$				50	200		20	200		V/mV
	$T_{AMIN} \leq T_A \leq T_{AMAX}$, $R_L \geq 2 \, k\Omega$,										
	$V_S = \pm 20V, V_O = \pm 15V$	32									V/mV
	$V_S = \pm 15V, V_O = \pm 10V$				25			15			V/mV
	$V_S = \pm 5V, V_O = \pm 2V$	10									V/mV
Output Voltage Swing	$V_S = \pm 20V$										
	$R_L \geq 10 \, k\Omega$	±16									V
	$R_L \geq 2 \, k\Omega$	±15									V
	$V_S = \pm 15V$										
	$R_L \geq 10 \, k\Omega$				±12	±14		±12	±14		V
	$R_L \geq 2 \, k\Omega$				±10	±13		±10	±13		V
Output Short Circuit Current	$T_A = 25°C$	10	25	35		25			25		mA
	$T_{AMIN} \leq T_A \leq T_{AMAX}$	10		40							mA
Common-Mode Rejection Ratio	$T_{AMIN} \leq T_A \leq T_{AMAX}$										
	$R_S \leq 10 \, k\Omega, V_{CM} = \pm 12V$				70	90		70	90		dB
	$R_S \leq 50 \, k\Omega, V_{CM} = \pm 12V$	80	95								dB

Electrical Characteristics (Continued)

PARAMETER	CONDITIONS	LM741A/LM741E MIN	LM741A/LM741E TYP	LM741A/LM741E MAX	LM741 MIN	LM741 TYP	LM741 MAX	LM741C MIN	LM741C TYP	LM741C MAX	UNITS
Supply Voltage Rejection Ratio	$T_{AMIN} \leq T_A \leq T_{AMAX}$, $V_S = \pm 20V$ to $V_S = \pm 5V$										
	$R_S \leq 50\Omega$	86	96								dB
	$R_S \leq 10 k\Omega$				77	96		77	96		dB
Transient Response	$T_A = 25°C$, Unity Gain										
Rise Time			0.25	0.8		0.3			0.3		μs
Overshoot			6.0	20		5			5		%
Bandwidth (Note 4)	$T_A = 25°C$	0.437	1.5								MHz
Slew Rate	$T_A = 25°C$, Unity Gain	0.3	0.7			0.5			0.5		V/μs
Supply Current	$T_A = 25°C$					1.7	2.8		1.7	2.8	mA
Power Consumption	$T_A = 25°C$										
	$V_S = \pm 20V$		80	150							mW
	$V_S = \pm 15V$					50	85		50	85	mW
LM741A	$V_S = \pm 20V$										
	$T_A = T_{AMIN}$			165							mW
	$T_A = T_{AMAX}$			135							mW
LM741E	$V_S = \pm 20V$			150							mW
	$T_A = T_{AMIN}$			150							mW
	$T_A = T_{AMAX}$			150							mW
LM741	$V_S = \pm 15V$										
	$T_A = T_{AMIN}$					60	100				mW
	$T_A = T_{AMAX}$					45	75				mW

Note 1: The maximum junction temperature of the LM741/LM741A is 150°C, while that of the LM741C/LM741E is 100°C. For operation at elevated temperatures, devices in the TO-5 package must be derated based on a thermal resistance of 150°C/W junction to ambient, or 45°C/W junction to case. The thermal resistance of the dual-in-line package is 100°C/W junction to ambient.

Note 2: For supply voltages less than ±15V, the absolute maximum input voltage is equal to the supply voltage.

Note 3: Unless otherwise specified, these specifications apply for $V_S = \pm 15V$, $-55°C \leq T_A \leq +125°C$ (LM741/LM741A). For the LM741C/LM741E, these specifications are limited to $0°C \leq T_A \leq +70°C$.

Note 4: Calculated value from: BW (MHz) = 0.35/Rise Time(μs).

Instrumentation Amplifiers

LH0036/LH0036C Instrumentation Amplifier

General Description

The LH0036/LH0036C is a true micro power instrumentation amplifier designed for precision differential signal processing. Extremely high accuracy can be obtained due to the 300 MΩ input impedance and excellent 100 dB common mode rejection ratio. It is packaged in a hermetic TO-8 package. Gain is programmable with one external resistor from 1 to 1000. Power supply operating range is between ±1V and ±18V. Input bias current and output bandwidth are both externally adjustable or can be set by internally set values. The LH0036 is specified for operation over the −55°C to +125°C temperature range and the LH0036C is specified for operation over the −25°C to +85°C temperature range.

Features

- High input impedance 300 MΩ
- High CMRR 100 dB
- Single resistor gain adjust 1 to 1000
- Low power 90µW
- Wide supply range ±1V to ±18V
- Adjustable input bias current
- Adjustable output bandwidth
- Guard drive output

Equivalent Circuit and Connection Diagrams

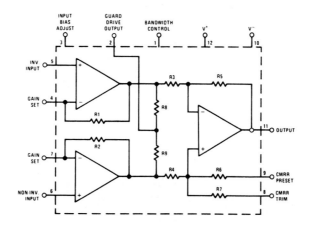

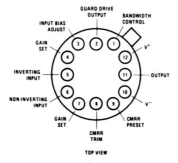

TOP VIEW

Order Number LH0036G or LH0036CG
See NS Package H12B

381

Absolute Maximum Ratings

Supply Voltage	±18V	Short Circuit Duration	Continuous
Differential Input Voltage	±30V	Operating Temperature Range	
Input Voltage Range	±V_S	LH0036	−55°C to +125°C
Shield Drive Voltage	±V_S	LH0036C	−25°C to +85°C
CMRR Preset Voltage	±V_S	Storage Temperature Range	−65°C to +150°C
CMRR Trim Voltage	±V_S	Lead Temperature, Soldering 10 seconds	300°C
Power Dissipation (Note 3)	1.5W		

Electrical Characteristics (Notes 1 and 2)

PARAMETER	CONDITIONS	LH0036 MIN	LH0036 TYP	LH0036 MAX	LH0036C MIN	LH0036C TYP	LH0036C MAX	UNITS
Input Offset Voltage (V_{IOS})	$R_S = 1.0k\Omega$, $T_A = 25°C$		0.5	1.0		1.0	2.0	mV
	$R_S = 1.0k\Omega$			2.0			3.0	mV
Output Offset Voltage (V_{OOS})	$R_S = 1.0k\Omega$, $T_A = 25°C$		2.0	5.0		5.0	10	mV
	$R_S = 1.0k\Omega$			6.0			12	mV
Input Offset Voltage Tempco ($\Delta V_{IOS}/\Delta T$)	$R_S \leq 1.0k\Omega$		10			10		μV/°C
Output Offset Voltage Tempco ($\Delta V_{OOS}/\Delta T$)			15			15		μV/°C
Overall Offset Referred to Input (V_{OS})	$A_V = 1.0$		2.5			6.0		mV
	$A_V = 10$		0.7			1.5		mV
	$A_V = 100$		0.52			1.05		mV
	$A_V = 1000$		0.502			1.005		mV
Input Bias Current (I_B)	$T_A = 25°C$		40	100		50	125	nA
				150			200	nA
Input Offset Current (I_{OS})	$T_A = 25°C$		10	40		20	50	nA
				80			100	nA
Small Signal Bandwidth	$A_V = 1.0$, $R_L = 10k\Omega$		350			350		kHz
	$A_V = 10$, $R_L = 10k\Omega$		35			35		kHz
	$A_V = 100$, $R_L = 10k\Omega$		3.5			3.5		kHz
	$A_V = 1000$, $R_L = 10k\Omega$		350			350		Hz
Full Power Bandwidth	$V_{IN} = \pm 10V$, $R_L = 10k$, $A_V = 1$		5.0			5.0		kHz
Input Voltage Range	Differential	±10	+12		±10	+12		V
	Common Mode	±10	±12		±10	±12		V
Gain Nonlinearity			0.03			0.03		%
Deviation From Gain Equation Formula	$A_V = 1$ to 1000		±0.3	±1.0		±1.0	±3.0	%
PSRR	±5.0V ≤ V_S ≤ ±15V, $A_V = 1.0$		1.0	2.5		1.0	5.0	mV/V
	±5.0V ≤ V_S ≤ ±15V, $A_V = 100$		0.05	0.25		0.10	0.50	mV/V
CMRR	$A_V = 1.0$ DC to		1.0	2.5		2.5	5.0	mV/V
	$A_V = 10$ 100 Hz		0.1	0.25		0.25	0.50	mV/V
	$A_V = 100$ $\Delta R_S = 1.0k$		50	100		50	100	μV/V
Output Voltage	$V_S = \pm 15V$, $R_L = 10k\Omega$	±10	±13.5		±10	±13.5		V
	$V_S = \pm 1.5V$, $R_L = 100k\Omega$	±0.6	±0.8		±0.6	±0.8		V
Output Resistance			0.5			0.5		Ω
Supply Current			300	400		400	600	μA
Equivalent Input Noise Voltage	0.1 Hz < f < 10 kHz, $R_S < 50\Omega$		20			20		μV/p-p
Slew Rate	$\Delta V_{IN} = \pm 10V$, $R_L = 10k\Omega$, $A_V = 1.0$		0.3			0.3		V/μs
Settling Time	To ±10 mV, $R_L = 10k\Omega$, $\Delta V_{OUT} = 1.0V$							
	$A_V = 1.0$		3.3			3.8		μs
	$A_V = 100$		180			180		μs

Note 1: Unless otherwise specified, all specifications apply for $V_S = \pm 15V$, Pins 1, 3, and 9 grounded, −25°C to +85°C for the LH0036C and −55°C to +125°C for the LH0036.

Note 2: All typical values are for $T_A = 25°C$.

Note 3: The maximum junction temperature is 150°C. For operation at elevated temperature derate the G package on a thermal resistance of 90°C/W, above 25°C.

LH0036/LH0036C

Typical Performance Characteristics

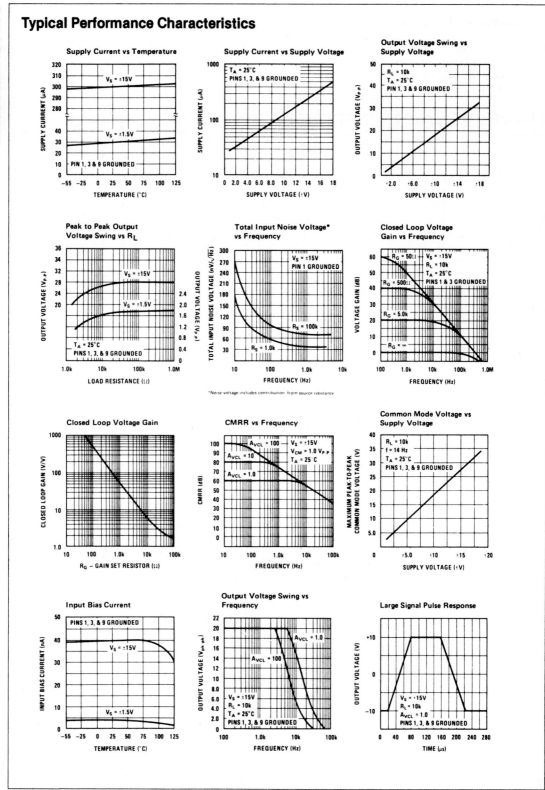

Typical Applications

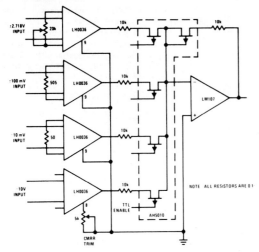

Pre MUX Signal Conditioning

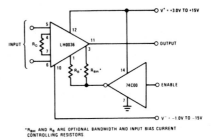

Instrumentation Amplifier with Logic Controlled Shut-Down

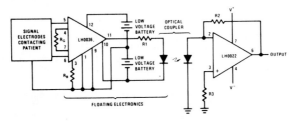

Isolation Amplifier for Medical Telemetry

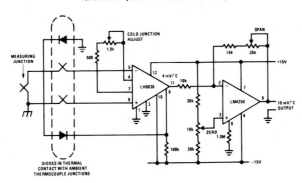

Thermocouple Amplifier with Cold Junction Compensation

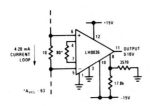

Process Control Interface

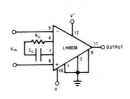

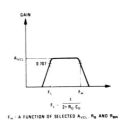

High Pass Filter

Applications Information

THEORY OF OPERATION

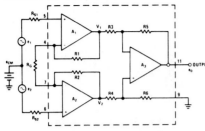

FIGURE 1. Simplified LH0036

The LH0036 is a 2 stage amplifier with a high input impedance gain stage comprised of A_1 and A_2 and a differential to single-ended unity gain stage, A_3. Operational amplifier, A_1, receives differential input signal, e_1, and amplifies it by a factor equal to $(R1 + R_G)/R_G$.

A_1 also receives input e_2 via A_2 and R2. e_2 is seen as an inverting signal with a gain of $R1/R_G$. A_1 also receives the common mode signal e_{CM} and processes it with a gain of +1. Hence:

$$V_1 = \frac{R1 + R_G}{R_G} e_1 - \frac{R1}{R_G} e_2 + e_{CM} \quad (1)$$

By similar analysis V_2 is seen to be:

$$V_2 = \frac{R2 + R_G}{R_G} e_2 - \frac{R2}{R_G} e_1 + e_{CM} \quad (2)$$

For R1 = R2:

$$V_2 - V_1 = \left[\left(\frac{2R1}{R_G}\right) + 1\right](e_2 - e_1) \quad (3)$$

Also, for R3 = R5 = R4 = R6, the gain of $A_3 = 1$, and:

$$e_0 = (1)(V_2 - V_1) = (e_2 - e_1)\left[1 + \left(\frac{2R1}{R_G}\right)\right] \quad (4)$$

As can be seen for identically matched resistors, e_{CM} is cancelled out, and the differential gain is dictated by equation (4).

For the LH0036, equation (4) reduces to:

$$A_{VCL} = \frac{e_0}{e_2 - e_1} = 1 + \frac{50k}{R_G} \quad (5a)$$

The closed loop gain may be set to any value from 1 ($R_G = \infty$) to 1000 ($R_G \cong 50\Omega$). Equation (5a) re-arranged in more convenient form may be used to select R_G for a desired gain:

$$R_G = \frac{50k}{A_{VCL} - 1} \quad (5b)$$

USE OF BANDWIDTH CONTROL (pin 1)

In the standard configuration, pin 1 of the LH0036 is simply grounded. The amplifier's slew rate in this configuration is typically 0.3V/μs and small signal bandwidth 350 kHz for $A_{VCL} = 1$. In some applications, particularly at low frequency, it may be desirable to limit bandwidth in order to minimize the overall noise bandwidth of the device. A resistor R_{BW} may be placed between pin 1 and ground to accomplish this purpose. Figure 2 shows typical small signal bandwidth versus R_{BW}.

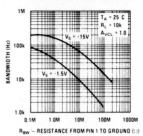

FIGURE 2. Bandwidth vs R_{BW}

It also should be noted that large signal bandwidth and slew rate may be adjusted down by use of R_{BW}. Figure 3 is plot of slew rate versus R_{BW}.

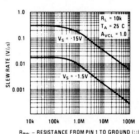

FIGURE 3. Output Slew Rate vs R_{BW}

CMRR CONSIDERATIONS

Use of Pin 9, CMRR Preset

Pin 9 should be grounded for nominal operation. An internal factory trimmed resistor, R6, will yield a CMRR in excess of 80 dB (for $A_{VCL} = 100$). Should a higher CMRR be desired, pin 9 should be left open and the procedure, in this section followed.

DC Off-set Voltage and Common Mode Rejection Adjustments

Off-set may be nulled using the circuit shown in Figure 4.

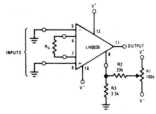

FIGURE 4. V_{OS} Adjustment Circuit

Pin 8 is also used to improve the common mode rejection ratio as shown in Figure 5. Null is

Applications Information (Cont'd)

achieved by alternately applying ±10V (for V^+ & V^- = 15V) to the inputs and adjusting R1 for minimum change at the output.

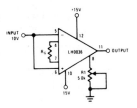

FIGURE 5. CMRR Adjustment Circuit

The circuits of Figure 4 and 5 may be combined as shown in Figure 6 to accomplish both V_{OS} and CMRR null. However, the V_{OS} and CMRR adjustment are interactive and several iterations are required. The procedure for null should start with the inputs grounded.

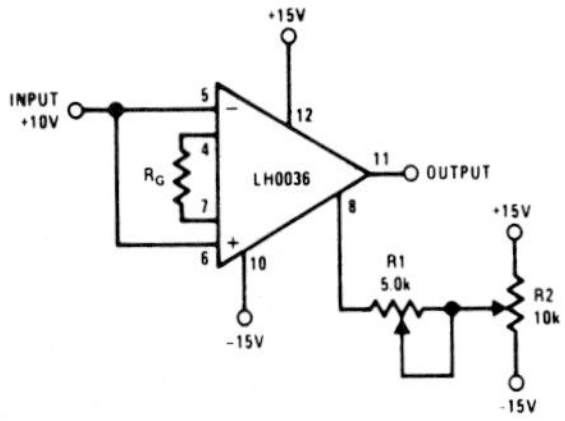

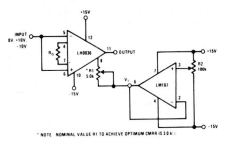

FIGURE 6. Combined CMRR, V_{OS} Adjustment Circuit

R2 is adjusted for V_{OS} null. An input of +10V is then applied and R1 is adjusted for CMRR null. The procedure is then repeated until the optimum is achieved.

A circuit which overcomes adjustment interaction is shown in Figure 7. In this case, R2 is adjusted first for output null of the LH0036. R1 is then adjusted for output null with +10V input. It is always a good idea to check CMRR null with a −10V input. The optimum null achievable will yield the highest CMRR over the amplifiers common mode range.

FIGURE 7. Improved V_{OS}, CMRR Nulling Circuit

AC CMRR Considerations

The ac CMRR may be improved using the circuit of Figure 8.

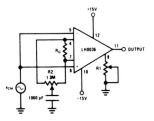

FIGURE 8. Improved AC CMRR Circuit

After adjusting R1 for best dc CMRR as before, R2 should be adjusted for minimum peak-to-peak voltage at the output while applying an ac common mode signal of the maximum amplitude and frequency of interest.

INPUT BIAS CURRENT CONTROL

Under nominal operating conditions (pin 3 grounded), the LH0036 requires input currents of 40 nA. The input current may be reduced by inserting a resistor (R_B) between 3 and ground or, alternatively, between 3 and V^-. For R_B returned to ground, the input bias current may be predicted by:

$$I_{BIAS} \cong \frac{V^+ - 0.5}{4 \times 10^8 + 800\, R_B} \quad (6a)$$

or

$$R_B = \frac{V^+ - 0.5 - (4 \times 10^8)(I_{BIAS})}{800\, I_{BIAS}} \quad (6b)$$

Where:

I_{BIAS} = Input Bias Current (nA)

R_B = External Resistor connected between pin 3 and ground (Ohms)

V^+ = Positive Supply Voltage (Volts)

Figure 9 is a plot of input bias current versus R_B.

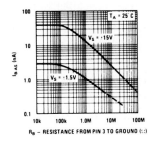

FIGURE 9. Input Bias Current as a Function of R_B

As indicated above, R_B may be returned to the negative supply voltage. Input bias current may then be predicted by:

$$I_{BIAS} \cong \frac{(V^+ - V^-) - 0.5}{4 \times 10^8 + 800\, R_B}$$

Applications Information (Cont'd)

or

$$R_B \cong \frac{(V^+ - V^-) - 0.5 - (4 \times 10^8)(I_{BIAS})}{800 \, I_{BIAS}} \quad (8)$$

Where:

I_{BIAS} = Input Bias Current (nA)

R_B = External resistor connected between pin 3 and V^- (Ohms)

V^+ = Positive Supply Voltage (Volts)

V^- = Negative Supply Voltage (Volts)

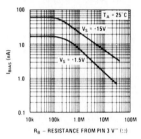

FIGURE 10. Input Bias Current as a Function of R_B

Figure 10 is a plot of input bias current versus R_B returned to V^- it should be noted that bandwidth is affected by changes in R_B. Figure 11 is a plot of bandwidth versus R_B.

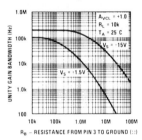

FIGURE 11. Unity Gain Bandwidth as a Function of R_B

BIAS CURRENT RETURN PATH CONSIDERATIONS

The LH0036 exhibits input bias currents typically in the 40 nA region in each input. This current must flow through R_{ISO} as shown in Figure 12.

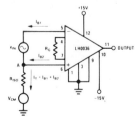

FIGURE 12. Bias Current Return Path

In a typical application, $V_S = \pm 15V$, $I_{B1} \cong I_{B2} \cong$ 40 nA, the total current, I_T, would flow through R_{ISO} causing a voltage rise at point A. For values of $R_{ISO} \geq 150 \, M\Omega$, the voltage at point A exceeds the +12V common range of the device. Clearly, for $R_{ISO} = \infty$, the LH0036 would be driven to positive saturation.

The implication is that a finite impedance must be supplied between the input and power supply ground. The value of the resistor is dictated by the maximum input bias current, and the common mode voltage. Under worst case conditions:

$$R_{ISO} \leq \frac{V_{CMR} - V_{CM}}{I_T} \quad (9)$$

Where:

V_{CMR} = Common Mode Range (10V for the LH0036)

V_{CM} = Common Mode Voltage

$I_T = I_{B1} + I_{B2}$

In applications in which the signal source is floating, such as a thermocouple, one end of the source may be grounded directly or through a resistor.

GUARD OUTPUT

Pin 2 of the LH0036 is provided as a guard drive pin in those stringent applications which require very low leakage and minimum input capacitance. Pin 2 will always be biased at the input common mode voltage. The source impedance looking into pin 2 is approximately 15 kΩ. Proper use of the guard/shield pin is shown in Figure 13.

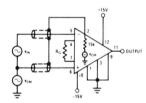

FIGURE 13. Use of Guard

For applications requiring a lower source impedance than 15 kΩ, a unity gain buffer, such as the LH0002 may be inserted between pin 2 and the input shields as shown in Figure 14.

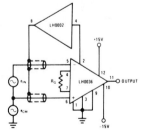

FIGURE 14. Guard Pin With Buffer

Definition of Terms

Bandwidth: The frequency at which the voltage gain is reduced to 0.707 of the low frequency (dc) value.

Closed Loop Gain, A_{VCL}: The ratio of the output voltage swing to the input voltage swing determined by $A_{VCL} = 1 + (50k/R_G)$. Where: R_G = Gain Set Resistor.

Common Mode Rejection Ratio: The ratio of input voltage range to the peak-to-peak change in offset voltage over this range.

Gain Equation Accuracy: The deviation of the actual closed loop gain from the predicted closed loop gain, $A_{VCL} = 1 + (50k/R_G)$ for the specified closed loop gain.

Input Bias Current: The current flowing at pin 5 and 6 under the specified operating conditions.

Input Offset Current: The difference between the input bias current at pins 5 and 6; i.e. $I_{OS} = |I_5 - I_6|$.

Input Stage Offset Voltage, V_{IOS}: The voltage which must be applied to the input pins to force the output to zero volts for $A_{VCL} = 100$.

Output Stage Offset Voltage, V_{OOS}: The voltage which must be applied to the input of the output stage to produce zero output voltage. It can be measured by measuring the overall offset at unity gain and subtracting V_{IOS}.

$$V_{OOS} = \left[V_{OS} \bigg|_{A_{VCL}=1} \right] - \left[V_{OS} \bigg|_{A_{VCL}=1000} \right]$$

Overall Offset Voltage:

$$V_{OS} = V_{IOS} + \frac{V_{OOS}}{A_{VCL}}$$

Power Supply Rejection Ratio: The ratio of the change in offset voltage, V_{OS}, to the change in supply voltage producing it.

Resistor, R_B: An optional resistor placed between pin 3 of the LH0036 and ground (or V^-) to reduce the input bias current.

Resistor, R_{BW}: An optional resistor placed between pin 1 of the LH0036 and ground (or V^-) to reduce the bandwidth of the output stage.

Resistor, R_G: A gain setting resistor connected between pins 4 and 7 of the LH0036 in order to program the gain from 1 to 1000.

Settling Time: The time between the initiation of an input step function and the time when the output voltage has settled to within a specified error band of the final output voltage.

Industrial Blocks

LM3080/LM3080A Operational Transconductance Amplifier

General Description

The LM3080 is a programmable transconductance block intended to fulfill a wide variety of variable gain applications. The LM3080 has differential inputs and high impedance push-pull outputs. The device has high input impedance and its transconductance (gm) is directly proportional to the amplifier bias current (I_{ABC}).

High slew rate together with programmable gain make the LM3080 an ideal choice for variable gain applications such as sample and hold, multiplexing, filtering, and multiplying.

The LM3080AH and LM3080AJ are guaranteed over the temperature range $-55°C$ to $+125°C$; the LM3080N, LM3080H, LM3080AN and LM3080J are guaranteed from $0°C$ to $+70°C$.

Features

- Slew Rate (unity gain compensated): 50 V/μs
- Fully Adjustable Gain: 0 to gm R_L limit
- Extended gm Linearity: 3 decades
- Flexible Supply Voltage Range: ±2V to ±18V
- Adjustable Power Consumption

Schematic and Connection Diagrams

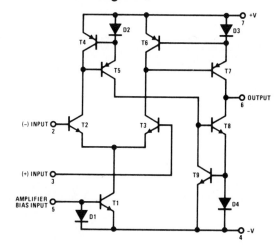

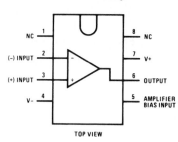

Dual-In-Line Package

TOP VIEW

Order Number LM3080AJ or LM3080J
See NS Package J08A

Order Number LM3080AN
See NS Package N08B

Absolute Maximum Ratings

Supply Voltage (Note 2)	
LM3080	±18 V
LM3080A	±22 V
Power Dissipation	250 mW
Differential Input Voltage	±5 V
Amplifier Bias Current (I_{ABC})	2 mA
DC Input Voltage	$+V_S$ to $-V_S$
Output Short Circuit Duration	Indefinite
Operating Temperature Range	
LM3080N, LM3080H, LM3080AN	
or LM3080J	0°C to +70°C
LM3080AH or LM3080AJ	−55°C to +125°C
Storage Temperature Range	−65°C to +150°C
Lead Temperature (Soldering, 10 seconds)	300°C

Electrical Characteristics (Note 1)

Parameter	Conditions	LM3080 Min.	LM3080 Typ.	LM3080 Max.	LM3080A Min.	LM3080A Typ.	LM3080A Max.	Units
Input Offset Voltage			0.4	5		0.4	2	mV
	Over Specified Temperature Range			6			5	mV
	$I_{ABC} = 5\,\mu A$		0.3			0.3	2	mV
Input Offset Voltage Change	$5\,\mu A \leq I_{ABC} \leq 500\,\mu A$		0.1			0.1	3	mV
Input Offset Current			0.1	0.6		0.1	0.6	μA
Input Bias Current			0.4	5		0.4	5	μA
	Over Specified Temperature Range		1	7		1	8	μA
Forward Transconductance (gm)		6700	9600	13000	7700	9600	12000	μmho
	Over Specified Temperature Range	5400			4000			μmho
Peak Output Current	$R_L = 0$, $I_{ABC} = 5\,\mu A$		5		3	5	7	μA
	$R_L = 0$	350	500	650	350	500	650	μA
	$R_L = 0$ Over Specified Temperature Range	300			300			μA
Peak Output Voltage								
Positive	$R_L = \infty$, $5\,\mu A \leq I_{ABC} \leq 500\,\mu A$	+12	+14.2		+12	+14.2		V
Negative	$R_L = \infty$, $5\,\mu A \leq I_{ABC} \leq 500\,\mu A$	−12	−14.4		−12	−14.4		V
Amplifier Supply Current			1.1			1.1		mA
Input Offset Voltage Sensitivity								
Positive	$\Delta V_{OFFSET}/\Delta V+$		20	150		20	150	μV/V
Negative	$\Delta V_{OFFSET}/\Delta V-$		20	150		20	150	μV/V
Common Mode Rejection Ratio		80	110		80	110		dB
Common Mode Range		±12	±14		±12	±14		V
Input Resistance		10	26		10	26		kΩ
Magnitude of Leakage Current	$I_{ABC} = 0$		0.2	100		0.2	5	nA
Differential Input Current	$I_{ABC} = 0$, Input = ±4 V		0.02	100		0.02	5	nA
Open Loop Bandwidth			2			2		MHz
Slew Rate	Unity Gain Compensated		50			50		V/μs

Note 1: These specifications apply for $V_S = \pm 15$ V and $T_A = 25°C$, amplifier bias current (I_{ABC}) = 500 μA, unless otherwise specified.

Note 2: Selections to supply voltage above ±22V, contact the factory.

Typical Performance Characteristics

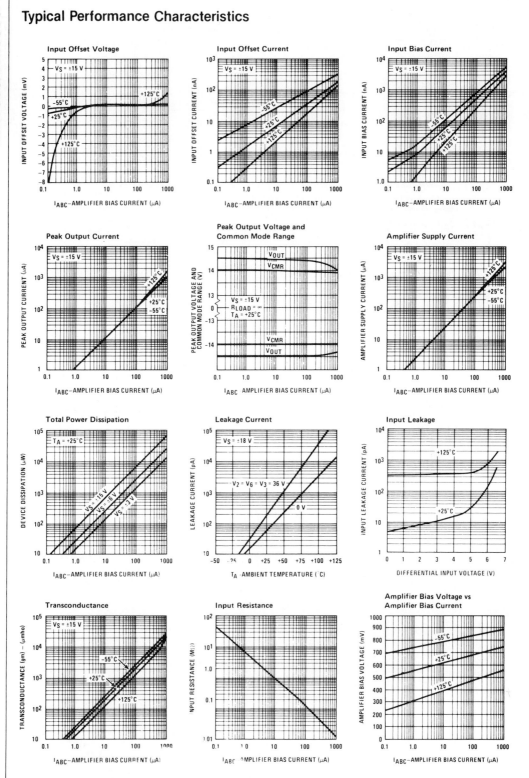

Typical Performance Characteristics (Continued)

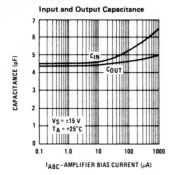

Input and Output Capacitance

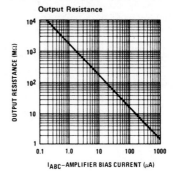

Output Resistance

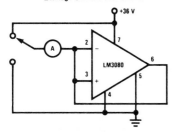

Leakage Current Test Circuit

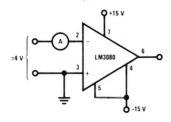

Differential Input Current Test Circuit

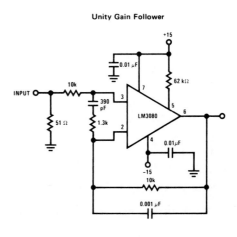

Unity Gain Follower

Audio/Radio Circuits

LM1596/LM1496 Balanced Modulator-Demodulator

General Description

The LM1596/LM1496 are double balanced modulator-demodulators which produce an output voltage proportional to the product of an input (signal) voltage and a switching (carrier) signal. Typical applications include suppressed carrier modulation, amplitude modulation, synchronous detection, FM or PM detection, broadband frequency doubling and chopping.

The LM1596 is specified for operation over the $-55°C$ to $+125°C$ military temperature range. The LM1496 is specified for operation over the $0°C$ to $+70°C$ temperature range.

Features

- Excellent carrier suppression

    65 dB typical at 0.5 MHz
    50 dB typical at 10 MHz

- Adjustable gain and signal handling

- Fully balanced inputs and outputs

- Low offset and drift

- Wide frequency response up to 100 MHz

Schematic and Connection Diagrams

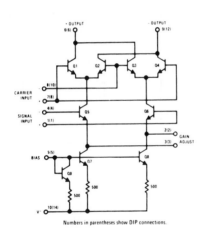

Numbers in parentheses show DIP connections.

Metal Can Package

Note: Pin 10 is connected electrically to the case through the device substrate.

Order Number LM1496H or LM1596H
See NS Package H08C

Dual-In-Line Package

Order Number LM1496N
See NS Package N14A

Typical Application and Test Circuit

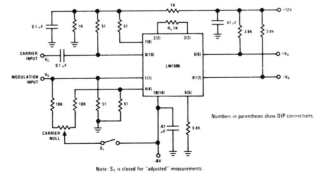

Numbers in parentheses show DIP connections.

Note: S_1 is closed for "adjusted" measurements.

Suppressed Carrier Modulator

LM1596/LM1496

Absolute Maximum Ratings

Internal Power Dissipation (Note 1)	500 mW
Applied Voltage (Note 2)	30V
Differential Input Signal ($V_7 - V_8$)	±5.0V
Differential Input Signal ($V_4 - V_1$)	±$(5+I_5 R_e)$ V
Input Signal ($V_2 - V_1$, $V_3 - V_4$)	5.0V
Bias Current (I_5)	12 mA
Operating Temperature Range LM1596	−55°C to +125°C
LM1496	0°C to +70°C
Storage Temperature Range	−65°C to +150°C
Lead Temperature (Soldering, 10 sec)	300°C

Electrical Characteristics (T_A = 25°C, unless otherwise specified, see test circuit)

PARAMETER	CONDITIONS	LM1596 MIN	LM1596 TYP	LM1596 MAX	LM1496 MIN	LM1496 TYP	LM1496 MAX	UNITS
Carrier Feedthrough	V_C = 60 mVrms sine wave, f_C = 1.0 kHz, offset adjusted		40			40		µVrms
	V_C = 60 mVrms sine wave, f_C = 10 MHz, offset adjusted		140			140		µVrms
	V_C = 300 mV$_{pp}$ square wave, f_C = 1.0 kHz, offset adjusted		0.04	0.2		0.04	0.2	mVrms
	V_C = 300 mV$_{pp}$ square wave, f_C = 1.0 kHz, offset not adjusted		20	100		20	150	mVrms
Carrier Suppression	f_S = 10 kHz, 300 mVrms, f_C = 500 kHz, 60 mVrms sine wave, offset adjusted	50	65		50	65		dB
	f_S = 10 kHz, 300 mVrms, f_C = 10 MHz, 60 mVrms sine wave, offset adjusted		50			50		dB
Transadmittance Bandwidth	R_L = 50Ω, Carrier Input Port, V_C = 60 mVrms sine wave, f_S = 1.0 kHz, 300 mVrms sine wave		300			300		MHz
	Signal Input Port, V_S = 300 mVrms sine wave, $V_7 - V_8$ = 0.5 Vdc		80			80		MHz
Voltage Gain, Signal Channel	V_S = 100 mVrms, f = 1.0 kHz, $V_7 - V_8$ = 0.5 Vdc	2.5	3.5		2.5	3.5		V/V
Input Resistance, Signal Port	f = 5.0 MHz, $V_7 - V_8$ = 0.5 Vdc		200			200		kΩ
Input Capacitance, Signal Port	f = 5.0 MHz, $V_7 - V_8$ = 0.5 Vdc		2.0			2.0		pF
Single Ended Output Resistance	f = 10 MHz		40			40		kΩ
Single Ended Output Capacitance	f = 10 MHz		5.0			5.0		pF
Input Bias Current	$(I_1 + I_4)/2$		12	25		12	30	µA
Input Bias Current	$(I_7 + I_8)/2$		12	25		12	30	µA
Input Offset Current	$(I_1 - I_4)$		0.7	5.0		0.7	5.0	µA
Input Offset Current	$(I_7 - I_8)$		0.7	5.0		5.0	5.0	µA
Average Temperature Coefficient of Input Offset Current	(−55°C < T_A < +125°C) (0°C < T_A < +70°C)		2.0			2.0		nA/°C nA/°C
Output Offset Current	$(I_6 - I_9)$		14	50		14	60	µA
Average Temperature Coefficient of Output Offset Current	(−55°C < T_A < +125°C) (0°C < T_A < +70°C)		90			90		nA/°C nA/°C
Signal Port Common Mode Input Voltage Range	f_S = 1.0 kHz		5.0			5.0		V$_{p-p}$
Signal Port Common Mode Rejection Ratio	$V_7 - V_8$ = 0.5 Vdc		−85			−85		dB
Common Mode Quiescent Output Voltage			8.0			8.0		Vdc
Differential Output Swing Capability			8.0			8.0		V$_{p-p}$
Positive Supply Current	$(I_6 + I_9)$		2.0	3.0		2.0	3.0	mA
Negative Supply Current	(I_{10})		3.0	4.0		3.0	4.0	mA
Power Dissipation			33			33		mW

Note 1: LM1596 rating applies to case temperatures to +125°C; derate linearly at 6.5 mW/°C for ambient temperature above 75°C. LM1496 rating applies to case temperatures to +70°C.

Note 2: Voltage applied between pins 6-7, 8-1, 9-7, 9-8, 7-4, 7-1, 8-4, 6-8, 2-5, 3-5.

Typical Performance Characteristics

Carrier Suppression vs Carrier Input Level
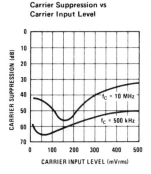

Carrier Suppression vs Frequency
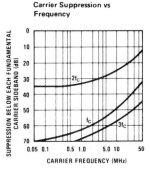

Carrier Feedthrough vs Frequency
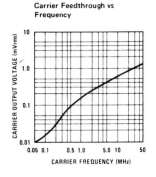

Sideband Output vs Carrier Levels
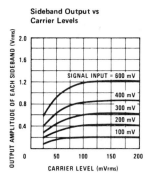

Sideband and Signal Port Transadmittances vs Frequency
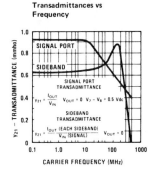

Signal-Port Frequency Response
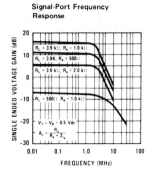

Typical Applications (Continued)

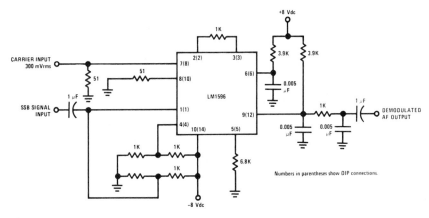

SSB Product Detector

This figure shows the LM1596 used as a single sideband (SSB) suppressed carrier demodulator (product detector). The carrier signal is applied to the carrier input port with sufficient amplitude for switching operation. A carrier input level of 300 mVrms is optimum. The composite SSB signal is applied to the signal input port with an amplitude of 5.0 to 500 mVrms. All output signal components except the desired demodulated audio are filtered out, so that an offset adjustment is not required. This circuit may also be used as an AM detector by applying composite and carrier signals in the same manner as described for product detector operation.

Typical Applications (Continued)

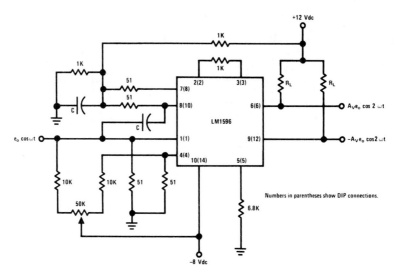

Broadband Frequency Doubler

The frequency doubler circuit shown will double low-level signals with low distortion. The value of C should be chosen for low reactance at the operating frequency.

Signal level at the carrier input must be less than 25 mV peak to maintain operation in the linear region of the switching differential amplifier. Levels to 50 mV peak may be used with some distortion of the output waveform. If a larger input signal is available a resistive divider may be used at the carrier input, with full signal applied to the signal input.

LM555/LM555C Timer

General Description

The LM555 is a highly stable device for generating accurate time delays or oscillation. Additional terminals are provided for triggering or resetting if desired. In the time delay mode of operation, the time is precisely controlled by one external resistor and capacitor. For astable operation as an oscillator, the free running frequency and duty cycle are accurately controlled with two external resistors and one capacitor. The circuit may be triggered and reset on falling waveforms, and the output circuit can source or sink up to 200 mA or drive TTL circuits.

Features

- Direct replacement for SE555/NE555
- Timing from microseconds through hours
- Operates in both astable and monostable modes
- Adjustable duty cycle
- Output can source or sink 200 mA
- Output and supply TTL compatible
- Temperature stability better than 0.005% per °C
- Normally on and normally off output

Applications

- Precision timing
- Pulse generation
- Sequential timing
- Time delay generation
- Pulse width modulation
- Pulse position modulation
- Linear ramp generator

Schematic Diagram

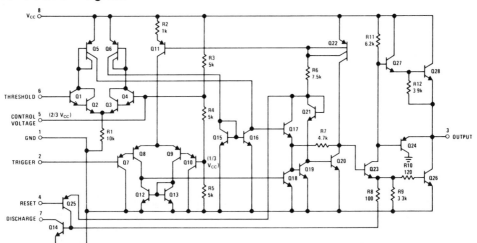

Connection Diagrams

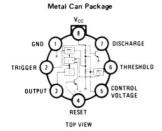

Metal Can Package
TOP VIEW

Order Number LM555H, LM555CH
See NS Package H08C

Dual-In-Line Package
TOP VIEW

Order Number LM555CN
See NS Package N08B
Order Number LM555J or LM555CJ
See NS Package J08A

Absolute Maximum Ratings

Supply Voltage	+18V
Power Dissipation (Note 1)	600 mW
Operating Temperature Ranges	
LM555C	0°C to +70°C
LM555	−55°C to +125°C
Storage Temperature Range	−65°C to +150°C
Lead Temperature (Soldering, 10 seconds)	300°C

Electrical Characteristics (T_A = 25°C, V_{CC} = +5V to +15V, unless otherwise specified)

PARAMETER	CONDITIONS	LM555 MIN	LM555 TYP	LM555 MAX	LM555C MIN	LM555C TYP	LM555C MAX	UNITS
Supply Voltage		4.5		18	4.5		16	V
Supply Current	V_{CC} = 5V, R_L = ∞		3	5		3	6	mA
	V_{CC} = 15V, R_L = ∞ (Low State) (Note 2)		10	12		10	15	mA
Timing Error, Monostable								
Initial Accuracy			0.5			1		%
Drift with Temperature	R_A, R_B = 1k to 100 k, C = 0.1μF, (Note 3)		30			50		ppm/°C
Accuracy over Temperature			1.5			1.5		%
Drift with Supply			0.05			0.1		%/V
Timing Error, Astable								
Initial Accuracy			1.5			2.25		%
Drift with Temperature			90			150		ppm/°C
Accuracy over Temperature			2.5			3.0		%
Drift with Supply			0.15			0.30		%/V
Threshold Voltage			0.667			0.667		x V_{CC}
Trigger Voltage	V_{CC} = 15V	4.8	5	5.2		5		V
	V_{CC} = 5V	1.45	1.67	1.9		1.67		V
Trigger Current			0.01	0.5		0.5	0.9	μA
Reset Voltage		0.4	0.5	1	0.4	0.5	1	V
Reset Current			0.1	0.4		0.1	0.4	mA
Threshold Current	(Note 4)		0.1	0.25		0.1	0.25	μA
Control Voltage Level	V_{CC} = 15V	9.6	10	10.4	9	10	11	V
	V_{CC} = 5V	2.9	3.33	3.8	2.6	3.33	4	V
Pin 7 Leakage Output High			1	100		1	100	nA
Pin 7 Sat (Note 5)								
Output Low	V_{CC} = 15V, I_7 = 15 mA		150			180		mV
Output Low	V_{CC} = 4.5V, I_7 = 4.5 mA		70	100		80	200	mV
Output Voltage Drop (Low)	V_{CC} = 15V							
	I_{SINK} = 10 mA		0.1	0.15		0.1	0.25	V
	I_{SINK} = 50 mA		0.4	0.5		0.4	0.75	V
	I_{SINK} = 100 mA		2	2.2		2	2.5	V
	I_{SINK} = 200 mA		2.5			2.5		V
	V_{CC} = 5V							
	I_{SINK} = 8 mA		0.1	0.25				V
	I_{SINK} = 5 mA					0.25	0.35	V
Output Voltage Drop (High)	I_{SOURCE} = 200 mA, V_{CC} = 15V		12.5			12.5		V
	I_{SOURCE} = 100 mA, V_{CC} = 15V	13	13.3		12.75	13.3		V
	V_{CC} = 5V	3	3.3		2.75	3.3		V
Rise Time of Output			100			100		ns
Fall Time of Output			100			100		ns

Note 1: For operating at elevated temperatures the device must be derated based on a +150°C maximum junction temperature and a thermal resistance of +45°C/W junction to case for TO-5 and +150°C/W junction to ambient for both packages.
Note 2: Supply current when output high typically 1 mA less at V_{CC} = 5V.
Note 3: Tested at V_{CC} = 5V and V_{CC} = 15V.
Note 4: This will determine the maximum value of $R_A + R_B$ for 15V operation. The maximum total ($R_A + R_B$) is 20 MΩ.
Note 5: No protection against excessive pin 7 current is necessary providing the package dissipation rating will not be exceeded.

Typical Performance Characteristics

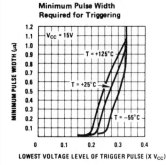
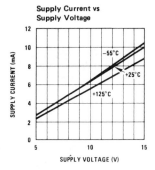
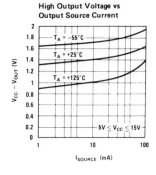

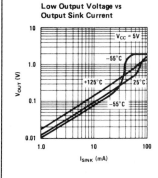
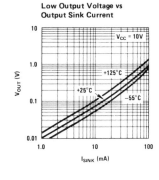
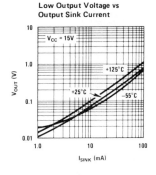

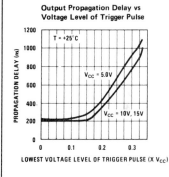
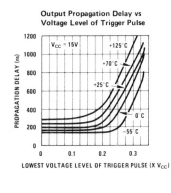
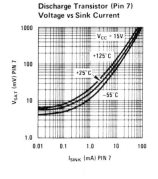

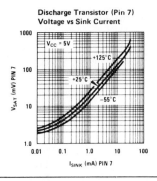

Applications Information

MONOSTABLE OPERATION

In this mode of operation, the timer functions as a one-shot (*Figure 1*). The external capacitor is initially held discharged by a transistor inside the timer. Upon application of a negative trigger pulse of less than 1/3 V_{CC} to pin 2, the flip-flop is set which both releases the short circuit across the capacitor and drives the output high.

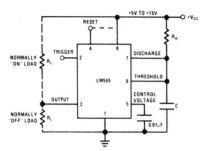

FIGURE 1. Monostable

The voltage across the capacitor then increases exponentially for a period of t = 1.1 $R_A C$, at the end of which time the voltage equals 2/3 V_{CC}. The comparator then resets the flip-flop which in turn discharges the capacitor and drives the output to its low state. *Figure 2* shows the waveforms generated in this mode of operation. Since the charge and the threshold level of the comparator are both directly proportional to supply voltage, the timing internal is independent of supply.

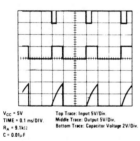

V_{CC} = 5V   Top Trace: Input 5V/Div.
TIME = 0.1 ms/DIV.   Middle Trace: Output 5V/Div.
R_A = 9.1 kΩ   Bottom Trace: Capacitor Voltage 2V/Div.
C = 0.01 μF

FIGURE 2. Monostable Waveforms

During the timing cycle when the output is high, the further application of a trigger pulse will not effect the circuit. However the circuit can be reset during this time by the application of a negative pulse to the reset terminal (pin 4). The output will then remain in the low state until a trigger pulse is again applied.

When the reset function is not in use, it is recommended that it be connected to V_{CC} to avoid any possibility of false triggering.

Figure 3 is a nomograph for easy determination of R, C values for various time delays.

NOTE: In monostable operation, the trigger should be driven high before the end of timing cycle.

ASTABLE OPERATION

If the circuit is connected as shown in *Figure 4* (pins 2 and 6 connected) it will trigger itself and free run as a

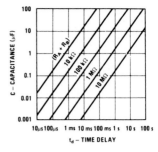

FIGURE 3. Time Delay

multivibrator. The external capacitor charges through $R_A + R_B$ and discharges through R_B. Thus the duty cycle may be precisely set by the ratio of these two resistors.

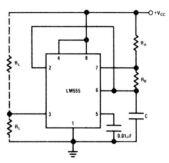

FIGURE 4. Astable

In this mode of operation, the capacitor charges and discharges between 1/3 V_{CC} and 2/3 V_{CC}. As in the triggered mode, the charge and discharge times, and therefore the frequency are independent of the supply voltage.

Figure 5 shows the waveforms generated in this mode of operation.

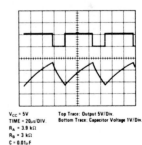

V_{CC} = 5V   Top Trace: Output 5V/Div.
TIME = 20μs/DIV.   Bottom Trace: Capacitor Voltage 1V/Div.
R_A = 3.9 kΩ
R_B = 3 kΩ
C = 0.01 μF

FIGURE 5. Astable Waveforms

The charge time (output high) is given by:
$$t_1 = 0.693 (R_A + R_B) C$$

And the discharge time (output low) by:
$$t_2 = 0.693 (R_B) C$$

Thus the total period is:
$$T = t_1 + t_2 = 0.693 (R_A + 2R_B) C$$

Applications Information (Continued)

The frequency of oscillation is:

$$f = \frac{1}{T} = \frac{1.44}{(R_A + 2R_B)C}$$

Figure 6 may be used for quick determination of these RC values.

The duty cycle is: $\quad D = \dfrac{R_B}{R_A + 2R_B}$

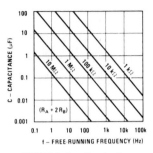

FIGURE 6. Free Running Frequency

FREQUENCY DIVIDER

The monostable circuit of *Figure 1* can be used as a frequency divider by adjusting the length of the timing cycle. *Figure 7* shows the waveforms generated in a divide by three circuit.

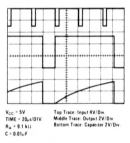

FIGURE 7. Frequency Divider

PULSE WIDTH MODULATOR

When the timer is connected in the monostable mode and triggered with a continuous pulse train, the output pulse width can be modulated by a signal applied to pin 5. *Figure 8* shows the circuit, and in *Figure 9* are some waveform examples.

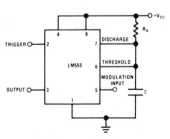

FIGURE 8. Pulse Width Modulator

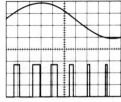

FIGURE 9. Pulse Width Modulator

PULSE POSITION MODULATOR

This application uses the timer connected for astable operation, as in *Figure 10*, with a modulating signal again applied to the control voltage terminal. The pulse position varies with the modulating signal, since the threshold voltage and hence the time delay is varied. *Figure 11* shows the waveforms generated for a triangle wave modulation signal.

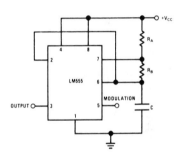

FIGURE 10. Pulse Position Modulator

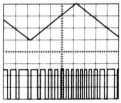

FIGURE 11. Pulse Position Modulator

LINEAR RAMP

When the pullup resistor, R_A, in the monostable circuit is replaced by a constant current source, a linear ramp is

Applications Information (Continued)

generated. *Figure 12* shows a circuit configuration that will perform this function.

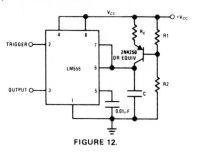

FIGURE 12.

Figure 13 shows waveforms generated by the linear ramp.

The time interval is given by:

$$T = \frac{2/3 \, V_{CC} \, R_E \, (R_1 + R_2) \, C}{R_1 \, V_{CC} - V_{BE} \, (R_1 + R_2)}$$

$$V_{BE} \approx 0.6V$$

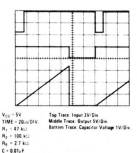

$V_{CC} = 5V$
TIME = 20μs/DIV.
$R_1 = 47\,k\Omega$
$R_2 = 100\,k\Omega$
$R_E = 2.7\,k\Omega$
$C = 0.01\,\mu F$

Top Trace: Input 3V/Div
Middle Trace: Output 5V/Div.
Bottom Trace: Capacitor Voltage 1V/Div.

FIGURE 13. Linear Ramp

50% DUTY CYCLE OSCILLATOR

For a 50% duty cycle, the resistors R_A and R_B may be connected as in *Figure 14*. The time period for the output high is the same as previous, $t_1 = 0.693 \, R_A \, C$. For the output low it is $t_2 =$

$$[(R_A \, R_B)/(R_A + R_B)] \, C \ln \left[\frac{R_B - 2R_A}{2R_B - R_A} \right]$$

Thus the frequency of oscillation is $f = \dfrac{1}{t_1 + t_2}$

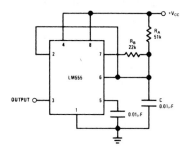

FIGURE 14. 50% Duty Cycle Oscillator

Note that this circuit will not oscillate if R_B is greater than $1/2 \, R_A$ because the junction of R_A and R_B cannot bring pin 2 down to $1/3 \, V_{CC}$ and trigger the lower comparator.

ADDITIONAL INFORMATION

Adequate power supply bypassing is necessary to protect associated circuitry. Minimum recommended is $0.1\mu F$ in parallel with $1\mu F$ electrolytic.

Lower comparator storage time can be as long as $10\mu s$ when pin 2 is driven fully to ground for triggering. This limits the monostable pulse width to $10\mu s$ minimum.

Delay time reset to output is $0.47\mu s$ typical. Minimum reset pulse width must be $0.3\mu s$, typical.

Pin 7 current switches within 30 ns of the output (pin 3) voltage.

Industrial Blocks

LM567/LM567C Tone Decoder

General Description

The LM567 and LM567C are general purpose tone decoders designed to provide a saturated transistor switch to ground when an input signal is present within the passband. The circuit consists of an I and Q detector driven by a voltage controlled oscillator which determines the center frequency of the decoder. External components are used to independently set center frequency, bandwidth and output delay.

Features

- 20 to 1 frequency range with an external resistor
- Logic compatible output with 100 mA current sinking capability
- Bandwidth adjustable from 0 to 14%

- High rejection of out of band signals and noise
- Immunity to false signals
- Highly stable center frequency
- Center frequency adjustable from 0.01 Hz to 500 kHz

Applications

- Touch tone decoding
- Precision oscillator
- Frequency monitoring and control
- Wide band FSK demodulation
- Ultrasonic controls
- Carrier current remote controls
- Communications paging decoders

Schematic and Connection Diagrams

Metal Can Package

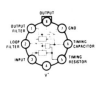

TOP VIEW

Order Number LM567H or LM567CH
See NS Package H08C

Dual-In-Line Package

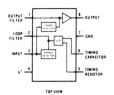

TOP VIEW

Order Number LM567CN
See NS Package N08B

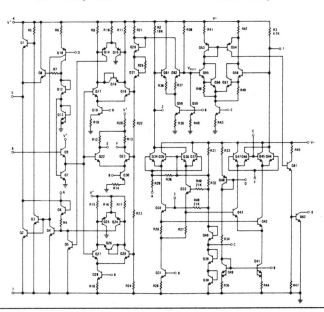

403

Absolute Maximum Ratings

Supply Voltage Pin	10V
Power Dissipation (Note 1)	300 mW
V_8	15V
V_3	−10V
V_3	$V_8 + 0.5V$
Storage Temperature Range	−65°C to +150°C

Electrical Characteristics (AC Test Circuit, $T_A = 25°C$, $V_C = 5V$)

PARAMETERS	CONDITIONS	LM567 MIN	LM567 TYP	LM567 MAX	LM567C/LM567CN MIN	LM567C/LM567CN TYP	LM567C/LM567CN MAX	UNITS
Power Supply Voltage Range		4.75	5.0	9.0	4.75	5.0	9.0	V
Power Supply Current	$R_L = 20k$							
Quiescent			6	8		7	10	mA
Power Supply Current	$R_L = 20k$							
Activated			11	13		12	15	mA
Input Resistance		18	20	22	15	20	25	kΩ
Smallest Detectable Input Voltage	$I_L = 100$ mA, $f_i = f_o$		20	25		20	25	mVrms
Largest No Output Input Voltage	$I_C = 100$ mA, $f_i = f_o$	10	15		10	15		mVrms
Largest Simultaneous Outband Signal to Inband Signal Ratio			6			6		dB
Minimum Input Signal to Wideband Noise Ratio	$B_n = 140$ kHz		−6			−6		dB
Largest Detection Bandwidth		12	14	16	10	14	18	% of f_o
Largest Detection Bandwidth Skew			1	2		2	3	% of f_o
Largest Detection Bandwidth Variation with Temperature			±0.1	0.25		±0.1	0.5	%/°C
Largest Detection Bandwidth Variation with Supply Voltage	4.75V − 6.75V		±1	±2		±1	±5	%V
Highest Center Frequency		100	500		100	500		kHz
Center Frequency Stability	$0 < T_A < 70$		35 ± 60			35 ± 60		ppm/°C
	$-55 < T_A < +125$		35 ± 140			35 ± 140		ppm/°C
Center Frequency Shift with Supply Voltage	4.75V − 6.75V		0.5	1.0		0.4	2.0	%/V
Fastest ON-OFF Cycling Rate			$f_o/20$			$f_o/20$		
Output Leakage Current	$V_8 = 15V$		0.01	25		0.01	25	μA
Output Saturation Voltage	$e_i = 25$ mV, $I_8 = 30$ mA		0.2	0.4		0.2	0.4	V
	$e_i = 25$ mV, $I_8 = 100$ mA		0.6	1.0		0.6	1.0	
Output Fall Time			30			30		ns
Output Rise Time			150			150		ns

Note 1: The maximum junction temperature of the LM567 is 150°C, while that of the LM567C and LM567CN is 100°C. For operating at elevated temperatures, devices in the TO-5 package must be derated based on a thermal resistance of 150°C/W, junction to ambient or 45°C/W, junction to case. For the DIP the device must be derated based on a thermal resistance of 187°C/W, junction to ambient.

Typical Performance Characteristics

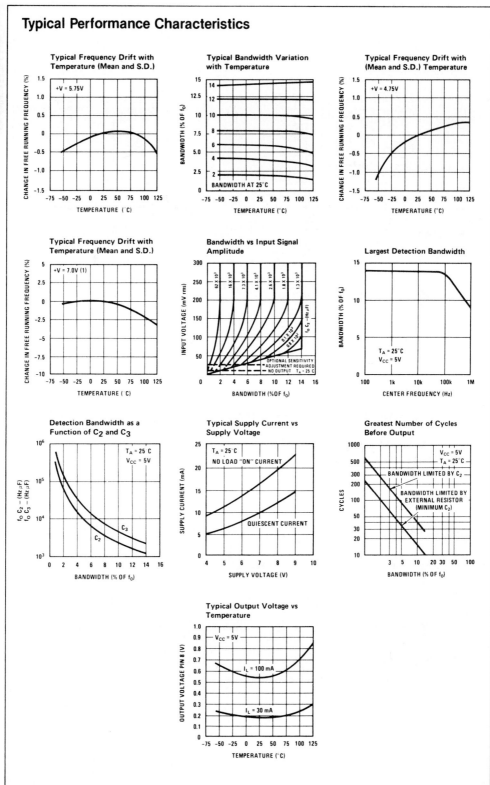

Typical Applications

Touch-Tone Decoder

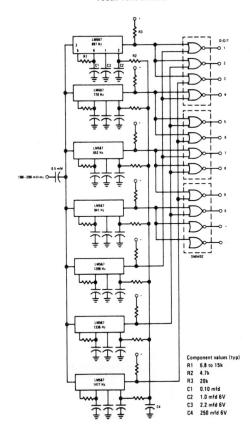

Component values (typ)
- R1 6.8 to 15k
- R2 4.7k
- R3 20k
- C1 0.10 mfd
- C2 1.0 mfd 6V
- C3 2.2 mfd 6V
- C4 250 mfd 6V

Oscillator with Quadrature Output

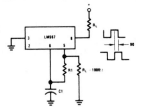

Connect pin 3 to 2.8V to invert output.

Oscillator with Double Frequency Output

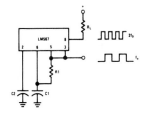

Precision Oscillator Drive 100 mA Loads

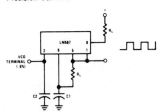

AC Test Circuit

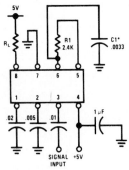

f_i = 100 kHz +5V
*Note: Adjust for f_O = 100 kHz.

Applications Information

The center frequency of the tone decoder is equal to the free running frequency of the VCO. This is given by

$$f_o \cong \frac{1}{1.1 R_1 C_1}$$

The bandwidth of the filter may be found from the approximation

$$BW = 1070 \sqrt{\frac{V_i}{f_o C_2}} \quad \text{in \% of } f_o$$

Where:

V_i = Input voltage (volts rms), $V_i \leq 200$ mV

C_2 = Capacitance at Pin 2 (μF)

APPENDIX B: SUGGESTED REFERENCES

Coughlin, Robert F., and Frederick F. Driscoll: *Operational Amplifiers and Linear Integrated Circuits*, 3d ed., Prentice-Hall, Englewood Cliffs, N.J., 1987.

Faulkenberry, Luces M.: *An Introduction to Operational Amplifiers with Linear IC Applications*, 2d ed., Wiley, New York, 1982.

Gayakwad, Ramakant A.: *Op Amps and Linear Integrated Circuit Technology*, Prentice-Hall, Englewood Cliffs, N.J., 1983.

Jung, Walter G.: *IC Op Amp Cookbook*, 2d ed., Howard W. Sams, Indianapolis, 1981.

Kuo, Benjamin C.: *Automatic Control Systems*, 5th ed., Prentice-Hall, Englewood Cliffs, N.J., 1987.

Lancaster, Don: *Active Filter Cookbook*, Howard W. Sams, Indianapolis, 1975.

Malvino, Albert P.: *Electronic Principles*, 3d ed., McGraw-Hill, New York, 1985.

Millman, Jacob, and Arvin Grabel: *Microelectronics*, 2d ed., McGraw-Hill, New York, 1987.

Roberge, J. K.: *Operational Amplifiers: Theory and Practice*, Wiley, New York, 1975.

Roddy, Dennis, and John Coolen: *Electronics: Theory, Circuits and Devices*, Reston, Reston, Va., 1982.

Schuler, Charles A., and William L. McNamee: *Industrial Electronics and Robotics*, McGraw-Hill, New York, 1986.

Sedra, Adel S., and Kenneth C. Smith: *Microelectronic Circuits*, Holt, Rinehart and Winston, New York, 1982.

Van Valkenburg, M. E.: *Analog Filter Design*, Holt, Rinehart and Winston, New York, 1982.

Young, Thomas: *Linear Integrated Circuits*, Wiley, New York, 1981.

⸻: *Linear Systems and Digital Signal Processing*, Prentice-Hall, Englewood Cliffs, N.J., 1985.

Zanger, Henry: *Semiconductor Devices and Circuits*, Wiley, New York, 1984.

Linear Databook. National Semiconductor Corp., Santa Clara, Calif., 1982.

APPENDIX C
ANSWERS TO ODD-NUMBERED PROBLEMS

Chapter 1

1-1. $I_T = 600 \ \mu A$, $I_{C1} = I_{C2} = 300 \ \mu A$, $V_{CE1} = V_{CE2} = 6.7$ V, $A_d = 115$, $r_{in} = 17.3$ kΩ.

1-3. $A_{CM} = 0.36$, CMRR = 50 dB.

1-5. $I_{C3} = 9$ mA, $I_{C1} = I_{C2} = 4.6$ mA, $V_{CE1} = V_{CE2} = 13.18$ V, $V_{C1} = V_{C2} = 11.48$ V, $r_{in} = 10.5$ kΩ, $A_d = 23$, $V_{CE3} = 15.46$ V.

1-7. $I_{C4} = 2$ mA, $I_{C1} = I_{C2} = 1$ mA, $V_{CE1} = V_{CE2} = 5.7$ V, $V_{CE4} = 9.3$ V, $r_{in} = 7.4$ kΩ, $A_d = 192.3$

1-9. $V_{in} = 6.75$ mV$_{P-P}$

1-11. $R_1 = 900 \ \Omega$, $R_2 = 583 \ \Omega$

1-13. $v_{o1} = 10.5 + 1.72 \sin 500t$ V, $v_{o2} = 10.5 - 1.72 \sin 500t$ V

1-15. $I_z = 8$ mA, $I_L = 8.6$ mA

1-17. $I_{C3} = 700 \ \mu A$, $I_{C1} = I_{C2} = 350 \ \mu A$, $I_{C6} = 3$ mA, $I_{C4} = I_{C5} = 1.5$ mA, $V_{CE1} = V_{CE2} = 2.7$ V, $V_{CE3} = 10.6$ V, $V_{CE4} = V_{CE5} = 5.7$ V, $V_{oQ} = 7.1$V, $r_{in2} = 3.47$ kΩ, $r_{in1} = 14.85$ kΩ, $A_{d1} = 41.7$, $A_{d2} = 95.2$, $A_{d(overall)} = 3969$, $V_{CE6} = 13.35$ V

1-19. $R_3 = 412.5 \ \Omega$, $A_v = 0.55$

1-21. $I_{C5} = 2$ mA, $I_{C1} = 10 \ \mu A$, $I_{C2} = 1$ mA, $I_{C3} = 1$ mA, $I_{C4} = 10 \ \mu A$, $V_{CE2} = 19.4$ V, $V_{CE3} = 9.4$ V, $V_{CE5} = 15.9$ V, $V_{oQ} = 8$ V, $r_{in} = 520$ kΩ, $A_d = 192$

Chapter 2

2-1. $A_v = 196.1$

2-3. $\beta = 0.0125$

APPENDIX C ANSWERS TO ODD-NUMBERED PROBLEMS **409**

2-5. $A_v = -2.52$, $i_{R1} = 22.2 \sin 10^3 t$ μA, $i_{RF} = -22.2 \sin 10^3 t$ μA, $v_o = -1.512 \sin 10^3 t$ μA, $i_L = -56 \sin 10^3 t$ μA

2-7. $R_1 = 17.37$ kΩ

2-9. $A_v = 100$

2-11. $A_v = 1.6$

2-13. $V_o = 0.999$ V, $V_{id} = 999$ μV, $I_F = 0.997$ μA

2-15. $I_1 = 50$ μA, $I_F = -50$ μA, $V_o = 0.9$ V

2-17. 312.5 kHz

2-19. $R_2/R_1 = R_4/R_3 = 7.07$

2-21. $A_v = 42.3$ dB

2-23. $R_1 = 1.28$ kΩ

Chapter 3

3-1. $I_{IO} = 4$ nA

3-3. $R_B = 16.88$ kΩ

3-5. $A_n = 3$, $V_{oB} = 1.5$ mV

3-7. $R_Y = 50.6$ kΩ, $R_X = 12$ kΩ, $R_B = 27$ Ω

3-9. $R_Z = 2$ kΩ, $R_Y = 1.33$ MΩ, $R_X = 100$ kΩ

3-11. $R_3 = 6.8$ kΩ, $R_1 = 1.5$ kΩ, $C_2 = 20$ pF, $C_1 = 500$ pF

3-13. $\theta_t = 188.8°$, $\theta_{pm} = -8.8°$, $A_{\text{loop}} = 80$ dB (unstable)

3-15. $f_{\max} = 26.53$ kHz

3-17. $f_{\max} = 122.4$ kHz

Chapter 4

4-1. (a) $I_F = 55.56$ μA, (b) $V_o = 3.78$ V, (c) $I_L = 1.72$ mA, (d) $I_o = 1.78$ mA, (e) 7.95 kΩ

4-3. $R_F = 33.3$ kΩ

4-5. $I_1 = -48.2$ μA, $I_2 = 22.2$ μA, $I_F = 26$ μA, $I_o = 2.056$ mA, $V_o = 2$V

4-7. $R_1 = 1.47$ kΩ, $R_B = 1.39$ kΩ

4-9. $R_G = 408.2$ Ω

4-11. $R = 2.5$ kΩ, $i_L = 3.2 \cos 100t$ mA

4-13.

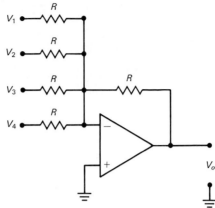

4-15. $BW_{max} = 63.9$ kHz, $BW_{min} = 2.08$ kHz

4-17. CMRR = 61.6 dB

Chapter 5

5-1. $v_o = -1.69 \sin(2\pi 1000t + 20.9°)$ V

5-3. $v_o = 1.1 \sin(10{,}000t - 90°)$ V $= -1.1 \cos 10{,}000t$ V

5-5.

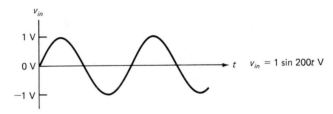

$v_{in} = 1 \sin 200t$ V

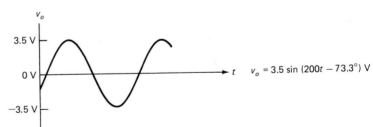

$v_o = 3.5 \sin(200t - 73.3°)$ V

5-7. $V_o = 2.2$ V

5-9.

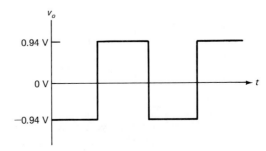

5-11.

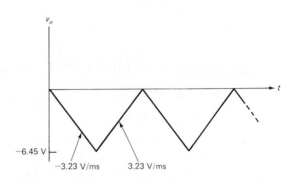

5-13. $v_1 = 640 \cos 2\pi 100t$ mV, $v_2 = -20.5 \sin 2\pi 100t$ mV

5-15. $V_{in} = 4$ V, $V_o = -0.453$ V; $V_{in} = 8$ V, $V_o = -0.471$ V

5-17. $V_{in} = 0.35$ V, $V_o = -0.70$ V; $V_{in} = 0.40$ V, $V_o = -4.80$ V

5-19.

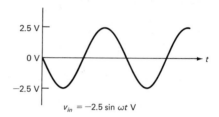

5-21.

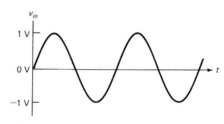

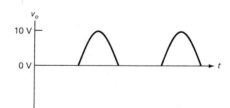

5-23.

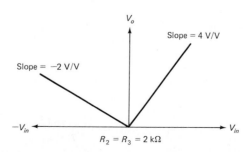

Chapter 6

6-1. $C = 0.0047\ \mu F$

6-3. R's remain unchanged. $C_1 = 0.067\ \mu F$, $C_2 = 0.029\ \mu F$

6-5. $R_1 = R_2 = 3.8\ k\Omega$, $R_B = 33.3\ k\Omega$

6-7. $R_1 = R_2 = 10\ k\Omega$, $C_1 = C_2 = 0.017\ \mu F$

6-9. $R_B = 50.3\ k\Omega$, $R_D = 85.6\ k\Omega$, $C_1 = C_2 = 0.0059\ \mu F$, $C_3 = C_4 = 0.0126\ \mu F$

6-11. $f_U = 4.32\ kHz$, $Q = 2.73$

6-13. $R_B = 16.1\ k\Omega$, $R_1 = R_2 = 7.96\ k\Omega$, $R_3 = R_4 = 31.83\ k\Omega$

6-15. $R_1 = 24.1\ k\Omega$, $R_2 = 96.5\ k\Omega$, $R_3 = 1\ k\Omega$

6-17. $R = 1\ k\Omega$

6-19. $R = 3.39\ k\Omega$, $R_\alpha = 505.1\ k\Omega$

6-21. $R_1 = 153\ \Omega$

Chapter 7

7-1. $T = 1.26\ ms$, $D = 70\%$

7-3. $T = 16.67\ ms$, $D = 50\%$

7-5. $V_{UT} = 4\ V$, $V_{LT} = 0\ V$

7-7.

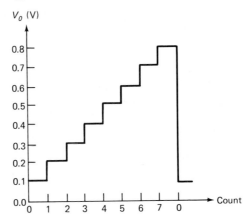

7-9. $R_M = 226\ k\Omega$, $R_F = 106\ k\Omega$, $R_1 = 21.2\ k\Omega$

7-11. $I_M = 115.3\ \mu A$, $V_{oQ} = 8\ V$, $A_v = 28.8$

7-13. $I_{ABC} = 226\ \mu A$, $g_m = 4.34\ mS$, $A_v = -434$

7-15. $R_{ABC} = 71.4\ k\Omega$, $R_L = 6\ k\Omega$

7-17. $ROC_{max} = SR_{max} = 0.173\ V/\mu s$

7-19. $v_o = 0.5 \sin 10^4 t$ V

7-21. $R_2 = 4.74$ kΩ

7-23. (a) $V_{dc} = 1$ V, (b) $V_{dc} = 1.5$ V, (c) $V_{dc} = 2.25$ V, (d) $V_{dc} = 2.25$ V

7-25. $V_R = 2.31$ V

7-27. $R_o = 2.88$ kΩ, $A_{loop} = 336$ kHz/V, lock range $= 80$ kHz

7-29. $R_1 = 4.54$ kHz, $C_2 = 0.156$ μF

Chapter 8

8-1. Resolution $= 0.024\%$, 1 LSB $= 1.221$ mV

8-3. (a) 2.5 V, (b) 5.0 V, (c) 5.156 V, (d) 4.884 V

8-5. $V_{ref} = -2.50$ V

8-7. $R_{(+)} = 12.5$ kΩ, $R_{(-)} = 25$ kΩ

8-9. $I_{FS} = 1.00$ mA, $R_L = 10$ kΩ

8-11. 136_{10} (1000 1000$_2$)

8-13. FSR $= 8.00$ V

8-15. $T_{C(min)} = 500$ μs, $T_{C(max)} = 254.5$ ms

8-17. $V_{in} = 6.875$ V

8-19. $T_C = 18$ ms, count $= 204_{10}$ (1100 1100$_2$)

8-21. $f_{s(max)} = 10$ kHz, $f_{alias} = 5$ kHz

8-23. -55.2 dB

Chapter 9

9-1. $I_{L1} = 24$ mA, $I_{L2} = 48$ mA, $I_{CT} = 24$ mA

9-3. $P_{reg} = 17.5$ W, $\eta = 41.7\%$

9-5. $P_C = 1.6$ W

9-7. $R_{SC} = 0.14$ Ω, $P_{RSC} = 3.5$ W

9-9. $I_{knee} = 4.85$ A, $I_{SC} = 718.7$ mA

9-11. $\beta_{min} = 15.4$

9-13. $R_1 = 54.7$ Ω

INDEX

Accuracy, 299
Active filter (*see* Filter)
Active loading, 44
A/D converter, 312–333
   (*See also* Converter, A/D)
Aliasing, 334–337
All-pass filter, 229–230
Amplifier:
   antilogarithmic, 174–176
   current (ICIS), 139
   current difference, 247–253
   differential (*see* Differential amplifier)
   instrumentation, 126–134
   logarithmic, 171–174
   Norton, 247
   operational (*see* Operational amplifier)
   operational transconductance (OTA), 253–264
   push-pull, 42
   summing, 111–123
   transconductance, 253
   transresistance (ICVS), 138, 310–311
Amplitude modulation, 258–260
Analog multiplexer, 176–179
Analog switch, 243–247
Antialiasing filter, 335–337
Antilogarithmic amplifier, 174–176
Averager, 117–119

Balanced modulator, 264–270
Band-reject filter, 222–226
Bandlimiting, 104
Bandpass filter, 213–222
   multiple feedback, 216–220
   wideband, 214–216
Bandstop filter, 222–226
Bandwidth, 76, 193
Bessel filter, 199
Bode plot, 75
Bounding, 239

Break frequency, 75
Butterworth filter, 198, 199, 339

Capacitor, sampling, 246
Capture range, 276–277
CDA (current difference amplifier), 247–253
Chebyshev filter, 198, 199, 211
Closed-loop gain, 60
Common-mode gain, 33
Common-mode input voltage, 32, 124
Common-mode rejection ratio, 124
Comparator, 234–243
Continuous converter, 328–330
Conversion gain, 278
Conversion time, 318–319
Converter:
   A/D, 312–333
      continuous, 328–330
      dual-slope, 324–327
      flash, 328–330
      integrating, 324–327
      parallel, 328–330
      ramp, 319–321
      successive approximation, 322–324
      tracking, 330–331
   current-to-voltage (I/V), 310
   D/A, 295–312
      R-2R ladder, 306–307
      weighted resistor, 303–306
   voltage-to-current (V/I), 253
   voltage-to-frequency (*see* Voltage-controlled oscillator)
Corner frequency, 75, 190, 211
Critical damping, 197
Critical frequency, 75
Current amplifier (ICIS), 139
Current difference amplifier (CDA), 247–253
Current mirror, 11, 248
Current-mode switch, 244

Current source, 8–17
   mirror, 11, 248
   zener-biased, 14

D/A converter, 295–312
   (*see also* Converter, D/A)
Damping coefficient, 192
Darlington pair, 47
Deadband, 241
Definite integral, 162
Demodulation:
   of DSB, 266–267
   of DSB-SC, 266–268
   of FM, 282
   synchronous, 282–283
Derivative, 156
Differential, 162
Differential amplifier:
   differential-input–differential-output, 25
   differential-input–single-ended-output, 29
   op amp–based, 123
   single-ended-input–differential-output, 22
   single-ended-input–single-ended-output, 17
Differential input voltage, 27
Differential nonlinearity, 317–318
Differential pair, 1
Differentiator, 155–162
DSB-SC (double-sideband suppressed-carrier)
      modulation, 266–268
Dual-slope converter, 324–327
Duty cycle, 236

Feedback factor, 60
Feedback gain, 60
Field-effect transistor, 47, 219–220, 362
Filter:
   all-pass, 229–230
   antialiasing, 335–337
   band-reject, 222–226
   bandpass, 213–222
   bandstop, 222–226
   bessel, 199
   Butterworth, 198, 199, 339
   Chebyshev, 198, 199, 211
   loops, 280–281
   notch, 222–226
   reconstruction, 337, 341
   Sallen-Key VCVS, 201–208
   state-variable, 226–228

First-order response, 76, 192, 193
Flash converter, 328–330
FM demodulation, 282
Foldback limiting, 356
Four-quadrant multiplier, 177
Frequency scaling, 202
Frequency synthesis, 283–284
Full-scale range, 313
Full-scale voltage, 297

Gain:
   closed-loop, 60
   common-mode, 33
   conversion, 278
   feedback, 60
   unity, 201–205
Gain-bandwidth product, 76
Gain error, 316
Geometric mean, 193
Graphic equalizer, 220–222
Guard drive, 133

Half-power point, 191
Hilbert transformer, 269
Howland current source, 137
Hybrid IC, 58
Hysteresis, 239
Hysteresis loop, 241

ICIS (current amplifier), 139
ICVS (transresistance amplifier), 138, 310–311
Impedance scaling, 202
Input bias current, 86
Input offset voltage, 86, 89
Input range division, 313
Integral, definite, 162
Instrumentation amplifier, 126–134
Integrand, 162
Integrating converter, 324–327
Integrator, 162–170
I/V converter, 310

Lever shifter, 39–42
Linearity error, 300
Logarithmic amplifier, 171–174
Loop filter, 280–281
Loop gain, 97

Midband, 74
Miller input resistance, 68
Modulation:
   amplitude, 258–260
   DSB, 264
   DSB-SC, 266–268
   pulse code, 275
   pulse-position, 274–275
   pulse width, 274–275, 362
   SSB-SC, 268–270
Modulation envelope, 258
Modulation index, 258, 262
Modulator:
   balanced, 264–270
   OTA, 260–264
Multiplier, 177

Noise gain, 78, 87
Nonlinearity error, 317
Norton amplifier, 247
Notch filter, 222–226
Null depth, 225
Nyquist sampling theorem, 334–335

Offset error, 300, 316
Offset null compensation, 90
Operational amplifier:
   differential amplifier, 123–126
   inverting, 63–69, 146–151
   noninverting, 59–63, 152–155
   summing, 111–123
Operational transconductance amplifier (OTA), 253–264
Oscillator, voltage-controlled, 274–275, 281, 312
Output offset voltage, 90
Overcurrent protection, 353–356
Overdamped response, 198
Oversampling, 334
Overshoot, 301

Parallel converter, 328–330
Passband, 190
PCM (pulse code modulation), 275
Phase comparator, 277–280
Phase detector, 277
Phase-locked loop (PLL), 275–289
Phase margin, 97
Photocoupler, 219

PLL (phase-locked loop), 275–289
Precision rectifier, 181–184
   full-wave, 182
   half-wave, 181
PRF (pulse repetition frequency), 273
PRR (pulse repetition rate), 273
Pulse code modulation (PCM), 275
Pulse-position modulation, 274–275
Pulse repetition frequency (PRF), 273
Pulse repetition rate (PRR), 273
Pulse width modulation (PWM), 274–275, 362
Push-pull amplifier, 42
PWM (pulse width modulation), 274–275, 362

Quantization error, 314–316
Quantization noise, 314, 337–339, 341

Ramp A/D converter, 319–321
Reconstruction filter, 337, 341
Regulator:
   series, 350–351
   shunt, 349
   switching, 362–364
   zener, 349–350
Resolution, 296
Ripple channel, 211
Rolloff, 192
R-2R ladder converter, 306–307

Sallen-Key VCVS:
   equal component, 205–208
   unity gain, 201–205
Sample and hold, 246
Sampling capacitor, 246
Schmitt trigger, 240–242
Schockley's equation, 171
Series-pass transistor, 350
Series regulator, 350–351
Servo, 311
Settling time, 300–301
Shunt regulator, 349
Side frequency, 258, 266
Sideband, 258, 266
Signal-to-quantization-noise ratio (SQNR), 316, 337
Single-sideband suppressed-carrier (SSB-SC) modulation, 268–270

Slew rate, 100–105
Small signal, 95
Spectrum analyzer, 220–221
SQNR (signal-to-quantization-noise ratio), 316, 337
SSB-SC (single-sideband suppressed-carrier) modulation, 268–270
State-variable filter, 226–228
Step function, 159
Stop band, 190
Strain gage, 129, 180
Successive approximation converter, 322–324
Summing amplifier, 111–123
Summing junction, 113, 251, 252
Summing node, 113
Summing point, 113
Switching regulator, 362–364
Synchronous demodulation, 282–283

Timer (IC), 271–275
Touch-tone frequencies, 287
Track-and-hold, 246–247, 324
Tracking converter, 330–331
Transconductance amplifier (VCIS), 253
Transistor:
    field-effect, 47, 219–220, 362
    series-pass, 350
Transresistance amplifier (ICVS), 138, 310–311
Two-quadrant multiplier, 177

Underdamped response, 198
Undersampling, 334

VCIS (voltage-controlled current source), 134, 253
VCO, 274–275, 281, 312
VCVS (voltage-controlled voltage source), 56, 71
V/I converter, 253
Virtual ground, 66, 113
Voltage:
    common-mode input, 32, 124
    differential input, 27
    full-scale, 297
    input offset, 86, 89
    output offset, 90
Voltage-controlled current source (VCIS), 134, 253
Voltage-controlled oscillator, 274–275, 281, 312
Voltage-controlled voltage source (VCVS), 56, 71
Voltage follower, 69
Voltage regulation, 346–348

Weighted averager, 119
Weighted resistor converter, 303–306
Window comparator, 242–243

Zener regulator, 349–350
Zero-crossing detector, 238